Bioactive Polyphenols for Health and Pathology Treatment

Bioactive Polyphenols for Health and Pathology Treatment

Edited by

Antonio Segura Carretero
Department of Analytical Chemistry, Faculty of Sciences, University of Granada, Granada, Spain

Maria de la Luz Cádiz Gurrea
Department of Analytical Chemistry, Faculty of Sciences, University of Granada, Granada, Spain

Javad Sharifi-Rad
Department of Medicine, College of Medicine, Korea University, Seoul, Republic of Korea

ELSEVIER

ACADEMIC PRESS
An imprint of Elsevier

Academic Press is an imprint of Elsevier
125 London Wall, London EC2Y 5AS, United Kingdom
525 B Street, Suite 1650, San Diego, CA 92101, United States
50 Hampshire Street, 5th Floor, Cambridge, MA 02139, United States

ISBN: 978-0-443-13821-8

For Information on all Academic Press publications
visit our website at https://www.elsevier.com/books-and-journals

Publisher: Mara Conner
Acquisitions Editor: Nina Bandeira
Editorial Project Manager: Lena Sparks
Production Project Manager: Sruthi Satheesh
Cover Designer: Mark Rogers

Typeset by MPS Limited, Chennai, India

Contents

Section 5 Polyphenols and cutaneous-related conditions

List of contributors

Óscar Beas
Department of Pharmacology and Physiology, Legal and Forensic Medicine, Faculty of Veterinary Medicine, Aragon Health Research Institute, University of Zaragoza, Zaragoza, Spain; Department of Molecular Medicine, School of Medicine, University of Colima, Colima, Mexico

Maria João Bessa
UNIPRO – Oral Pathology and Rehabilitation Research Unit, University Institute of Health Sciences (IUCS), CESPU, Gandra, Portugal

Guillermo Cásedas
Agri-Food Institute of Aragon IA2, CITA, University of Zaragoza, Zaragoza, Spain; Department of Pharmacy, Faculty of Health Sciences, Universidad San Jorge, Villanueva de Gállego, Zaragoza, Spain

Yaiza María Yáñez Castillo
Department of Urology, San Cecilio University Hospital, Granada, Spain

Paulo C. Costa
REQUIMTE/UCIBIO, MedTech-Laboratory of Pharmaceutical Technology, Department of Drug Sciences, Faculty of Pharmacy, University of Porto, Porto, Portugal; Associate Laboratory i4HB – Institute for Health and Bioeconomy, Faculty of Pharmacy, University of Porto, Porto, Portugal

Carmen del Río
Institute of Biomedicine of Seville (IBiS), Hospital Universitario Virgen del Rocío, CSIC, Universidad de Sevilla, Seville, Spain

Cristina Delerue-Matos
REQUIMTE/LAQV, ISEP, Polytechnic of Porto, Porto, Portugal

Marina Delgado-Machuca
Department of Pharmacology and Physiology, Legal and Forensic Medicine, Faculty of Veterinary Medicine, Aragon Health Research Institute, University of Zaragoza, Zaragoza, Spain; Department of Molecular Medicine, School of Medicine, University of Colima, Colima, Mexico

Patricia Fernández-Moreno
Department of Analytical Chemistry, University of Granada, Granada, Granada, Spain

Álvaro Fernández-Ochoa
Department of Analytical Chemistry, University of Granada, Granada, Granada, Spain

Julio Galvez
Department of Pharmacology, Center for Biomedical Research (CIBM), University of Granada, Granada, Spain; Instituto de Investigación Biosanitaria de Granada (ibs. GRANADA), Granada, Spain; Centro de Investigación Biomédica En Red para Enfermedades Hepáticas y Digestivas (CIBER-EHD), School of Pharmacy, University of Granada, Granada, Spain

David Garcia-Burgos
Department of Psychobiology, Institute of Neurosciences, University of Granada, Granada, Spain

Ángela González-Díaz
Institute of Biomedicine of Seville (IBiS), Hospital Universitario Virgen del Rocío, CSIC, Universidad de Sevilla, Seville, Spain

Carlos Hernández-Martínez
Department of Physical Education and Sports, Faculty of Sports Science, Sport and Health University Research Institute (iMUDS), University of Granada, Granada, Spain

Felipe Huertas-Camarasa
Department of Analytical Chemistry, University of Granada, Granada, Granada, Spain

Lucas Jurado-Fasoli
Department of Physical Education and Sports, Faculty of Sports Science, Sport and Health University Research Institute (iMUDS), University of Granada, Granada, Spain; Department of Physiology, Faculty of Medicine, University of Granada, Granada, Andalucía, Spain

Francisco Les
Agri-Food Institute of Aragon IA2, CITA, University of Zaragoza, Zaragoza, Spain; Department of Pharmacy, Faculty of Health Sciences, Universidad San Jorge, Villanueva de Gállego, Zaragoza, Spain

Francisco Javier Leyva-Jimenez
Regional Institute for Applied Scientific Research (IRICA), University of Castilla-La Mancha, Ciudad Real, Spain; Area of Food Science and Technology, Faculty of Chemical Sciences and Technologies, University of Castilla-La Mancha, Ciudad Real, Spain

Víctor López
Agri-Food Institute of Aragon IA2, CITA, University of Zaragoza, Zaragoza, Spain; Department of Pharmacy, Faculty of Health Sciences, Universidad San Jorge, Villanueva de Gállego, Zaragoza, Spain

Vikky Kumar Mahto
University Department of Botany, Lalit Narayan Mithila University, Darbhanga, Bihar, India

Juan J. Martin-Olmedo
Department of Physical Education and Sports, Faculty of Sports Science, Sport and Health University Research Institute (iMUDS), University of Granada, Granada, Spain; Department of Physiology, Faculty of Medicine, University of Granada, Granada, Andalucía, Spain

Arjun Kumar Mehara
University Department of Botany, Lalit Narayan Mithila University, Darbhanga, Bihar, India

Joan Montaner
Institute of Biomedicine of Seville (IBiS), Hospital Universitario Virgen del Rocío, CSIC, Universidad de Sevilla, Seville, Spain; Department of Neurology, Hospital Universitario Virgen Macarena, Seville, Spain

Ana M Najar
Institute of Biomedicine of Seville (IBiS), Hospital Universitario Virgen del Rocío, CSIC, Universidad de Sevilla, Seville, Spain; Department of Neurology, Hospital Universitario Virgen Macarena, Seville, Spain

Francisco J. Osuna-Prieto
Department of Physical Education and Sports, Faculty of Sports Science, Sport and Health University Research Institute (iMUDS), University of Granada, Granada, Spain; Hospital Universitari Joan XXIII of Tarragona, Pere Virgili Health Research Institute (IISPV), Tarragona, Spain; CIBER of Diabetes and Associated Metabolic Diseases (CIBERDEM)-Institute of Health Carlos III (ISCIII), Madrid, Spain

Javier Quero
Department of Pharmacology and Physiology, Legal and Forensic Medicine, Faculty of Veterinary Medicine, Aragon Health Research Institute, University of Zaragoza, Zaragoza, Spain

Francisca Rodrigues
REQUIMTE/LAQV, ISEP, Polytechnic of Porto, Porto, Portugal

Alba Rodriguez-Nogales
Department of Pharmacology, Center for Biomedical Research (CIBM), University of Granada, Granada, Spain; Instituto de Investigación Biosanitaria de Granada (ibs. GRANADA), Granada, Spain

María Jesús Rodríguez-Yoldi
Department of Pharmacology and Physiology, Legal and Forensic Medicine, Faculty of Veterinary Medicine, Aragon Health Research Institute, University of Zaragoza, Zaragoza, Spain; Agri-Food Institute of Aragon IA2, CITA, University of Zaragoza, Zaragoza, Spain; Centro de Investigación Biomédica en Red de Fisiopatología de la Obesidad y Nutrición (CIBEROBN), Instituto de Salud Carlos III, Madrid, Spain

Marina Romero-Bernal
Institute of Biomedicine of Seville (IBiS), Hospital Universitario Virgen del Rocío, CSIC, Universidad de Sevilla, Seville, Spain

Jonatan R. Ruiz
Department of Physical Education and Sports, Faculty of Sports Science, Sport and Health University Research Institute (iMUDS), University of Granada, Granada, Spain; CIBER of Obesity and Nutrition Physiopathology (CIBEROBN), Institute of Health Carlos III, Granada, Spain; Biosanitary Research Institute, Ibs.Granada, Granada, Spain

María Teresa Melgarejo Segura
Department of Urology, San Cecilio University Hospital, Granada, Spain

Chaminda Jaya Seneviratne
The University of Queensland, School of Dentistry, Brisbane, QLD, Australia

Ana Margarida Silva
REQUIMTE/LAQV, ISEP, Polytechnic of Porto, Porto, Portugal

Ankit Kumar Singh
University Department of Botany, Lalit Narayan Mithila University, Darbhanga, Bihar, India

Vikas Kumar Singh
Department of Science & Humanities, Supaul College of Engineering, Supaul, Bihar, India

Vipendra Kumar Singh
School of Biosciences and Bioengineering, Indian Institute of Technology Mandi, VPO Kamand, Mandi, Himachal Pradesh, India

Filipa Teixeira
REQUIMTE/LAQV, ISEP, Polytechnic of Porto, Porto, Portugal

María del Carmen Villegas-Aguilar
Department of Analytical Chemistry, University of Granada, Granada, Granada, Spain

Laurence James Walsh
The University of Queensland, School of Dentistry, Brisbane, QLD, Australia

SECTION 1

Overview of the role of polyphenols from herbal extracts to nutraceuticals

CHAPTER ONE

Historical insights of polyphenols in human health

Javier Quero[1], Víctor López[2,3], Francisco Les[2,3], Guillermo Cásedas[2,3], Marina Delgado-Machuca[1,4], Óscar Beas[1,4] and María Jesús Rodríguez-Yoldi[1,2,5]

[1]Department of Pharmacology and Physiology, Legal and Forensic Medicine, Faculty of Veterinary Medicine, Aragon Health Research Institute, University of Zaragoza, Zaragoza, Spain
[2]Agri-Food Institute of Aragon IA2, CITA, University of Zaragoza, Zaragoza, Spain
[3]Department of Pharmacy, Faculty of Health Sciences, Universidad San Jorge, Villanueva de Gállego, Zaragoza, Spain
[4]Department of Molecular Medicine, School of Medicine, University of Colima, Colima, Mexico
[5]Centro de Investigación Biomédica en Red de Fisiopatología de la Obesidad y Nutrición (CIBEROBN), Instituto de Salud Carlos III, Madrid, Spain

Introduction

Natural products, due to their extraordinary chemical diversity, have been considered an important source of bioactive substances with therapeutic potential since ancient times. These natural products, among which the most used for their biological properties are polyphenols, are mainly extracted from plants (Habli et al., 2017; Simoben et al., 2015).

Polyphenols are secondary metabolites, also known as phytochemicals, produced by plants or fungi that represent the largest group of non energy compounds in plant foods from our daily diet such as fruits, vegetables, cereals, cocoa, tea, coffee, wine, etc. (Caponio et al., 2020; Lippolis et al., 2023; Montagna et al., 2019; Quero et al., 2020).

Different epidemiological studies have correlated the consumption of cereals, fruits and vegetables, characteristic of the Mediterranean diet, with the lower risk of suffering diseases associated to oxidative stress such as cancer, inflammation, osteoarthritis, cardiovascular diseases (CVDs), diabetes, etc. (Adolphe et al., 2010; Asensi et al., 2011; Basu & Maier, 2018; Costea et al., 2018; Hardman, 2014; Herrera-Marcos et al., 2017; Paesa Ancín-Azpilicueta et al., 2022; Paesa Nogueira et al., 2022; Quero et al., 2022, 2021; St-Onge & Zuraikat, 2019; Toledo et al., 2015; Velderrain-Rodríguez et al., 2021). These protective effects could be related to the

Bioactive Polyphenols for Health and Pathology Treatment.
DOI: https://doi.org/10.1016/B978-0-443-13821-8.00001-9

antioxidant and antiinflammatory properties of polyphenols (Caponio et al., 2020; Fischer et al., 2018; García-Lafuente et al., 2009; Pandey & Rizvi, 2009; Singh et al., 2014; Tutino et al., 2021; Yao et al., 2004).

The antioxidant effects of polyphenols can be exerted directly, by eliminating free radicals, or indirectly, by modulating the expression of genes and activity of enzymes involved in redox homeostasis (Arbonés-Mainar et al., 2008). Therefore polyphenols could help endogenous antioxidant systems control oxidative homeostasis by reducing excess reactive oxygen species (ROS) and reactive nitrogen species (RNS).

On this line, in vitro studies have shown that polyphenols are capable of donating an electron or a hydrogen atom to neutralize free radicals. In lipid peroxidation chain reactions, polyphenols act as chain breakers converting free radicals into stable radicals by donating an electron (Guo et al., 2009). Polyphenols can also reduce the rate of oxidation by inhibiting or deactivating the precursors of free radicals and consequently suppressing their generation (Tsao, 2010). Through different interactions with enzymes, polyphenols can decrease levels of hydrogen peroxide, superoxide, and hydroperoxide anions by inducing antioxidant enzymes such as catalase, superoxide dismutase (SOD), and glutathione peroxidase, as well as they are found to inhibit the expression of pro oxidant enzymes such as xanthine oxidase (Tsao, 2010). However, despite their different antioxidant capacities, depending on cell type, intracellular pH and polyphenol concentrations, and chemical structure, polyphenols have also been found to show a pro oxidant behavior (Chikara et al., 2018; Kim & Kim, 2018; León-González et al., 2015; Roleira et al., 2015).

The antiinflammatory effect of dietary polyphenols could be related to the regulation or modulation of inflammatory pathways, inhibiting the release of metabolites or cytokines such as TNF-alpha and different interleukins (Vauzour, 2012; Yang et al., 2001). Polyphenols also contribute to the antiinflammatory effect due to their ability to reduce ROS and RNS along with their iron and copper chelating activity (Crascì et al., 2018).

Polyphenols, which are mainly found as glycosylated derivatives in plants, must undergo various intestinal transformations by digestive enzymes and the colon microbiota, to be hydrolyzed to aglycones and other bioactive metabolites and absorbed by enterocytes (Lippolis et al., 2023; Marin et al., 2015; Silva et al., 2019). Aglycones are metabolized again in the enterocytes before being carried to the liver, where these products undergo final enzymatic transformations becoming conjugated metabolites, hydrophilic molecules that enter the bloodstream and are distributed to tissues and organs or eventually excreted in the urine (Marin et al., 2015). Thus the beneficial effect of polyphenols on human health is often not direct but is due to the interaction of conjugated metabolites with genes and enzymes that modulate intracellular signaling cascades involved in cell growth, proliferation, and death, as well as in antioxidant and antiinflammatory responses (Crozier et al., 2009).

Although phenols were probably isolated in crude form from plants at the end of the 18th century, it was the active ingredient of willow bark that was discovered and isolated from willow bark into yellow crystals and named salicin by Johann Buchner in 1828; this could be considered the first phenolic compound of plant origin isolated for a therapeutic purpose. Later, in 1838, Raffaele Piria obtained a compound from the crystals isolated from willow bark, which he named salicylic acid. Subsequently, quercetin, which is one of the most important and ubiquitous flavonoids, might have also been one of the first flavonoids isolated from plants; its name was used since 1857 derived from "*quercetum*" (oak forest) (Wick, 2012).

Classification

Polyphenols are characterized by the presence of one or more hydroxyl groups on an aromatic ring. These molecules can be classified into soluble and insoluble compounds according to their location in the plant. The first is characterized by low and medium molecular weight compounds free in the soluble fraction of cell, while insoluble polyphenols include molecules linked to polysaccharides or proteins of the cell wall (Reis-Giada, 2013).

Polyphenols can also be classified into two main groups with different subclasses depending on their molecular weight, chemical structure, and complexity. These two groups are flavonoids (chalcones, flavanones, flavones, flavonols, flavanonols, isoflavonoids, flavanols, catechins, and anthocyanidins) and non-flavonoids (phenolic acids, curcuminoids, stilbenes, coumarins, lignans, and tannins). Chalcones are synthesized by the addition of three molecules of malonyl-CoA to 4-coumaryl-CoA, which in turn is derived from L-phenylalanine. Finally, isomerases, reductases, hydrolases, and dioxygenases modify chalcones leading to the different flavonoid subgroups (Falcone Ferreyra et al., 2012; Martens et al., 2010; Saito et al., 2013).

Effect of polyphenols on human health

In this chapter, we are going over the main uses of polyphenols along history based on their biological activities, as well as describing the main polyphenol compounds used in modern medicine to prevent and treat diseases, some of which are presented in Fig. 1.1.

Phenols and phenolic acids

Phenols and phenolic acids are the most basic and simple structures of what the scientific knowledge consider within the phytochemical group of polyphenols. Simple phenols are organic compounds that have at least one hydroxyl linked to a benzene, while phenolic acids must also contain a carboxylic function. Unlike what happens with phenolic acids, simple phenols are rare in nature. The simple phenols most described for their pharmacological and therapeutic properties are probably hydroquinone and a phenol of monoterpene origin known as thymol (Fig. 1.1).

Hydroquinone is a simple phenol with antiseptic and antimicrobial activity but is also an inhibitor of melanin synthesis with important

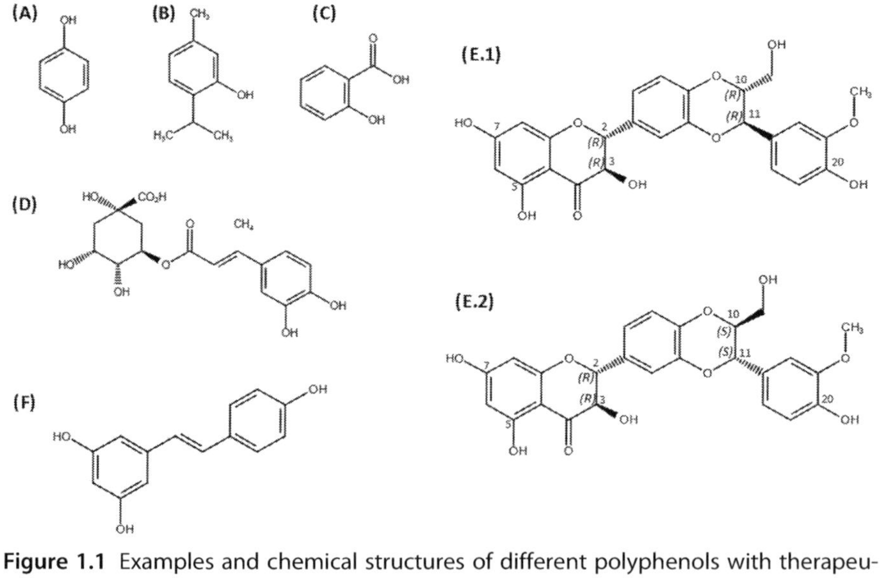

Figure 1.1 Examples and chemical structures of different polyphenols with therapeutic interests: hydroquinone (A), thymol (B), salicylic acid (C), chlorogenic acid (D), silybin A (E.1) and silybin B (E.2) and trans-resveratrol (F).

cosmetic applications. In nature, hydroquinone can be found in the form of arbutoside (also called arbutin) in the plant known as bearberry, scientifically named *Arctostaphylos uva-ursi*, from the Ericaceae family. This plant species is a shrub with evergreen leaves and edible red berry like fruits. The leaves are rich in arbutin, which in the body is transformed into hydroquinone, leading to an antiseptic action on the urinary tract (Schindler). Medicines and other products containing bearberry leaf or bearberry extracts are used for inflammation and urinary tract infections (Gágyor et al., 2021; Schindler et al., 2002). Furthermore, hydroquinone as an isolated active ingredient has been very useful in cosmetic products to treat skin blemishes and hyperpigmentation (Tung et al., 2023).

Thymol is another simple phenolic structure but with a particular terpenic origin; thymol is found in aromatic plants such as thyme (*Thymus vulgaris* L.), oregano (*Origanum vulgare* L.) or savory (*Satureja montana* L.). It is a compound with antioxidant and free radical scavenging activity but also antibacterial action. The plants mentioned above (thyme, oregano, and savory) have traditionally been used in phytotherapy as plants with antibacterial and digestive properties and in fact leaves from these species are used in cooking and gastronomy for their power to preserve foods and dishes. This compound, thymol, can be obtained by distillation of the aerial parts, giving rise to an essential oil with thymol and its isomer carvacrol, which also has antioxidant and antibacterial activity. Thymol is an irritating substance and is not recommended for ingestion by children or pregnant women, nor is it recommended through the skin. Recent works show that the essential oils with thymol have antibacterial, antifungal, or antiparasitic properties (Kowalczyk et al., 2020); thymol has shown activity against biofilms of *Escherichia coli*, *Listeria monocytogenes*, *Pseudomonas putida*, and *Staphylococcus aureus* but also against fungi like *Cryptococcus or* parasites like *Leishmania infantum* or *Anisakis* (López et al., 2018); thymol can be used as an isolated compound but it is generally used as the bioactive compound from oregano, thyme, or savory extracts.

According to phenolic acids, it is probable that salicylic acid (Fig. 1.1) may be the most important compound in terms if pharmacological properties and medicinal applications. We could say that the history of the non-steroidal antiinflammatory drugs (NSAIDs) goes back thousands of years because willow bark was already described as an analgesic remedy by Hippocrates (5th century BC). However, the first NSAID isolated was salicin extracted from the bark of *Salix* by Johannes Buchner in Germany. Salicin is a phenolic compound that is metabolized and biotransformed in the liver into salicylic acid. In 1838, salicin was transformed into salicylic acid by the Frenchman Raffaele Piria and later in 1897–1899, Felix Hoffman, who worked for Bayer, obtained acetylsalicylic acid from salicylic acid isolated from *Spiraea ulmaria*, from which it is derived, the name Aspirin (Wick, 2012). Thanks to the discovery of Aspirin, the pathophysiological phenomena related to pain, inflammation, eicosanoids, and the cyclooxygenase enzyme began to be known. From this moment on, the rest of the NSAIDs used today such as ibuprofen, naproxen, diclofenac, etc. were synthesized. Although the current clinical applications of acetylsalicylic acid as an analgesic, antipyretic, and antiinflammatory are well known, we must not forget the essential role it plays as an antiplatelet agent in the prophylaxis of ischemic attacks.

Phenolic acids are better represented in nature by compounds derived from benzoic or cinnamic acid. One of the most abundant phenolic acids in plants is chlorogenic and neochlorogenic acid (Fig. 1.1). These caffeic acid derivates are widely distributed in edible and medicinal plants, even in mushrooms. Both compounds exert interesting antioxidant properties, but different animal and human studies have demonstrated effects on metabolic diseases like obesity and diabetes as well as cardiovascular disorders. A recent systematic review has shown that chlorogenic acids have hepatoprotective effects against liver diseases through different molecular and signaling targets. These phenolic compounds may activate the erythroid 2-related factor 2 (Nrf2) and inhibit the toll-like receptor 4 (TLR4)/nuclear factor-κB (NF-κB) among others; nevertheless it has also been proposed that these compounds may exert effects on the microbiota (Xue et al., 2023; Zeng et al., 2022). Coffee is a source of chlorogenic acid and some studies reveal that green bean coffee extract with 500 mg of chlorogenic acid per day lowers body weight (Kanchanasurakit et al., 2023). These effects could be induced by a synergism between phenolic acids and caffeine, which has been already described as a lipolytic agent. Green coffee extract has also been shown to regulate glycemic indices, lipid profile and cardiometabolic risk factors (Asbaghi et al., 2020; Morvaridi et al., 2020; Naylor et al., 2021).

Flavonoids

Flavonoids constitute a diverse group of phytochemicals found in numerous fruits, vegetables, seeds, and beverages such as tea, beer, and wine. They are synthesized from a molecule of phenylalanine and three of malonyl-CoA through what is known as the "flavonoid biosynthetic pathway." The basic chemical structure of flavonoids consists of a C6-C3-C6 carbon skeleton, where the C6 components are aromatic rings joined by three carbon atoms that can form a third pyran or pyrone ring (A-C rings). Flavonoids are diaryl-propanoid phenolic compounds, meaning that their structure is of the C_6-C_3-C_6 type, with two aromatic (benzene) rings joined together by a three-carbon chain cyclized through an oxygen (Fig. 1.2). All flavonoids are hydroxylated structures on the aromatic ring and are therefore polyphenolic. They have a carbonyl in position 4, and variations occur at positions 1, 2, and 3 of the C_3 unit and in the B ring. Over 6,000 distinct types of flavonoids have been identified and categorized into six primary families based on their structural variations: flavanones, flavanols, flavones, flavonols, anthocyanidins, and isoflavonoids (Chang et al., 2018; Panche et al., 2016).

In plants, flavonoids serve various functions, including protection against oxidative damage from solar radiation, promoting plant growth, aiding in pollination, and offering defense against predators, among others (Havsteen, 2002; Luna et al., 2020).

In humans, flavonoids are notable for their antioxidant properties, while their pharmacological actions are diverse, encompassing anticancer, cardioprotective, neuroprotective, antiinflammatory, anti infectious, and other effects. Numerous studies have explored the relationship between flavonoid consumption and cancer prevention. There is evidence indicating an inverse association, particularly with quercetin, in the risk of cancer (Neuhouser, 2004).

A meta-analysis based on 12 studies suggests that flavonoid intake has a protective effect on the risk of developing lung cancer in some

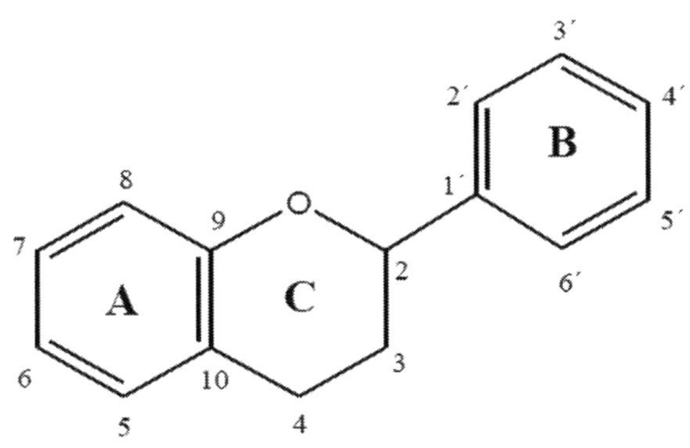

Figure 1.2 Flavonoid skeleton.

populations but not in others (Tang et al., 2009). Additionally, flavonoids may protect against smoking-related cancer risks (Woo et al., 2013). The consumption of anthocyanidins, flavanones, and flavones is inversely associated with the risk of esophageal cancer (Cui et al., 2016).

Concerning endocrine cancers, a meta-analysis suggests that higher consumption of flavonols, flavones, and isoflavones is associated with a reduced risk of specific women's cancers such as breast, ovarian, and endometrial cancer (Liu et al., 2022). Isoflavones, a subgroup of flavonoids, are widely used by women for their phytoestrogenic activity. Prospective studies on soy isoflavone consumption indicate a clear decrease in the risk of breast cancer in pre- and postmenopausal women (Boutas et al., 2022). However, in the context of specific cancers in men, a higher intake of total flavonoids might be significantly associated with an elevated risk of prostate cancer, according to the meta-analysis by Liu et al. (Liu et al., 2022). Nevertheless, a previous meta-analysis did not find a significant association between flavonoid intake and prostate cancer risk (Guo et al., 2016). Other studies suggest that green tea catechins or phytoestrogens tend to reduce the incidence of prostate cancer (Guo et al., 2017; Zhang et al., 2017).

Flavonoids have also been studied in gastrointestinal cancers. A meta-analysis has demonstrated that the consumption of various subclasses of flavonoids, such as flavonols, flavones, and anthocyanidins, could potentially reduce the risk of colorectal cancer. Notably, quercetin, a flavonol subgroup, and apigenin, a flavone, when consumed in high quantities, may reduce the risk of colon and rectal cancer, respectively (Chang et al., 2018). High consumption of phytosterols like soy isoflavones, epicatechin, and procyanidin has also been significantly associated with a reduced risk of colorectal cancer (Jin et al., 2012; Yu et al., 2016). A slight reduction in gastrointestinal cancer risk has also been observed (Tse & Eslick, 2016).

Polyphenols have also been investigated in CVDs. Increased dietary intake of total flavonoids is associated with a lower risk of CVD compared to those with low intake (Kim & Je, 2017; Liu et al., 2017; Micek et al., 2021). Intake of catechins, anthocyanins, and flavan-3-ols is more closely associated with CVD prevention, while catechins, flavonols, and flavones are linked to coronary disease. An increase in flavanones only showed improvement regarding stroke risk. Among individual compounds, quercetin and kaempferol intake is associated with reduced risk of coronary and CVDs, respectively (Micek et al., 2021). Soy isoflavones are known to potentially reduce CVD-related mortality (Nachvak et al., 2019).

Moreover, flavonoids are recognized for their vitamin P-like action, possessing venoactive and phlebotonic properties, which are important in the treatment of chronic venous insufficiency. A well-known polyphenolic extract is Daflon, containing a micronized purified fraction of flavonoids. Two meta-analyses of more than seven trials with this flavonoid extract demonstrated significant improvement in nine leg symptoms: pain, heaviness, swelling, cramps, paresthesia, burning, itching, leg redness, functional discomfort, and quality of life compared to the placebo group (Kakkos & Nicolaides, 2018; Sheikh et al., 2020). This micronized purified flavonoid fraction and hydroxyethylrutosides have shown the capacity to reduce and heal venous leg ulcers (Scallon & Bell-Syer, 2013).

Other flavonoids and extracts rich in these compounds with phlebotonic activity include rutinosides, hidrosmin, diosmin, Centella asiatica, French maritime pine bark extract, grape seed extract, etc. These have the potential to slightly reduce leg edema and likely reduce ankle circumference compared to the placebo. However, the meta-analysis evaluating phlebotonics showed little to no difference in quality of life and ulcer healing compared to the placebo. Additionally, phlebotonics may slightly increase adverse events, with gastrointestinal disorders being the most frequently reported adverse events (Martinez-Zapata et al., 2020).

Flavonoids can also regulate blood pressure. A review on chocolate and cocoa products rich in flavanols showed a small hypotensive effect (2 mmHg) in primarily healthy adults in the short term. These results were more significant when comparing (pre)hypertensive participants to normotensive participants (Ried et al., 2017). Quercetin has also demonstrated the ability to reduce blood pressure in normotensive and (pre) hypertensive patients (Popiolek-Kalisz & Fornal, 2022). This significantly lowered systolic blood pressure in the mixed population and in the normotensive subgroup and diastolic blood pressure in the (pre)hypertensive subgroup.

Flavonoids have also been demonstrated to possess antidiabetic and antiinflammatory properties. The relationship between hyperglycemia, inflammation, and vascular complications is now well established. A meta-analysis determined that the chemical structure of these compounds is crucial in determining their pharmacological action (Shamsudin et al., n.d.). To exert antiinflammatory and antidiabetic effects, a double bond C2-C3 (C ring) and hydroxyl groups at positions C3', C4', C5, and C7 in rings A and B are required. Furthermore, the substitution at C3 of the C ring diminishes the antiinflammatory action but enhances the antidiabetic activity of flavonoids. In this regard, another meta-analysis supports the hypothesis that quercetin reduces serum glucose levels within a dosage range of 10–50 mg/kg (Bule et al., 2019). Furthermore, two reviews concluded that quercetin has the ability to prevent diabetic nephropathy and improve glycemic control, dyslipidemias, inflammation, fibrotic lesions, and oxidative stress associated with diabetes mellitus (Hu et al., 2022; Li et al., 2022). These effects are more pronounced when high and sustained doses are administered over time.

Other flavonoids such as naringenin also demonstrate the ability to inhibit inflammation and oxidative stress related to the autoimmune system (Alimohammadi et al., 2022). Another study using orange juice, rich in flavonoids and ascorbic acid, demonstrated that it could reduce the concentrations of IL-6, C-reactive protein, and malondialdehyde in the participants (Cara et al., 2022).

Flavonoids can also exhibit valuable biological activities in the treatment of sepsis and associated inflammation. Flavonoids can strongly inhibit the inflammatory response by reducing the levels of previously mentioned proinflammatory mediators, such as tumor necrosis factor-alpha, and interleukins, as well as oxidative stress. In addition, they have antibacterial and antiapoptotic activities, being capable of increasing the survival rate by over 50% according to a meta-analysis (De Oliveira Rodrigues Junior et al., 2023).

Among the mentioned flavonoids, luteolin has demonstrated the capability to reduce inflammation and oxidative stress, modulate the immune response, and prevent organ damage during sepsis (Vajdi et al., 2023). This flavone can downregulate the gene expression of proinflammatory cytokines, such as Nod-3 receptor protein and the nuclear factor kappa B. Additionally, luteolin decreases the hyperactivity of macrophages, neutrophils, and lymphocytes.

Another pharmacological activity traditionally studied in flavonoids is neuroprotection, although there is a smaller number of clinical trials and meta-analyses on this topic. One of the most studied flavonoids for this activity is Epigallocatechin-3-O-gallate. This catechin has demonstrated a positive effect in Alzheimer's disease (Zhang et al., 2020). It has shown the ability to reduce the level of Aβ42, regulate the activity of α-, β-, γ-secretases, inhibit tau phosphorylation and the activity of the enzyme AChE, and possess antioxidant, antiinflammatory, and antiapoptotic capabilities.

Therefore foods rich in flavonoids, including fruits, vegetables, seeds, flowers, tea, cocoa, and wine, or extracts rich in these bioactive compounds, may play a crucial role in maintaining human health and preventing and treating various diseases. Nevertheless, these results are constrained by trial heterogeneity, and further human studies with improved conditions and larger sample sizes are needed. Additionally, due to the wide array of flavonoids, the individual active principles of each subgroup and their therapeutic potential for each health issue should be analyzed separately.

Anthocyanins

Anthocyanins are polyphenols responsible for the red to blue-purple colors in fruits and plants and can be considered as a subgroup within flavonoids. The most representative sources of anthocyanins are quite diverse, ranging from elderberry, aronia, and bilberries to black grapes and cherries (Kozłowska & Szostak-Wegierek, 2014). Around 500 anthocyanins and 23 anthocyanidins (as they mainly exist as glycosides) have been described but the most explored are identified in Fig. 1.3: cyanidin, delphinidin,

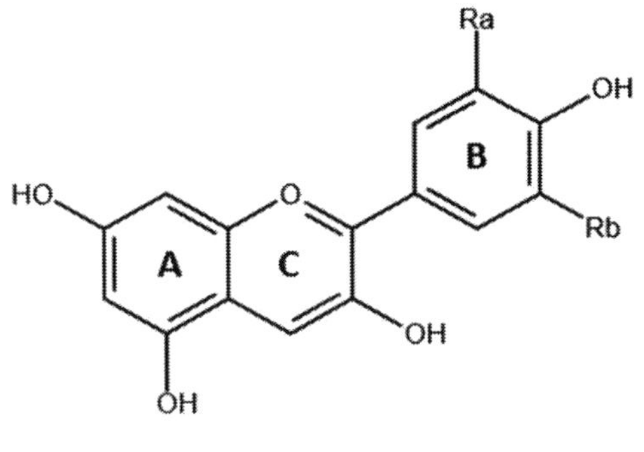

Name	Ra	Rb
Pelargonidin (Pg)	H	H
Cyanidin (Cy)	OH	H
Delphinidin (De)	OH	OH
Peonidin (Pn)	OCH_3	H
Petunidin (Pt)	OH	OCH_3
Malvidin (Ma)	OCH_3	OCH_3

Figure 1.3 Anthocyanidins skeleton and most relevant examples.

malvidin, pelargonidin, peonidin, and petunidin (Castañeda-Ovando et al., 2009; He & Giusti, 2010; Khoo et al., 2017). The distribution of these anthocyanidins in fruits and vegetables is 50%, 12%, 7%, 12%, 12%, and 7%, respectively (Khoo et al., 2017). The sugar moieties mostly connect to the 3-position on the C-ring or the 5, 7-position on the A-ring. These monosaccharides involve xylose (Xyl), galactose (Gal), glucose (Glu), arabinose (Ara), rhamnose (Rha), or glucuronic acid (GlcA) (Li et al., 2013).

Anthocyanins have significant content in promoting lifelong health maintenance due to their composition of secondary metabolites. Traditionally, high-content anthocyanin plants have been used for their properties on venous insufficiency, gut disorders, diabetes, wounds and inflammation, and urinary tract infections (EM, 2014; Chew et al., 2019; De Mello et al., 2017; Di Martino et al., 2006; Habanova et al., 2016). Cyanidin, peonidin, delphinidin, petunidin, and malvidin decreased the plasma concentrations of several NF-kB-regulated pro-inflammatory chemokines and immunoregulatory cytokines (Karlsen et al., 2007). For the last 15years, there has been an increase interest in this research area, as the average publications related to anthocyanins in PubMed augmented more than 5five times (390 manuscripts in 2009 compared to 1728 in 2023). In this way, new insights are focused on the benefits of anthocyanins in human health.

Besides traditional uses, last publications confirmed that anthocyanins have an essential role the regulation of other pathologies such as cardiovascular and neurodegenerative diseases, visual and immune system disorders, and cancer as they are capable of modulating oxidative stress. The field of oxidative stress research spans various disciplines (cell biology, physiology, chemistry) and plays a crucial role in understanding health, disease as mentioned above, and the importance of antioxidant enzymes in maintaining redox balance (Balsano & Alisi, 2009; Sies, 2015).

Anthocyanins have earned recent attention from researchers for their potential preventive and therapeutic effects on human health, primarily attributed to their in vitro antioxidant activity. Anthocyanins are considered real free radical scavengers due to their capacity to counteract ROS by two different mechanisms: hydrogen atom transfer (HAT) and single electron transfer (SET) (Fukumoto & Mazza, 2000; Jhin & Hwang, 2014). Various methods have been employed to evaluate this activity, which is influenced by the molecular structure of anthocyanins in cellular, animal, or in vivo models (Pojer et al., 2013; Wallace & Monica Giusti, 2015).

Some clinical trials indicate that anthocyanins not only have an antioxidant effect but have also been shown antiinflammatory results in blood biomarkers related to CVDs (Nemzer et al., 2011). The ANTHONIA research consortium conducted a study involving 30 healthy women who consumed anthocyanin-rich beverages. The participants were divided into three groups and consumed different drinks for 14 days. The study found that the enzymatic activities of plasma SOD and catalase significantly increased in the groups that consumed anthocyanin-rich beverages, but not in the placebo group. However, other enzymes and antiinflammatory parameters were not affected by the intervention (Kuntz et al., 2014). A randomized, double-blinded, placebo controlled, crossover trial demonstrated that, the consumption of a full anthocyanin juice blend (cyanidin 3-rutoside, cyanidin 3-diglycoside, and cyanidin 3-glucoside) in 12 healthy subjects, rapid increase in serum antioxidants and exerted protecting effect in cells from oxidative damage. Juice administration also resulted in a statistically significant decrease in serum lipid peroxidation within 2 hours of consumption (Jensen et al., 2008). One of the most popular sources of anthocyanins is blueberries and its consumption during six weeks in eighteen male participants reduced the levels of oxidized DNA bases and increased the resistance to oxidatively induced DNA damage (Riso et al., 2013). On the other hand, American researchers examined the combined effect of cyanidin and delphinidin on postprandial dysmetabolism and inflammation in healthy individuals consuming high-fat meal. This extract reduced endotoxemia, glucose and triglyceride levels, and inflammatory markers, demonstrating its potential benefits in mitigating the negative effects of unhealthy diets (Cremonini et al., 2022). Following these cardiovascular protective effects, a purified anthocyanin supplement was able to improve HDL-cholesterol levels in prehypertensive and non-dyslipidemic men (Hassellund et al., 2013). While regular consumption of bilberries was studied to determine its effects on CVD risk reduction. The research involved women and men who consumed frozen bilberries three times a week for 6 weeks. The study found that the consumption of bilberries led to a decrease in parameters such as total cholesterol, LDL-C, triglycerides, glucose, albumin, and γ-glutamyltransferase, while increasing HDL-C. The findings suggest that regular intake of bilberries can be important in reducing the risk of CVDs by improving lipid profile and increasing HDL-C levels (Habanova et al., 2016). Patients who suffered from acute myocardial infarction (AMI) were also selected to test anthocyanin potential benefits. Treatment group received standard medical therapy combined to bilberry powder. In comparison to the control group (standard medical therapy), the intervention group demonstrated a reduction in oxidized low-density lipoprotein after 8 weeks supplementation (Arevström et al., 2019). The effects of purified anthocyanins (320 mg/day) on platelet granule secretion in individuals with hypercholesterolemia were investigated in a randomized, double-blind clinical trial with 150 hypercholesterolemic participants for 24 weeks (Song et al., 2014). The consumption of anthocyanins resulted in significant reductions in plasma levels of β-thromboglobulin (β-TG), soluble P-selectin, and Regulated on Activation Normal T cell Expressed and Secreted (RANTES) compared to the placebo. These findings provide evidence that anthocyanins, by inhibiting platelet granule secretion, have the potential to be a potent cardioprotective agent.

Diabetes is another cardiovascular disorder that can be ameliorated by anthocyanin consumption as de Mello et al. proposed (de Mello et al., 2017). In this study, fifteen Finnish participants at high risk of developing type 2 diabetes received a rich bilberry diet which increased fasting serum hippuric acid. Authors associated this improvement with a better glycaemic control and β-cell function. A delphinidin-rich maqui berry extract commercialized as Delphinol was tested as a potential antidiabetic agent. Three different clinical trials demonstrated that Delphinol significantly and dose dependently lowered basal glycemia and insulinemia, controlled post prandial blood glucose and lowered blood glucose, and improved blood lipid profile in prediabetic individuals (Alvarado Schoenlau et al., 2016; Alvarado Leschot et al., 2016; Hidalgo et al., 2014).

Anthocyanins from blueberries have been associated with the improvement of visual disorders as an adjuvant treatment in age-related macular degeneration (AMD), eye cataracts, glaucoma and eye fatigue. 43 participants were supplemented with an anthocyanin-rich bilberry extract for 8 weeks. Intervention group obtained better results in objective (functional visual acuity, keratoconjunctival epithelial damage, and fluorescein tear film break-up time) and subjective (evaluated by a questionnaire) parameters than placebo group in eye fatigue clinical trial (Ozawa et al., 2015). Another related comparison trial established that oral consumption of 240 mg standardized bilberry extract for 12 weeks alleviates the tonic accommodation of the ciliary muscle that is caused by tasks involving visual display terminals (VDT) and near vision tasks (Kosehira et al., 2020). In a study carried out in 332 patients with normal tension glaucoma, the administration of 60 mg of anthocyanins ($n = 132$) resulted in the improvement of some parameters such as the Humphrey Visual Field (HVF) test, the logarithm of the minimum angle resolution of corrected visual acuity (logMAR BCVA) and intraocular pressure compared to placebo group values. This means that long-term use of anthocyanins for at least a year can help improve visual function in glaucoma patients (Shim et al., 2012).

Based on the clinical trials carried out to date, it can be concluded that consuming foods or supplements rich in anthocyanins show promising potential as an adjuvant and/or preventive therapeutic tool in various ocular pathologies (visual acuity, intraocular pressure, and ciliary muscle function) and CVDs (diabetes, hypercholesterolemia, AMI, thrombosis). Additionally, an increase in antioxidant enzymes and a reduction in inflammation have been observed in response to anthocyanin consumption. However, it is important to note that more research is required to fully understand the mechanisms of action and determine appropriate doses to obtain maximum therapeutic benefits. Overall, the results obtained so far support the potential role of anthocyanins in the treatment and prevention of various diseases, although more studies are needed to confirm and expand these findings.

Tannins

Historically, the term tannin comes from the English word tanning, which denotes the method of using different organic substances to change animal skins into leather (Khanbabaee & Van Ree, 2001).

Chemically, tannins are polymeric forms derived from phenolic, non-nitrogenous, water soluble compounds found in food products and plant materials, typically obtained from plants and algae (Sharifi-Rad et al., 2022). These substances harbor medicinal properties with potential for nutraceutical intervention in treating diseases (Moilanen et al., 2016). Numerous recent studies have demonstrated the medicinal potential of these compounds in managing chronic and degenerative illnesses, owing to their antioxidants, antimicrobial, antimutagenic, and antiproliferative properties, among others (Koleckar et al., 2008; Olivas-Aguirre et al., 2015; Sharifi-Rad et al., 2022). The physiological functions are attributed to conjugated metabolites that interact with genes and enzymes, which modulate intracellular signaling pathways involved in cell growth, multiplication, and apoptosis. Moreover, they exhibit antioxidant and antiinflammatory responses (Olivas-Aguirre et al., 2015; Quero et al., 2020), offering a multitude of health benefits to consumers.

Tannins are classified based on their structure into four two main groups: ***hydrolyzable tannins*** (HT; gallotannins, ellagitannins, complex tannins) and **condensed tannins (CT)** (Khanbabaee & Van Ree, 2001).

Gallotannins are those tannins in which galloyl units or their metadespid derivatives are linked to various polyol, catechin units, or triterpenoids. Ellagitannins are those tannins in which at least two of the galloyl units are C-C coupled to each other and do not contain a glycosidically bound catechin unit. The condensed tannins are all oligomeric and polymeric. Proanthocyanidins are formed by linking the C-4 of one catechin to the C-8 or C-6 of the next monomeric catechin. Complex tannins are those in which a catechin unit is glycosidically linked to a gallotannin or ellagitannin unit so they can be considered derived from gallic or ellagic acids (Khanbabaee & Van Ree, 2001).

Gallotanins (gallic acid base) and ellagitanins (ellagic acid base) are known as HT due to their ability to be easily hydrolyzed in the presence of acids, bases, or enzymes both in vitro and in vivo. These compounds are assembled into polymeric structures of high molecular weight with different binding capacities to proteins or other macromolecules (Olivas-Aguirre et al., 2015).

Hydrolyzable tannins

HTs have generated great interest in the scientific community, being natural compounds derived from plants, seeds, fruits, and vegetables. They exhibit multiple biological activities that are beneficial to the health of the organisms that consume them. Their high antioxidant and free radical scavenging capacity is the main reason for the study of these compounds in the field of health, due to their possible effects in the prevention of oxidation related diseases (Larrosa et al., 2010). In recent years, the generation and application of new knowledge about these compounds have been promoted, describing multiple beneficial effects in the treatment of various diseases (Olivas-Aguirre et al., 2015).

One of the relevant aspects of the antiinflammatory activity of products rich in HT is their modulating effect on the composition of the intestinal microbiota. Studies suggest that extracts and juices of *Punica granatum* exhibit beneficial mechanisms in the composition of the microbiota, improving the symptoms of intestinal inflammation (Kiss & Piwowarski, 2018). Elagitannins (ETs) have also been shown in recent research to have functional capacity for cardiovascular risk, with relevant antiatherogenic, antithrombotic, antiinflammatory, and antiangiogenic effects (Larrosa et al., 2010). They also have vasodilatory, bactericidal, hypolipidemic, hypoglycemic capacity and an effect on improving muscle strength (Koleckar et al., 2008; Sharifi-Rad et al., 2022). Supporting the benefits of its application in the treatment of chronic degenerative diseases and vascular (Koleckar et al., 2008). The mechanisms by which these effects are obtained are still being studied but are attributed to the oxidative capacity of their hydrolysis and bioconversion products (Olivas-Aguirre et al., 2015; Sharifi-Rad et al., 2022). Other research describes the involvement of these molecules in systemic health. The ellagitannins punicalagin, chebulinic acid, geraniin, and corillagin from the fruits of *Terminalia chebula* have been described as potential natural inhibitors of acetylcholinesterase for the treatment of Alzheimer's disease (Chen et al., 2023). Melastomadodecandrum (elagitanino) is also used clinically for the treatment of metrorrhagic bleeding (Lu et al., 2023). The favorable results of various studies highlight its effective potential in abnormal uterine bleeding, promoting the use of this natural agent in the treatment of gynecological hemorrhagic diseases (Lu et al., 2023).

On the other hand, galitanins (GT) propose a potentially beneficial effect in preventing and treating various cancers. These anti-cancer and antimutagenic capabilities may be attributed to the compound's antioxidant properties (Chung et al., 1998). Recent in vitro and in vivo studies suggest that pentagalloyl glucose (a hydrolyzable gallotan) has demonstrated cytotoxic effects on cancers including those of the prostate, breast, lung, head and neck, liver, leukemia, cervix, colorectal, and pancreas (Wen et al., 2023). This compound can inhibit tumor growth by modulating the transduction of signals responsible for controlling cell growth, division, apoptosis, angiogenesis, and metastasis (Wen et al., 2023). Additionally, recent studies have demonstrated the antimicrobial properties of gallotanins against a variety of gram-positive and gram-negative bacterial strains, such as those extracted from *Galla chinensis* (gallotannins) that show antimicrobial activity against *Salmonella typhimurium* and *Bacillus cereus* (Fraga-Corral et al., 2020).

HTs have predominantly been utilized in natural traditional medicine. Their abundance in nature has facilitated their use in various historical contexts, leading to their subsequent modification and current application in the medical, pharmaceutical, and veterinary fields (Fraga-Corral et al., 2020). The bioactivities of tannins are considered by the European Medicines Agency (EMA), who permit the use of extracts that contain particular compounds, like those extracted from *Myricaria bracteata*. These extracts possess very significant antiinflammatory activity, which is useful for therapeutic purposes in humans and animals suffering from ear edema and arthritis (Fraga-Corral et al., 2020).

Condensed tannins

Condensed tannins are the most widely found in nature and represent an important component of the human diet, they have functional capacity as physiological antioxidants, preventing oxidative damage (Gourlay et al., 2022; Koleckar et al., 2008). In recent decades, different studies have corroborated the direct relationship between the consumption of these substances and the health benefits (Zeng et al., 2020). They can be found in nuts, fruits, seeds, stems, and bark of a wide variety of tree species such as mimosa, quebracho, and oak, as well as in the leaves, branches, and stems of some legumes. High levels of these tannins have also been reported in cocoa, grape seeds, bananas, apples, and persimmons (Panzella & Napolitano, 2022).

Condensed tannins are also called proanthocyanidins because they can release anthocyanidins upon depolymerization, which occurs only under strongly acidic conditions (Olivas-Aguirre et al., 2015). The three representative groups of condensed tannins are procyanidins, proanthocyanidins, prodelphinidins, and profisetinines. These phytochemicals are considered beneficial to human health and exhibit various pharmacological properties. Its antioxidant, anticancer, antidiabetic, neuroprotective, and antimicrobial effects have been validated (Rauf et al., 2019). In recent years, in vitro and in vivo studies have been carried out, reporting an important antiparasitic efficacy, postulating itself as a therapeutic option to reduce the costs of gastrointestinal nematode infections (Greiffer et al., 2022). Recent studies have shown that grape seed proanthocyanidin extract (GSPE) has an antiobesity effect (Liu et al., 2020). Oligomeric proanthocyanidins (OPCs) and punicalagin have also been found to be potent inhibitors against severe acute respiratory syndrome coronavirus 2 (SARS-CoV-2) infection with different mechanisms. showing inhibitory activity against omicron variants of SARS-CoV-2 infection (Chen et al., 2023). Other epidemiological research has linked the consumption of these molecules to a reduction in the occurrence of some diseases such as hypertension and CVDs. The European Prospective Investigation into Cancer and Nutrition (EPIC) study, conducted in eight European countries, revealed an inverse association between PsA intake and the incidence of type 2 diabetes (Bladé et al., 2016).

There are multiple effects generally attributed to tannins that over time have consolidated solid evidence-based knowledge, recognizing its therapeutic and biological applications within medical practice. However, the application of tannins within the clinic is limited due to particular characteristics of each compound such as solubility, bioavailability, half life, lack of precision of some mechanisms of action in the human body that affect the effectiveness in pharmacological delivery systems (Chung et al., 1998). There are currently multiple investigations working to improve this problem, including new forms of administration with the integration of some types of organic and metallic nanoparticles.

Coumarins

Coumarins are phenolic compounds derived from cinnamic acid known as colorless benzopyrones found in edible, medicinal plants, fungi, bacteria, and species (Di Stasi, 2021). They can be classified into simple coumarins, simple prenylates, simple geranilates, furan, pyranus, sesquiterpenyl, and oligomeric (Sarker & Nahar, 2017).

Coumarins are of great interest in the area of phytochemicals, especially due to their structural diversity and medicinal properties (Sarker & Nahar, 2017). Coumarins and their derivatives exert a wide range of bioactive properties such as anticoagulants, analgesics, antibacterial, antiinflammatory, antioxidant, antitumor, antiviral, and enzyme inhibitors whose beneficial action is attributed to free radical scavenging effects (Garg et al., 2020; Sarker & Nahar, 2017). Clinical trials with coumarin compounds have shown antitumor activity against several types of cancer, including breast cancer, lung cancer, malignant melanoma, prostate cancer, and metastatic renal cell carcinoma (Venkata Sairam et al., 2016).

Research has shown that coumarins can be used as drugs for various diseases, mainly in those in which they are needed to reduce edema and inflammation, the effect of which is by inhibiting the biosynthesis of prostaglandins (Garg et al., 2020; Rostom et al., 2022). Especially umbelyferone and esculetin how antioxidant properties and protect cellular DNA from oxidative damage (Cai & Cai, 2023; Garg et al., 2020). There are also coumarin derivatives reported as intestinal antiinflammatory products and their available pharmacodynamic data supporting the protective effects of these products in the acute and subchronic phase of intestinal inflammation (Garg et al., 2020; Di Stasi, 2021).

Another coumarin derivative is esculetin, which has been widely used not only in expectorant and antitussive aspects but also as antiinflammatory, antioxidant, antibacterial, and antitumor therapies. Research conducted on rats/mice and cell lines reports antiinflammatory and antioxidant effects in models of induced gut inflammation (Garg et al., 2020; Venkata Sairam et al., 2016).

There are currently drugs with coumarin compounds approved by the FDA as drugs for use in clinical practice. Such is the case with warfarin (dicumarol). This drug is used as an anticoagulant due to its action of blocking the enzyme vitamin K reductase, thus altering the coagulation mechanism (Gao et al., 2021). Warfarin was discovered because when the plant species *Melilotus officinalis* was badly preserved the simple coumarin melilotoside, found in its aerial parts, can be transformed and dimerized into dicoumarol, a dimeric coumarin with anticoagulant properties. Another example is Microfleg ®, which contains a synergistic and complementary combination of bioflavonoids and coumarins, used as a food supplement to improve circulation (Garg et al., 2020). In this way, many coumarin derivatives are found in the pharmaceutical industry. Over time, sufficient evidence has been collected from both experimental and clinical studies to validate the beneficial effects of these compounds on different diseases (Flores-Morales et al., 2023). These lines of research are of interest to public health due to their potential health benefits. However, due to their potent bioactive molecules, it is important to apply and use these substances properly. Therefore its use is recommended only in preparations of pharmaceutical formulations (Garg et al., 2020).

Lignans

Lignans have received considerable attention in both chemical and pharmacological sciences, and in recent decades, they have played a significant role due to their complex structures and biological activities, especially their antiinflammatory and antitumor properties. They are natural compounds polymerized by two phenylpropanoid (C_6-C_3) derivatives and can be found in a wide variety of plants and seeds. Most of them are free in nature, while some exist as glucosides. Due to their chemical structure and compound bioavailability, along with advancements in extraction, separation, and compound identification technologies, lignans have become a crucial starting point in cancer drug development in the last decade (Mukhija et al., 2022). Lignans are also attributed with antioxidant effects as polyphenolic substances, aiding in combating the effects of harmful free radicals (Zhou et al., 2022).

Among the variety of plants species containing these compounds, a special mention should be for milk thistle (*Silybum marianum*), a plant native to Southern Europe, Southern Russia, and Asia, also adapted to other countries. It contains between 65% and 80% flavonolignans, primarily constituted by silimarin, which is a mixture of silibinins A and B (Fig. 1.1), silichristin and others like the flavonoid taxifolin. Nowadays, special formulations are developed considering the compound's bioavailability. Due to the lipophilic nature of silimarin, it is commercially sold in the form of capsules with oil carriers for better systemic bioavailability, rather than as an herbal tea, which is common with other types of plants (Camini & Costa, 2020).

Silymarin are not only antioxidant phenolic compounds but also possess a variety of beneficial properties for use in the treatment of various pathologies. They have hepatoprotective activities traditionally used in alternative medicine to treat conditions such as hepatitis, cirrhosis, jaundice, and as a remedy for snake, insect, fungus, and alcohol poisoning (Hariyanti et al., 2023; Mastron et al., 2014; Rambaldi et al., 2005). The pharmacokinetics of silymarin from a standardized product in both sick and healthy patients have been shown to be rapidly absorbed and eliminated (Abenavoli et al., 2018; Xie et al., n.d.). Other studies in healthy volunteers found high bioavailability of silymarin, metabolizing quickly into its conjugates (Calani et al., 2012).

Studies conducted with silymarin in humans have shown effects on non-alcoholic fatty liver, where a dose of 140 mg per day of orally administered silymarin significantly reduced transaminase levels compared to the placebo (Hajaghamohammadi et al., 2008; Hashemi et al., 2009). More recent studies confirmed the effectiveness of silymarin with a dose of 210 mg/day (Solhi et al., 2014). As silymarin is considered a hepatoprotective agent, it can also be used for the treatment of viral diseases, given that the liver is a site of viral replication. Additionally, it is antioxidant and antiinflammatory, suggesting that silymarin may reduce virus-related liver damage (Federico et al., 2017).

In addition, preclinical studies discussed further have also set the potential of silymarin in the management of different disorders. Silymarin was found to prevent damage to pancreatic beta cells, which are crucial for insulin synthesis, and at the same time, modulate glucose homeostasis (Malekinejad et al., 2012). Diabetic nephropathy, a condition developed in patients with diabetes, is characterized by impaired glomerular filtration barrier due to high blood glucose levels, inflammation, and oxidative stress. It is a major cause of end stage renal failure (Conserva et al., 2016). Animal models induced with diabetes have provided evidence of the beneficial effects of silymarin (Gong et al., 2020).

A study in mice using an ischemia-reperfusion kidney injury model, where they were treated with 100 mg/kg of silymarin for 7 days before induction, showed lower levels of serum creatinine and urea. Simultaneously, it attenuated damage to the renal system, tubular cells, and the number of apoptotic cells (Tan et al., 2015, 19C.E./198C.E.). Another study conducted with rats induced with type 2 diabetes using streptozotocin and nicotinamide received doses of 60 and 120 mg/kg of silymarin for 60 days. The mice showed reductions in blood glucose, glycated hemoglobin, albumin, serum creatinine, and uric acid (Nasri, 2013).

Finally, recent studies demonstrated that silymarin significantly reduces the viability and cell migration of human gastric cancer cells in a dose dependent manner after administration (Zhang et al., 2018). A study in mice with gastric cancer xenografts observed increased apoptosis in the tumor with a dose of 100 mg/kg, reducing tumor volume (Kim et al., 2019). Therefore silymarin could be an adjuvant or chemopreventive treatment for cancer due to its ability to induce apoptosis and inhibit tumor growth (Kim et al., 2019). Another study in nude mice with prostate cancer xenografts showed that dietary silymarin reduces advanced prostate cancer growth compared to the control group, associated with a strong and significant decrease in tumor cell proliferation, an increase in apoptosis, and a decrease in tumor microvessel density (Singh et al., 2003).

Stilbenes: resveratrol

Stilbenes are phenolic compounds that are part of the nature of many plants and foods we consume, such as blackberries, grapes, red wine, and some nuts. A well known compound of stilbenes is resveratrol, mostly found in red wine, which over the years has been associated with many health benefits due to its antioxidant, anticancer and cardiovascular protection activities (Lin et al., 2018; Wu et al., 2018; Zhao et al., 2019). Resveratrol has two phenol rings (monophenol and diphenol) joined by a styrene double bond and exists in both cis and trans isomeric forms (Fig. 1.1). Trans-resveratrol appears to be the most abundant and stable natural form (Gambini et al., 2015). Resveratrol is naturally synthesized by more than 70 plant species in response to stress, bacterial or fungal infections, and ultraviolet irradiation (Hasan & Bae, 2017).

The overproduction of ROS in the immune system is a major factor in oxidative stress, which damages macromolecules, altering their functions and forming the basis for many age-related diseases such as cancer, diabetes, chronic kidney disease, cardiovascular, and neurodegenerative diseases (Liguori et al., 2018; Lin et al., 2018). Different in vitro and in vivo studies have confirmed that resveratrol possess strong antioxidant activity and primarily inhibits excess of ROS. In fibroblasts exposed to rotenone, a plant-origin insecticide that causes psychomotor impairment, resveratrol showed a decrease in mitochondrial fragmentation, maintaining mitochondrial membrane potential and preventing attenuation of oxidative phosphorylation, thus acting as a protector against the harmful impact of ROS (Sgarbi et al., 2018).

In addition, resveratrol has been shown to decrease advanced glycation end products (AGEs) (Shen et al., 2017). The accumulation of these AGEs, resulting from the absence of enzymatic reactions between proteins and sugars, causes damage at the tissue and cellular levels, including lipid peroxidation, changes in protein structures, endothelial dysfunction, and stimulation of inappropriate cellular activity (Galiniak et al., 2019). Recent reports demonstrate that increased glycation in the human body increases oxidative stress, which is associated with a higher prevalence of breast cancer, diabetes, and degenerative diseases (Adamska et al., 2019; Pinkas & Aschner, 2016; Rhee & Kim, 2018). In line with this, administration of resveratrol in drinking water to rats treated with chronic methylglyoxal, significantly reduced the level of advanced protein oxidation products, AGEs, and protein carbonyl in plasma, as well as oxidative stress markers in the liver (Yılmaz et al., 2018).

Due to this antioxidant activity, resveratrol has been associated to anticancer properties. Resveratrol regulates proapoptotic protein expression, reducing the development of different types of cancer, such as pancreatic (Zhou et al., 2019), cervical (Zhao et al., 2019), gastric (Wu et al., 2018), and thyroid cancer (Zheng et al., 2018). Besides its anticancer properties, resveratrol was also observed to have beneficial effects in the management of diabetes. Patients with type II diabetes and diabetic nephropathy received resveratrol at a dose of 500 mg per day, undergoing a significant decrease in urine albumin and creatinine, along with an increase in serum antioxidant enzymes (Sattarinezhad et al., 2019). Because of its significant impact on human health as an antioxidant, resveratrol is available for sale as a dietary supplement. However, resveratrol is also formulated with other phytochemicals such as piperine (Johnson et al., 2011) and/or encapsulated through nanotechnology for controlled release and improvement of its bioavailability (Gambini et al., 2015; Pangeni et al., 2014; Summerlin et al., 2015).

Preclinical studies with animal models have shown that resveratrol also has an effect on neuronal damage and improvement in the functioning of the central nervous system. It reduces neurodegeneration in the cerebral cortex and enhances memory in a murine model exposed to fluoride (Sharma et al., 2018). A study conducted in mice with vascular dementia revealed that resveratrol administration improves cognition, learning, and memory (Ma et al., 2013). It has also been demonstrated that in mice with Alzheimer's, it enhances cognition and induces neuroprotection in amyloid pathologies (Corpas et al., 2019). There is also evidence that the neuroprotective effect can be an effective treatment for depression, chronic neuropathic pain, and cerebral ischemia/reperfusion injury in a murine model, owing to the antioxidant and antiinflammatory properties of resveratrol (Hou et al., 2018; Moore et al., 2018; Tao et al., 2016; Xie et al., 2017).

Conclusions

As shown in this chapter, different in vitro, in vivo, and epidemiological studies have shown effective results in preventing and/or improving symptoms of inflammatory, oxidative, chronic degenerative diseases and metabolic. More research is needed due to inter-individual variability in most of these studies. However, it can be concluded that most of these studies support beneficial health effects. Particularly recommended is the promotion of these polyphenolic compounds and their intervention in the treatment of degenerative diseases through dietary supplementation rich in polyphenols, both as single agents and adjuvants of well established drugs.

Acknowledgments

The authors acknowledge The Ministerio de Ciencia e Innovación-Fondo Europeo de Desarrollo Regional (PID2022−136414OB-I00), Fondo Social Europeo-Gobierno de Aragón (B16-R1), Fondo Social Europeo de Desarrollo Rural-Gobierno de Aragón

(GCP2021002300) and CIBER Fisiopatología de la Obesidad y la Nutrición as an initiative of FEDER-ICCIII (CIBEROBN, CB06/03/1012) for financial support. Javier Quero was recipient of a contract from the Universidad de Zaragoza within Investigo Program subsidized by the European Economic Community (Plan Nacional de Recuperación, Transformación y Resiliencia). Marina Delgado-Machuca was recipient of a Programa de becas de Movilidad CUMex (Consorcio de Universidades Mexicanas)-AUIP 2023 fellowship.

References

Abenavoli, L., Izzo, A. A., Milić, N., Cicala, C., Santini, A., & Capasso, R. (2018). Milk thistle (Silybum marianum): A concise overview on its chemistry, pharmacological, and nutraceutical uses in liver diseases. *Phytotherapy Research, 32*(11), 2202–2213. Available from https://doi.org/10.1002/ptr.6171, http://onlinelibrary.wiley.com/journal/10.1002/(ISSN)1099-1573.

Adamska, A., Araszkiewicz, A., Pilacinski, S., Gandecka, A., Grzelka, A., Kowalska, K., Malinska, A., Nowicki, M., & Zozulinska-Ziolkiewicz, D. (2019). Dermal microvessel density and maturity is closely associated with atherogenic dyslipidemia and accumulation of advanced glycation end products in adult patients with type 1 diabetes. *Microvascular Research, 121*, 46–51. Available from https://doi.org/10.1016/j.mvr.2018.10.002.

Adolphe, J. L., Whiting, S. J., Juurlink, B. H. J., Thorpe, L. U., & Alcorn, J. (2010). Health effects with consumption of the flax lignan secoisolariciresinol diglucoside. *British Journal of Nutrition, 103*(7), 929–938. Available from https://doi.org/10.1017/S0007114509992753.

Alimohammadi, M., Mohammad, R. N., Rahimi, A., Faramarzi, F., Alizadeh-Navaei, R., & Rafiei, A. (2022). The effect of immunomodulatory properties of naringenin on the inhibition of inflammation and oxidative stress in autoimmune disease models: a systematic review and meta-analysis of preclinical evidence. *Inflammation Research, 71* (10-11), 1127–1142. Available from https://doi.org/10.1007/s00011-022-01599-7.

Alvarado, J., Schoenlau, F., Leschot, A., Salgado, A. M., & Portales, P. V. (2016). Delphinol® standardized maqui berry extract significantly lowers blood glucose and improves blood lipid profile in prediabetic individuals in three-month clinical trial. *Panminerva Medica, 58*(3), 1–6. Available from http://www.minervamedica.it/en/journals/panminerva-medica/archive.php.

Alvarado, J. L., Leschot, A., Olivera-Nappa, Á., Salgado, A. M., Rioseco, H., Lyon, C., & Vigil, P. (2016). Delphinidin-rich maqui berry extract (Delphinol®) lowers fasting and postprandial glycemia and insulinemia in prediabetic individuals during oral glucose tolerance tests. *BioMed Research International, 2016*. Available from https://doi.org/10.1155/2016/9070537, http://www.hindawi.com/journals/biomed/.

Arbonés-Mainar, J. M., Navarro, M. A., Lou-Bonafonte, J. M., Martínez-Gracia, M. V., & Osada, J. (2008). *Olive oil phenolics as potential therapeutical agents polyphenols and health: New and recent advances* (pp. 31–75). Spain: Nova Science Publishers, Inc. Available from https://www.novapublishers.com/catalog/product_info.php?products_id = 6901.

Arevström, L., Bergh, C., Landberg, R., Wu, H., Rodriguez-Mateos, A., Waldenborg, M., Magnuson, A., Blanc, S., & Fröbert, O. (2019). Freeze-dried bilberry (*Vaccinium myrtillus*) dietary supplement improves walking distance and lipids after myocardial infarction: an open-label randomized clinical trial. *Nutrition Research, 62*, 13–22. Available from https://doi.org/10.1016/j.nutres.2018.11.008.

Asbaghi, O., Sadeghian, M., Nasiri, M., Khodadost, M., Shokri, A., Panahande, B., Pirouzi, A., & Sadeghi, O. (2020). The effects of green coffee extract supplementation on glycemic indices and lipid profile in adults: a systematic review and dose-response meta-analysis of clinical trials. *Nutrition Journal, 19*(1). Available from https://doi.org/10.1186/s12937-020-00587-z.

Asensi, M., Ortega, A., Mena, S., Feddi, F., & Estrela, J. M. (2011). Natural polyphenols in cancer therapy. *Critical Reviews in Clinical Laboratory Sciences, 48*(5-6), 197–216. Available from https://doi.org/10.3109/10408363.2011.631268.

Balsano, C., & Alisi, A. (2009). Antioxidant effects of natural bioactive compounds. *Current Pharmaceutical Design, 15*(26), 3063–3073. Available from https://doi.org/10.2174/138161209789058084.

Basu, P., & Maier, C. (2018). Phytoestrogens and breast cancer: In vitro anticancer activities of isoflavones, lignans, coumestans, stilbenes and their analogs and derivatives. *Biomedicine & Pharmacotherapy, 107*, 1648–1666. Available from https://doi.org/10.1016/j.biopha.2018.08.100.

Bladé, C., Aragonès, G., Arola-Arnal, A., Muguerza, B., Bravo, F. I., Salvadó, M. J., Arola, L., & Suárez, M. (2016). Proanthocyanidins in health and disease. *BioFactors, 42*(1), 5–12. Available from https://doi.org/10.1002/biof.1249, http://onlinelibrary.wiley.com/journal/10.1002/(ISSN)1872-8081.

Boutas, I., Kontogeorgi, A., Dimitrakakis, C., & Kalantaridou, S. N. (2022). Soy isoflavones and breast cancer risk: A meta-analysis. *In Vivo, 36*(2), 556–562. Available from https://doi.org/10.21873/INVIVO.12737, http://iv.iiarjournals.org/lookup/doi/10.21873/invivo.12737.

Bule, M., Abdurahman, A., Nikfar, S., Abdollahi, M., & Amini, M. (2019). Antidiabetic effect of quercetin: A systematic review and meta-analysis of animal studies. *Food and Chemical Toxicology, 125*, 494–502. Available from https://doi.org/10.1016/j.fct.2019.01.037.

Cai, T., & Cai, B. (2023). Pharmacological activities of esculin and esculetin: A review. *Medicine, 102*(40), e35306. Available from https://doi.org/10.1097/md.0000000000035306.

Calani, L., Brighenti, F., Bruni, R., & Del Rio, D. (2012). Absorption and metabolism of milk thistle flavanolignans in humans. *Phytomedicine, 20*(1), 40–46. Available from https://doi.org/10.1016/j.phymed.2012.09.004.

Camini, F. C., & Costa, D. C. (2020). Silymarin: Not just another antioxidant. *Journal of Basic and Clinical Physiology and Pharmacology, 31*(4). Available from https://doi.org/10.1515/jbcpp-2019-0206, http://www.reference-global.com/loi/jbcpp.

Caponio, G. R., Cofano, M., Lippolis, T., Gigante, I., De Nunzio, V., Difonzo, G., Noviello, M., Tarricone, L., Gambacorta, G., Giannelli, G., De Angelis, M., & Notarnicola, M. (2020). Anti-proliferative and pro-apoptotic effects of digested aglianico grape pomace extract in human colorectal cancer cells. *Molecules, 27*(20), 6791. Available from https://doi.org/10.3390/molecules27206791.

Caponio, G. R., Lorusso, M. P., Sorrenti, G. T., Marcotrigiano, V., Difonzo, G., De Angelis, E., Guagnano, R., Ciaula, A. D., Diella, G., Logrieco, A. F., Montagna, M. T., Monaci, L., De Angelis, M., & Portincasa, P. (2020). Chemical characterization, gastrointestinal motility and sensory evaluation of dark chocolate: A nutraceutical boosting consumers' health. *Nutrients, 12*(4). Available from https://doi.org/10.3390/nu12040939, https://www.mdpi.com/2072-6643/12/4/939/pdf.

Cara, K. C., Beauchesne, A. R., Wallace, T. C., & Chung, M. (2022). Effects of 100% orange juice on markers of inflammation and oxidation in healthy and at-risk adult populations: A scoping review. *Advances in Nutrition, 13*(1), 116–137. Available from https://doi.org/10.1093/advances/nmab101, http://advances.nutrition.org/.

Castañeda-Ovando, A., Pacheco-Hernández, M. D. L., Páez-Hernández, M. E., Rodríguez, J. A., & Galán-Vidal, C. A. (2009). Chemical studies of anthocyanins: A review. *Food Chemistry, 113*(4), 859–871. Available from https://doi.org/10.1016/j.foodchem.2008.09.001.

Chang, H., Lei, L., Zhou, Y., Ye, F., & Zhao, G. (2018). Dietary flavonoids and the risk of colorectal cancer: an updated meta-analysis of epidemiological studies. *Nutrients, 10*(7), 950. Available from https://doi.org/10.3390/nu10070950.

Chen, H. F., Wang, W. J., Chen, C. Y., Chang, W. C., Hsueh, P. R., Peng, S. L., Wu, C. S., Chen, Y., Huang, H. Y., Shen, W. J., Wang, S. C., & Hung, M. C. (2023). The natural tannins oligomeric proanthocyanidins and punicalagin are potent inhibitors of infection by SARS-CoV-2 in vitro. *Taiwan bioRxiv*. Available from https://doi.org/10.1101/2023.01.12.523465, https://www.biorxiv.org.

Chew, B., Mathison, B., Kimble, L., McKay, D., Kaspar, K., Khoo, C., Chen, C.-Y. O., & Blumberg, J. (2019). Chronic consumption of a low calorie, high polyphenol cranberry beverage attenuates inflammation and improves glucoregulation and HDL cholesterol in healthy overweight humans: a randomized controlled trial. *European Journal of Nutrition, 58*(3), 1223–1235. Available from https://doi.org/10.1007/s00394-018-1643-z.

Chikara, S., Nagaprashantha, L. D., Singhal, J., Horne, D., Awasthi, S., & Singhal, S. S. (2018). Oxidative stress and dietary phytochemicals: Role in cancer chemoprevention and treatment. *Cancer Letters, 413*, 122–134. Available from https://doi.org/10.1016/j.canlet.2017.11.002, http://www.elsevier.com/locate/canlet.

Chung, K. T., Wong, T. Y., Wei, C. I., Huang, Y. W., & Lin, Y. (1998). Tannins and human health: A review. *Critical Reviews in Food Science and Nutrition, 38*(6), 421–464. Available from https://doi.org/10.1080/10408699891274273, http://www.tandf.co.uk/journals/titles/10408398.asp.

Conserva, F., Gesualdo, L., & Papale, M. (2016). A systems biology overview on human diabetic nephropathy: From genetic susceptibility to post-transcriptional and post-translational modifications. *Journal of Diabetes Research, 2016*, 1–23. Available from https://doi.org/10.1155/2016/7934504.

Corpas, R., Griñán-Ferré, C., Rodríguez-Farré, E., Pallàs, M., & Sanfeliu, C. (2019). Resveratrol induces brain resilience against Alzheimer neurodegeneration through proteostasis enhancement. *Molecular Neurobiology, 56*(2), 1502–1516. Available from https://doi.org/10.1007/s12035-018-1157-y.

Costea, T., Hudiţă, A., Ciolac, O. A., Gălăţeanu, B., Ginghină, O., Costache, M., Ganea, C., & Mocanu, M. M. (2018). Chemoprevention of colorectal cancer by dietary compounds. MDPI AG, Romania. *International Journal of Molecular Sciences, 19*(12). Available from https://doi.org/10.3390/ijms19123787, https://www.mdpi.com/1422-0067/19/12/3787/pdf.

Crascì, L., Lauro, M. R., Puglisi, G., & Panico, A. (2018). Natural antioxidant polyphenols on inflammation management: Anti-glycation activity vs metalloproteinases inhibition. Taylor and Francis Inc., Italy. *Critical Reviews in Food Science and Nutrition, 58*(6), 893–904. Available from https://doi.org/10.1080/10408398.2016.1229657, http://www.tandf.co.uk/journals/titles/10408398.asp.

Cremonini, E., Daveri, E., Iglesias, D. E., Kang, J., Wang, Z., Gray, R., Mastaloudis, A., Kay, C. D., Hester, S. N., Wood, S. M., Fraga, C. G., & Oteiza, P. I. (2022). A randomized placebo-controlled cross-over study on the effects of anthocyanins on inflammatory and metabolic responses to a high-fat meal in healthy subjects. *Redox Biology, 51*, 102273. Available from https://doi.org/10.1016/j.redox.2022.102273.

Crozier, A., Jaganath, I. B., & Clifford, M. N. (2009). Dietary phenolics: Chemistry, bioavailability and effects on health. *Natural Product Reports, 26*(8), 1001–1043. Available from https://doi.org/10.1039/b802662a.

Cui., Liu, X., Tian, Y., Xie, C., Li, Q., Cui, H., et al. (2016). Flavonoid Subclasses, and Esophageal Cancer Risk: A Meta-Analysis of Epidemiologic Studies. *Nutrients, 8*(6), l.

EM (2014). European Union herbal monograph on *Vaccinium myrtillus* L., fructus recens., 2014.

Falcone Ferreyra, M. L., Rius, S. P., & Casati, P. (2012). Flavonoids: Biosynthesis, biological functions, and biotechnological applications. Frontiers Research Foundation, Argentina. *Frontiers in Plant Science*, *3*. Available from https://doi.org/10.3389/fpls.2012.00222, http://journal.frontiersin.org/Journal/10.3389/fpls.2012.00222/pdf.

Federico, A., Dallio, M., & Loguercio, C. (2017). Silymarin/Silybin and chronic liver disease: A marriage of many years. *Molecules*, *22*(2), 191. Available from https://doi.org/10.3390/molecules22020191.

Fischer, N., Seo, E. J., & Efferth, T. (2018). Prevention from radiation damage by natural products. Elsevier GmbH, Germany. *Phytomedicine*, *47*, 192–200. Available from https://doi.org/10.1016/j.phymed.2017.11.005, http://www.urbanfischer.de/journals/phytomed.

Flores-Morales, V., Villasana-Ruíz, A. P., Garza-Veloz, I., González-Delgado, S., & Martinez-Fierro, M. L. (2023). Therapeutic effects of coumarins with different substitution patterns. MDPI, Mexico. *Molecules*, *28*(5). Available from https://doi.org/10.3390/molecules28052413, http://www.mdpi.com/journal/molecules.

Fraga-Corral, M., García-Oliveira, P., Pereira, A. G., Lourenço-Lopes, C., Jimenez-Lopez, C., Prieto, M. A., & Simal-Gandara, J. (2020). Technological application of tannin-based extracts. *Molecules*, *25*(3). Available from https://doi.org/10.3390/molecules25030614, https://www.mdpi.com/1420-3049/25/3/614/pdf.

Fukumoto, L. R., & Mazza, G. (2000). Assessing antioxidant and prooxidant activities of phenolic compounds. *Journal of Agricultural and Food Chemistry*, *48*(8), 3597–3604. Available from https://doi.org/10.1021/jf000220w.

Galiniak, S., Aebisher, D., & Bartusik-Aebisher, D. (2019). Health benefits of resveratrol administration. *Poland Acta Biochimica Polonica*, *66*(1), 13–21. Available from https://doi.org/10.18388/abp.2018_2749, https://ojs.ptbioch.edu.pl/index.php/abp/article/download/2749/1734/.

Gambini, J., Inglés, M., Olaso, G., Lopez-Grueso, R., Bonet-Costa, V., Gimeno-Mallench, L., Mas-Bargues, C., Abdelaziz, K. M., Gomez-Cabrera, M. C., Vina, J., & Borras, C. (2015). Properties of resveratrol: In vitro and in vivo studies about metabolism, bioavailability, and biological effects in animal models and humans. *Oxidative Medicine and Cellular Longevity*, *2015*, 1–13. Available from https://doi.org/10.1155/2015/837042.

Gao, L., Wang, F., Chen, Y., Li, F., Han, B., & Liu, D. (2021). The antithrombotic activity of natural and synthetic coumarins. *Fitoterapia*, *154*, 104947. Available from https://doi.org/10.1016/j.fitote.2021.104947.

García-Lafuente, A., Guillamón, E., Villares, A., Rostagno, M. A., & Martínez, J. A. (2009). Flavonoids as anti-inflammatory agents: Implications in cancer and cardiovascular disease. *Inflammation Research*, *58*(9), 537–552. Available from https://doi.org/10.1007/s00011-009-0037-3.

Garg, S. S., Gupta, J., Sharma, S., & Sahu, D. (2020). An insight into the therapeutic applications of coumarin compounds and their mechanisms of action. *European Journal of Pharmaceutical Sciences*, *152*, 105424. Available from https://doi.org/10.1016/j.ejps.2020.105424.

Gong, L., Guo, S., & Zou, Z. (2020). Resveratrol ameliorates metabolic disorders and insulin resistance in high-fat diet-fed mice. *Life Sciences*, *242*, 117212. Available from https://doi.org/10.1016/j.lfs.2019.117212.

Gourlay, G., Hawkins, B. J., Albert, A., Schnitzler, J. P., & Peter Constabel, C. (2022). Condensed tannins as antioxidants that protect poplar against oxidative stress from drought and UV-B. *Canada Plant Cell and Environment*, *45*(2),

362–377. Available from https://doi.org/10.1111/pce.14242, http://onlinelibrary.wiley.com/journal/10.1111/(ISSN)1365-3040.

Greiffer, L., Liebau, E., Herrmann, F. C., & Spiegler, V. (2022). Condensed tannins act as anthelmintics by increasing the rigidity of the nematode cuticle. *Scientific Reports*, *12*(1). Available from https://doi.org/10.1038/s41598-022-23566-2, https://www.nature.com/srep/.

Guo, J. J., Hsieh, H. Y., & Hu, C. H. (2009). Chain-breaking activity of carotenes in lipid peroxidation: A theoretical study. *Journal of Physical Chemistry B*, *113*(47), 15699–15708. Available from https://doi.org/10.1021/jp907822h, http://pubs.acs.org/doi/pdfplus/10.1021/jp907822h.

Guo, K., Liang, Z., Liu, L., Li, F., & Wang, H. (2016). Flavonoids intake and risk of prostate cancer: a meta-analysis of observational studies. *Andrologia*, *48*(10), 1175–1182. Available from https://doi.org/10.1111/and.12556.

Guo, Y., Zhi, F., Chen, P., Zhao, K., Xiang, H., Mao, Q., Wang, X., & Zhang, X. (2017). Green tea and the risk of prostate cancer: A systematic review and meta-analysis. *Medicine*, *96*(13). Available from https://doi.org/10.1097/MD.0000000000006426, http://journals.lww.com/md-journal.

Gágyor, I., Hummers, E., Schmiemann, G., Friede, T., Pfeiffer, S., Afshar, K., & Bleidorn, J. (2021). Herbal treatment with uva ursi extract versus fosfomycin in women with uncomplicated urinary tract infection in primary care: a randomized controlled trial. *Clinical Microbiology and Infection*, *27*(10), 1441–1447. Available from https://doi.org/10.1016/j.cmi.2021.05.032.

Habanova, M., Saraiva, J. A., Haban, M., Schwarzova, M., Chlebo, P., Predna, L., Gažo, J., & Wyka, J. (2016). Intake of bilberries (*Vaccinium myrtillus* L.) reduced risk factors for cardiovascular disease by inducing favorable changes in lipoprotein profiles. *Nutrition Research*, *36*(12), 1415–1422. Available from https://doi.org/10.1016/j.nutres.2016.11.010, http://www.elsevier.com/locate/nutres.

Habli, Z., Toumieh, G., Fatfat, M., Rahal, O. N., & Gali-Muhtasib, H. (2017). Emerging cytotoxic alkaloids in the battle against cancer: Overview of molecular mechanisms. *Lebanon Molecules*, *22*(2). Available from https://doi.org/10.3390/molecules22020250, http://www.mdpi.com/1420-3049/22/2/250/pdf.

Hajaghamohammadi, A. A., Ziaee, A., & Rafiei, R. (2008). The efficacy of silymarin in decreasing transaminase activities in non-alcoholic fatty liver disease: A randomized controlled clinical trial. *Brieflands, Iran Hepatitis Monthly*, *8*(3), 191–195. Available from http://hepmon.com/pdf/The_Efficacy_of_Silymarin_in_Decreasing_.pdf.

Hardman, W. E. (2014). Diet components can suppress inflammation and reduce cancer risk. *Nutrition Research and Practice*, *8*(3), 233–240. Available from https://doi.org/10.4162/nrp.2014.8.3.233, http://www.ncbi.nlm.nih.gov/pmc/articles/PMC4058555/pdf/nrp-8-233.pdf.

Hariyanti, T., Margiana, R., Al-Gazally, M. E., Patra, I., Lateef Al-Awsi, G. R., Hameed, N., Kayumova, D., Ansari, M. J., Torres-Criollo, L. M., Mustafa, Y. F., Abedi-Firouzjah, R., & Farhood, B. (2023). The protective effects of silymarin on the reproductive toxicity: A comprehensive review. *Current Medicinal Chemistry*, *30*(39), 4421–4449. Available from https://doi.org/10.2174/0929867330666230130115332.

Hasan, M., & Bae, H. (2017). An overview of stress-induced resveratrol synthesis in grapes: Perspectives for resveratrol-enriched grape products. *Molecules*, *22*(2), 294. Available from https://doi.org/10.3390/molecules22020294.

Hashemi, S. J., Hajiani, E., & Sardabi, E. H. (2009). A placebo-controlled trial of silymarin in patients with nonalcoholic fatty liver disease. *Brieflands, Iran Hepatitis Monthly*, *9*(4), 265–270. Available from http://hepmon.com/pdf/A_Placebo-Controlled_Trial_of_Silymarin_.pdf.

Hassellund, S. S., Flaa, A., Kjeldsen, S. E., Seljeflot, I., Karlsen, A., Erlund, I., & Rostrup, M. (2013). Effects of anthocyanins on cardiovascular risk factors and inflammation in pre-hypertensive men: A double-blind randomized placebo-controlled crossover study. *Journal of Human Hypertension*, *27*(2), 100–106. Available from https://doi.org/10.1038/jhh.2012.4.

Havsteen, B. H. (2002). The biochemistry and medical significance of the flavonoids. Elsevier Inc., Germany. *Pharmacology and Therapeutics*, *96*(2-3), 67–202. Available from https://doi.org/10.1016/S0163-7258(02)00298-X, http://www.elsevier.com/locate/pharmthera.

He, J., & Giusti, M. M. (2010). Anthocyanins: Natural colorants with health-promoting properties. *Annual Review of Food Science and Technology*, *1*(1), 163–187. Available from https://doi.org/10.1146/annurev.food.080708.100754.

Herrera-Marcos, L. V., Lou-Bonafonte, J. M., Arnal, C., Navarro, M. A., & Osada, J. (2017). Transcriptomics and the mediterranean diet: A systematic review. *Nutrients*, *9* (5). Available from https://doi.org/10.3390/nu9050472, http://www.mdpi.com/2072-6643/9/5/472/pdf.

Hidalgo, J., Flores, C., Hidalgo, M. A., Perez, M., Yañez, A., Quiñones, L., Caceres, D. D., & Burgos, R. A. (2014). Delphinol® standardized maqui berry extract reduces postprandial blood glucose increase in individuals with impaired glucose regulation by novel mechanism of sodium glucose cotransporter inhibition. *Panminerva Medica*, *56*(2), 1–7.

Hou, Y., Wang, K., Wan, W., Cheng, Y., Pu, X., & Ye, X. (2018). Resveratrol provides neuroprotection by regulating the JAK2/STAT3/PI3K/AKT/mTOR pathway after stroke in rats. *Genes & Diseases*, *5*(3), 245–255. Available from https://doi.org/10.1016/j.gendis.2018.06.001.

Hu, T., Yue, J., Tang, Q., Cheng, K.-W., Chen, F., Peng, M., Zhou, Q., & Wang, M. (2022). The effect of quercetin on diabetic nephropathy (DN): A systematic review and meta-analysis of animal studies. *Food & Function*, *13*(9), 4789–4803. Available from https://doi.org/10.1039/d1fo03958j.

Jensen, G. S., Wu, X., Patterson, K. M., Barnes, J., Carter, S. G., Scherwitz, L., Beaman, R., Endres, J. R., & Schauss, A. G. (2008). In vitro and in vivo antioxidant and anti-inflammatory capacities of an antioxidant-rich fruit and berry juice blend. Results of a pilot and randomized, double-blinded, placebo-controlled, crossover study. *Journal of Agricultural and Food Chemistry*, *56*(18), 8326–8333. Available from https://doi.org/10.1021/jf8016157.

Jhin, C., & Hwang, K. T. (2014). Prediction of radical scavenging activities of anthocyanins applying adaptive Neuro-Fuzzy inference system (ANFIS) with quantum chemical descriptors. *International Journal of Molecular Sciences*, *15*(8), 14715–14727. Available from https://doi.org/10.3390/ijms150814715, http://www.mdpi.com/1422-0067/15/8/14715/pdf.

Jin, H., Leng, Q., & Li, C. (2012). Dietary flavonoid for preventing colorectal neoplasms. *Cochrane Database Syst Rev*, *8*.

Johnson, J. J., Nihal, M., Siddiqui, I. A., Scarlett, C. O., Bailey, H. H., Mukhtar, H., & Ahmad, N. (2011). Enhancing the bioavailability of resveratrol by combining it with piperine. *Molecular Nutrition & Food Research*, *55*(8), 1169–1176. Available from https://doi.org/10.1002/mnfr.201100117.

Kakkos, S. K., & Nicolaides, A. N. (2018). Efficacy of micronized purified flavonoid fraction (Daflon®) on improving individual symptoms, signs and quality of life in patients with chronic venous disease: A systematic review and meta-analysis of randomized double-blind placebo-controlled trials. Edizioni Minerva Medica, Greece. *International Angiology*, *37*(2), 143–154. Available from https://doi.org/10.23736/S0392-9590.18.03975-5, https://www.minervamedica.it/en/getfreepdf/SkXoJuSL%252F0VGyL8P0U9kIflEMLq%252B5C1hcC4DjIPoVi9Mo6opInlSIXIpSHbtUaJ60L47poeHhRkpDysYV00cAw%253D%253D/R34Y2018N02A0143.pdf.

Kanchanasurakit, S., Saokaew, S., Phisalprapa, P., & Duangjai, A. (2023). Chlorogenic acid in green bean coffee on body weight: A systematic review and meta-analysis of randomized controlled trials. *Systematic Reviews*, *12*(1). Available from https://doi.org/10.1186/s13643-023-02311-4.

Karlsen, A., Retterstøl, L., Laake, P., Paur, I., Kjølsrud-Bøhn, S., Sandvik, L., & Blomhoff, R. (2007). Anthocyanins inhibit nuclear factor-κB activation in monocytes and reduce plasma concentrations of pro-inflammatory mediators in healthy adults, 3. *The Journal of Nutrition*, *137*(8), 1951−1954. Available from https://doi.org/10.1093/jn/137.8.1951.

Khanbabaee, K., & Van Ree, T. (2001). Tannins: Classification and definition. *Natural Product Reports*, *18*(6), 641−649. Available from https://doi.org/10.1039/b101061l.

Khoo, H. E., Azlan, A., Tang, S. T., & Lim, S. M. (2017). Anthocyanidins and anthocyanins: Colored pigments as food, pharmaceutical ingredients, and the potential health benefits. *Food and Nutrition Research*, *61*. Available from https://doi.org/10.1080/16546628.2017.1361779, https://foodandnutritionresearch.net/index.php/fnr/article/download/1257/4592.

Kim, C., & Kim, B. (2018). Anti-cancer natural products and their bioactive compounds inducing ER stress-mediated apoptosis: A review. *Nutrients*, *10*(8), 1021. Available from https://doi.org/10.3390/nu10081021.

Kim, S. H., Choo, G. S., Yoo, E. S., Woo, J. S., Han, S. H., Lee, J. H., & Jung, J. Y. (2019). Silymarin induces inhibition of growth and apoptosis through modulation of the MAPK signaling pathway in AGS human gastric cancer cells. *Oncology Reports*, *42*(5), 1904−1914. Available from https://doi.org/10.3892/or.2019.7295, https://www.spandidos-publications.com/or/42/5/1904.

Kim, Y., & Je, Y. (2017). Flavonoid intake and mortality from cardiovascular disease and all causes: A meta-analysis of prospective cohort studies. *Clinical Nutrition ESPEN*, *20*, 68−77. Available from https://doi.org/10.1016/j.clnesp.2017.03.004.

Kiss, A. K., & Piwowarski, J. P. (2018). Ellagitannins, gallotannins and their metabolites-the contribution to the anti-inflammatory effect of food products and medicinal plants. *Current Medicinal Chemistry*, *25*(37), 4946−4967. Available from https://doi.org/10.2174/0929867323666160919111559, http://www.eurekaselect.com/CDN/download.php?param = Sk9V0Uk5pBTF2MvQEkVOgL0NBNQyq8yNcS8zjNy89wMDoAxQdy5wnZGZn8fG5FwcDGxptY2Fn0aW79uLh3Bk6Znx48ZDpEzNkDIypZTJhlZWuUwNh2RiENGZahNWgJhZmTJhqODkm3ZGlEzMc2MTcVYi&key = VWlxhRWleuZECUtQaVNEDc2RvZZknVJQ6TMyjRDM71Rj4c1NcDV2bLTQp3Mz8ItM5jM16NjQbzMjyNERs2VEbeATcVYlTcVY.

Koleckar, V., Kubikova, K., Rehakova, Z., Kuca, K., Jun, D., Jahodar, L., & Opletal, L. (2008). Condensed and hydrolysable tannins as antioxidants influencing the health. *Mini-Reviews in Medicinal Chemistry*, *8*(5), 436−447. Available from https://doi.org/10.2174/138955708784223486.

Kosehira, M., Machida, N., & Kitaichi, N. (2020). A 12-week-long intake of bilberry extract (*Vaccinium myrtillus* L.) improved objective findings of ciliary muscle contraction of the eye: A randomized, double-blind, placebo-controlled, parallel-group comparison trial. *Nutrients*, *12*(3), 600. Available from https://doi.org/10.3390/nu12030600.

Kowalczyk, A., Przychodna, M., Sopata, S., Bodalska, A., & Fecka, I. (2020). Thymol and thyme essential oil—New insights into selected therapeutic applications. *Molecules*, *25*(18), 4125. Available from https://doi.org/10.3390/molecules25184125.

Kozłowska, A., & Szostak-Wegierek, D. (2014). Flavonoids—food sources and health benefits. *Roczniki Państwowego Zakładu Higieny.*, *65*(2), 79—85.

Kuntz, S., Kunz, C., Herrmann, J., Borsch, C. H., Abel, G., Fröhling, B., Dietrich, H., & Rudloff, S. (2014). Anthocyanins from fruit juices improve the antioxidant status of healthy young female volunteers without affecting anti-inflammatory parameters: Results from the randomised, double-blind, placebo-controlled, cross-over ANTHONIA (ANTHOcyanins in Nutrition Investigation Alliance) study. *British Journal of Nutrition*, *112*(6), 925—936. Available from https://doi.org/10.1017/S0007114514001482, http://journals.cambridge.org/BJN.

Larrosa, M., García-Conesa, M. T., Espín, J. C., & Tomás-Barberán, F. A. (2010). Ellagitannins, ellagic acid and vascular health. *Molecular Aspects of Medicine*, *31*(6), 513—539. Available from https://doi.org/10.1016/j.mam.2010.09.005.

León-González, A. J., Auger, C., & Schini-Kerth, V. B. (2015). Pro-oxidant activity of polyphenols and its implication on cancer chemoprevention and chemotherapy. *Biochemical Pharmacology*, *98*(3), 371—380. Available from https://doi.org/10.1016/j.bcp.2015.07.017, http://www.elsevier.com/locate/biochempharm.

Li, X., Ma, H., Huang, H., Li, D., & Yao, S. (2013). Natural anthocyanins from phytoresources and their chemical researches. *Natural Product Research*, *27*(4-5), 456—469. Available from https://doi.org/10.1080/14786419.2012.706299.

Li, Z., Deng, H., Guo, X., Yan, S., Lu, C., Zhao, Z., Feng, X., Li, Q., Wang, J., Zeng, J., & Ma, X. (2022). Effective dose/duration of natural flavonoid quercetin for treatment of diabetic nephropathy: A systematic review and meta-analysis of rodent data. *Phytomedicine*, *105*, 154348. Available from https://doi.org/10.1016/j.phymed.2022.154348.

Liguori, I., Russo, G., Curcio, F., Bulli, G., Aran, L., Della-Morte, D., Gargiulo, G., Testa, G., Cacciatore, F., Bonaduce, D., & Abete, P. (2018). Oxidative stress, aging, and diseases. Dove Medical Press Ltd., Italy. *Clinical Interventions in Aging*, *13*, 757—772. Available from https://doi.org/10.2147/CIA.S158513, http://www.dovepress.com/clinical-interventions-in-aging-journal.

Lin, Y. T., Wu, Y. C., Sun, G. C., Ho, C. Y., Wong, T. Y., Lin, C. H., Chen, H. H., Yeh, T. C., Li, C. J., Tseng, C. J., & Cheng, P. W. (2018). Effect of resveratrol on reactive oxygen species-induced cognitive impairment in rats with angiotensin II-induced early Alzheimer's disease. *Journal of Clinical Medicine*, 7(10). Available from https://doi.org/10.3390/JCM7100329, http://www.mdpi.com/2077-0383/7/10/329/pdf.

Lippolis, T., Cofano, M., Caponio, G. R., De Nunzio, V., & Notarnicola, M. (2023). Bioaccessibility and bioavailability of diet polyphenols and their modulation of gut microbiota. *International Journal of Molecular Sciences*, *24*(4). Available from https://doi.org/10.3390/ijms24043813, http://www.mdpi.com/journal/ijms.

Liu, F., Peng, Y., Qiao, Y., Huang, Y., Song, F., Zhang, M., & Song, F. (2022). Consumption of flavonoids and risk of hormone-related cancers: A systematic review and meta-analysis of observational studies. *Nutrition Journal*, *21*(1). Available from https://doi.org/10.1186/s12937-022-00778-w.

Liu, M., Yun, P., Hu, Y., Yang, J., Khadka, R. B., & Peng, X. (2020). Effects of grape seed proanthocyanidin extract on obesity. *Obesity Facts*, *13*(2), 279—291. Available from https://doi.org/10.1159/000502235, http://content.karger.com/ProdukteDB/produkte.asp?Aktion = JournalHome&ProduktNr = 233731.

Liu, X., Liu, Y. -J., Huang, Y., Yu, H. -J., Yuan, S., Tang, B., Wang, P., & He, Q. -Q. (2017). Dietary total flavonoids intake and risk of mortality from all causes and cardiovascular disease in the general population: A systematic review and meta-analysis of cohort studies. *Molecular Nutrition & Food Research*, *61*(6). Available from https://doi.org/10.1002/mnfr.201601003.

Lu, Q. P., Wu, M. L., Li, H. L., Zhou, Y., Xian, M. H., Huang, W. Z., Piao, X. H., & Ge, Y. W. (2023). Combined metabolite analysis and network pharmacology to elucidate the mechanisms of therapeutic effect of melastoma dodecandrum ellagitannins on abnormal uterine bleeding. *Chemistry and Biodiversity*, *20*(8). Available from https://doi.org/10.1002/cbdv.202300646, http://onlinelibrary.wiley.com/journal/10.1002/(ISSN)1612-1880.

Luna, R., R.-Garza, E., & Saldivar, S. (2020). Environmentally friendly methods for flavonoid extraction from plant material: impact of their operating conditions on yield and antioxidant properties. *ScientificWorldJournal*, *2020*, 2020.

López, V., Cascella, M., Benelli, G., Maggi, F., & Gómez-Rincón, C. (2018). Green drugs in the fight against Anisakis simplex—larvicidal activity and acetylcholinesterase inhibition of Origanum compactum essential oil. *Parasitology Research*, *117*(3), 861–867. Available from https://doi.org/10.1007/s00436-018-5764-3, link.springer.de/link/service/journals/00436/index.htm.

Ma, X. R., Sun, Z. K., Liu, Y. R., Jia, Y. J., Zhang, B. A., & Zhang, J. W. (2013). Resveratrol improves cognition and reduces oxidative stress in rats with vascular dementia. *Neural Regeneration Research*, *8*(22), 2050–2059. Available from https://doi.org/10.3969/j.issn.1673-5374.2013.22.004, http://www.nrronline.org/temp/NeuralRegenRes8222050-3035427_082554.pdf.

Malekinejad, H., Rezabakhsh, A., Rahmani, F., & Hobbenaghi, R. (2012). Silymarin regulates the cytochrome P450 3A2 and glutathione peroxides in the liver of streptozotocin-induced diabetic rats. *Phytomedicine*, *19*(7), 583–590. Available from https://doi.org/10.1016/j.phymed.2012.02.009.

Marin, L., Miguelez, E. M., Villar, C. J., & Lombo, F. (2015). Bioavailability of dietary polyphenols and gut microbiota metabolism: antimicrobial properties. *Biomed Res Int*, *2015*.

Martens, S., Preuß, A., & Matern, U. (2010). Multifunctional flavonoid dioxygenases: Flavonol and anthocyanin biosynthesis in Arabidopsis thaliana L. *Phytochemistry*, *71*(10), 1040–1049. Available from https://doi.org/10.1016/j.phytochem.2010.04.016.

Martinez-Zapata, M. J., Vernooij, R. W. M., Simancas-Racines, D., Uriona Tuma, S. M., Stein, A. T., Moreno Carriles, R. M. M., Vargas, E., & Bonfill Cosp, X. (2020). Phlebotonics for venous insufficiency. *Cochrane Database of Systematic Reviews*, *2020*(11). Available from https://doi.org/10.1002/14651858.CD003229.pub4, https://www.cochranelibrary.com/cdsr/table-of-contents.

Di Martino, P., Agniel, R., David, K., Templer, C., Gaillard, J. L., Denys, P., & Botto, H. (2006). Reduction of Escherichia coli adherence to uroepithelial bladder cells after consumption of cranberry juice: A double-blind randomized placebo-controlled cross-over trial. *World Journal of Urology*, *24*(1), 21–27. Available from https://doi.org/10.1007/s00345-005-0045-z.

Mastron, J. K., Siveen, K. S., Sethi, G., & Bishayee, A. (2014). Silymarin and hepatocellular carcinoma: A systematic, comprehensive, and critical review. *Anti-Cancer Drugs*, *26*(5), 475–486. Available from https://doi.org/10.1097/CAD.0000000000000211, http://journals.lww.com/anti-cancerdrugs/pages/default.aspx.

de Mello, V. D. F., Lankinen, M. A., Lindström, J., Puupponen-Pimiä, R., Laaksonen, D. E., Pihlajamäki, J., Lehtonen, M., Uusitupa, M., Tuomilehto, J., Kolehmainen, M., Törrönen, R., & Hanhineva, K. (2017). Fasting serum hippuric acid is elevated after bilberry (*Vaccinium myrtillus*) consumption and associates with improvement of fasting glucose levels and insulin secretion in persons at high risk of developing type 2 diabetes. *Molecular Nutrition & Food Research*, *61*(9), 1700019. Available from https://doi.org/10.1002/mnfr.201700019.

Micek, A., Godos, J., Del Rio, D., Galvano, F., & Grosso, G. (2021). Dietary flavonoids and cardiovascular disease: A comprehensive dose–response meta-analysis.

Molecular Nutrition & Food Research, 65(6). Available from https://doi.org/10.1002/mnfr.202001019.

Moilanen, J., Karonen, M., Tähtinen, P., Jacquet, R., Quideau, S., & Salminen, J. P. (2016). Biological activity of ellagitannins: Effects as anti-oxidants, pro-oxidants and metal chelators. *Phytochemistry, 125*, 65–72. Available from https://doi.org/10.1016/j.phytochem.2016.02.008, http://www.journals.elsevier.com/phytochemistry/.

Montagna, M. T., Diella, G., Triggiano, F., Caponio, G. R., Giglio, D., Caggiano, O., et al. (2019). Food of the Gods: History, science, and human health. *International Journal of Environmental Research and Public Health*. (24).

Moore, A., Beidler, J., & Hong, M. Y. (2018). Resveratrol and depression in animal models: A systematic review of the biological mechanisms. *Molecules, 23*(9). Available from https://doi.org/10.3390/molecules23092197, http://www.mdpi.com/1420-3049/23/9/2197/pdf.

Morvaridi, M., Rayyani, E., Jaafari, M., Khiabani, A., & Rahimlou, M. (2020). The effect of green coffee extract supplementation on cardio metabolic risk factors: A systematic review and meta-analysis of randomized controlled trials. *Journal of Diabetes & Metabolic Disorders, 19*(1), 645–660. Available from https://doi.org/10.1007/s40200-020-00536-x.

Mukhija, M., Joshi, B. C., Bairy, P. S., Bhargava, A., & Sah, A. N. (2022). Lignans: A versatile source of anticancer drugs. Springer Science and Business Media Deutschland GmbH, India. *Journal of Basic and Applied Sciences, 11*(1). Available from https://doi.org/10.1186/s43088-022-00256-6, bjbas.springeropen.com/.

Nachvak, S. M., Moradi, S., Anjom-Shoae, J., Rahmani, J., Nasiri, M., Maleki, V., et al. (2019). Soy isoflavones, and protein intake in relation to mortality from all causes, cancers, and cardiovascular diseases: A systematic review and dose-response meta-analysis of prospective cohort studies. *Journal of the Academy of Nutrition and Dietetics, 119*(9), 1483–500.

Nasri, H. (2013). Re: Effect of silymarin on streptozotocin-nicotinamide-induced type 2 diabetic nephropathy in rats. *Iranian Journal of Kidney Diseases*, 7(5), 414–415. Available from http://www.ijkd.org/index.php/ijkd/article/download/1155/586.

Naylor, L. H., Zimmermann, D., Guitard-Uldry, M., Poquet, L., Leveques, A., Eriksen, B., Bel Rhlid, R., Galaffu, N., D'urzo, C., De Castro, A., Van Schaick, E., Green, D. J., & Actis-Goretta, L. (2021). Acute dose-response effect of coffee-derived chlorogenic acids on the human vasculature in healthy volunteers: A randomized controlled trial. *American Journal of Clinical Nutrition, 113*(2), 370–379. Available from https://doi.org/10.1093/ajcn/nqaa312, http://www.ajcn.org/contents-by-date.2005.shtml.

Nemzer, B. V., Rodriguez, L. C., Hammond, L., DiSilvestro, R., Hunter, J. M., & Pietrzkowski, Z. (2011). Acute reduction of serum 8-iso-PGF2-alpha and advanced oxidation protein products in vivo by a polyphenol-rich beverage; a pilot clinical study with phytochemical and in vitro antioxidant characterization. *Nutrition Journal, 10*(1). Available from https://doi.org/10.1186/1475-2891-10-67.

Neuhouser, M. L. (2004). Dietary flavonoids and cancer risk: Evidence from human population studies. *Nutrition and Cancer, 50*(1), 1–7. Available from https://doi.org/10.1207/s15327914nc5001_1, http://www.tandfonline.com/toc/hnuc20/current.

Olivas-Aguirre, F. J., Wall-Medrano, A., González-Aguilar, G. A., López-Díaz, J. A., Álvarez-Parrilla, E., De La Rosa, L. A., & Ramos-Jimenez, A. (2015). Taninos hidrolizables; bioquímica, aspectos nutricionales y analíticos y efectos en la salud. *Nutricion Hospitalaria., 31*(1), 55–66. Available from https://doi.org/10.3305/nh.2015.31.1.7699, http://www.aulamedica.es/gdcr/index.php/nh/article/download/7699/pdf_7694.

De Oliveira Rodrigues Junior, E., de Santana, I. R., Durço, A. O., Conceição, L. S. R., Barreto, A. S., Menezes, I. A. C., Roman-Campos, D., & dos Santos, M. R. V.

(2023). The effects of flavonoids in experimental sepsis: A systematic review and meta-analysis. John Wiley and Sons Ltd, Brazil. *Phytotherapy Research, 37*(6), 2531–2551. Available from https://doi.org/10.1002/ptr.7846, http://onlinelibrary.wiley.com/journal/10.1002/(ISSN)1099-1573.

Ozawa, Y., Kawashima, M., Inoue, S., Inagaki, E., Suzuki, A., Ooe, E., Kobayashi, S., & Tsubota, K. (2015). Bilberry extract supplementation for preventing eye fatigue in video display terminal workers. *The Journal of Nutrition, Health & Aging, 19*(5), 548–554. Available from https://doi.org/10.1007/s12603-014-0573-6.

Paesa, M., Ancín-Azpilicueta, C., Velderrain-Rodríguez, G., Martin-Belloso, O., Gualillo, O., Osada, J., Rodríguez-Yoldi, M. J., & Mendoza, G. (2022). Anti-inflammatory and chondroprotective effects induced by phenolic compounds from onion waste extracts in ATDC-5 chondrogenic cell line. *Antioxidants, 11*(12), 2381. Available from https://doi.org/10.3390/antiox11122381.

Paesa, M., Nogueira, D. P., Velderrain-Rodríguez, G., Esparza, I., Jiménez-Moreno, N., Mendoza, G., Osada, J., Martin-Belloso, O., Rodríguez-Yoldi, M. J., & Ancín-Azpilicueta, C. (2022). Valorization of onion waste by obtaining extracts rich in phenolic compounds and feasibility of its therapeutic use on colon cancer. *Antioxidants, 11*(4). Available from https://doi.org/10.3390/antiox11040733, https://www.mdpi.com/2076-3921/11/4/733/pdf.

Panche, A. N., Diwan, A. D., & Chandra, S. R. (2016). Flavonoids: An overview. Cambridge University Press, India. *Journal of Nutritional Science, 5*. Available from https://doi.org/10.1017/jns.2016.41, http://journals.cambridge.org/action/displayJournal?jid = JNS.

Pandey, K. B., & Rizvi, S. I. (2009). Plant polyphenols as dietary antioxidants in human health and disease. *Oxidative Medicine and Cellular Longevity, 2*(5), 270–278. Available from https://doi.org/10.4161/oxim.2.5.9498India, http://www.landesbioscience.com/journals/oximed/article/PandeyOMCL2-5.pdf.

Pangeni, R., Sahni, J. K., Ali, J., Sharma, S., & Baboota, S. (2014). Resveratrol: Review on therapeutic potential and recent advances in drug delivery. *Expert Opinion on Drug Delivery, 11*(8), 1285–1298. Available from https://doi.org/10.1517/17425247.2014.919253, http://informahealthcare.com/loi/edd.

Panzella, L., & Napolitano, A. (2022). Condensed tannins, a viable solution to meet the need for sustainable and effective multifunctionality in food packaging: Structure, sources, and properties. *Journal of Agricultural and Food Chemistry, 70*(3), 751–758. Available from https://doi.org/10.1021/acs.jafc.1c07229.

Pinkas, A., & Aschner, M. (2016). Advanced glycation end-products and their receptors: Related pathologies, recent therapeutic strategies, and a potential model for future neurodegeneration studies. *Chemical Research in Toxicology, 29*(5), 707–714. Available from https://doi.org/10.1021/acs.chemrestox.6b00034, 0893-228X.

Pojer, E., Mattivi, F., Johnson, D., & Stockley, C. S. (2013). The case for anthocyanin consumption to promote human health: A review. *Comprehensive Reviews in Food Science and Food Safety, 12*(5), 483–508. Available from https://doi.org/10.1111/1541-4337.12024.

Popiolek-Kalisz, J., & Fornal, E. (2022). The effects of quercetin supplementation on blood pressure – meta-analysis. *Current Problems in Cardiology, 47*(11), 101350. Available from https://doi.org/10.1016/j.cpcardiol.2022.101350.

Quero, J., Ballesteros, L. F., Ferreira-Santos, P., Velderrain-Rodriguez, G. R., Rocha, C. M. R., Pereira, R. N., Teixeira, J. A., Martin-Belloso, O., Osada, J., & Rodríguez-Yoldi, M. J. (2022). Unveiling the antioxidant therapeutic functionality of sustainable olive pomace active ingredients. *Antioxidants, 11*(5). Available from https://doi.org/10.3390/antiox11050828, https://www.mdpi.com/2076-3921/11/5/828/pdf.

Quero, J., Jiménez-Moreno, N., Esparza, I., Osada, J., Cerrada, E., Ancín-Azpilicueta, C., & Rodríguez-Yoldi, M. J. (2021). Grape stem extracts with potential anticancer and antioxidant properties. *Antioxidants, 10*(2), 1–17. Available from https://doi.org/10.3390/antiox10020243, https://www.mdpi.com/2076-3921/10/2/243/pdf.

Quero, J., Mármol, I., Cerrada, E., & Rodríguez-Yoldi, M. J. (2020). Insight into the potential application of polyphenol-rich dietary intervention in degenerative disease management. *Food and Function, 11*(4), 2805–2825. Available from https://doi.org/10.1039/d0fo00216j, http://pubs.rsc.org/en/journals/journal/fo.

Rambaldi, A., Jacobs, B. P., Iaquinto, G., & Gluud, C. (2005). Milk thistle for alcoholic and/or hepatitis B or C liver diseases — A systematic Cochrane hepato-biliary group review with meta-analyses of randomized clinical trials. *American Journal of Gastroenterology, 100*(11), 2583–2591. Available from https://doi.org/10.1111/j.1572-0241.2005.00262.x.

Rauf, A., Imran, M., Abu-Izneid, T., Haq, I.-U., Patel, S., Pan, X., Naz, S., Sanches Silva, A., Saeed, F., & Ansar Rasul Suleria, H. (2019). Proanthocyanidins: A comprehensive review. *Biomedicine & Pharmacotherapy, 116*, 108999. Available from https://doi.org/10.1016/j.biopha.2019.108999.

Reis-Giada, M. L. (2013). Oxidative stress and chronic degenerative diseases — A role for antioxidants. *Intech, 2013*.

Rhee, S. Y., & Kim, Y. S. (2018). The role of advanced glycation end products in diabetic vascular complications. *Diabetes and Metabolism Journal, 42*(3), 188–195. Available from https://doi.org/10.4093/dmj.2017.0105, https://e-dmj.org/Synapse/Data/PDFData/2004DMJ/dmj-42-188.pdf.

Ried, K., Fakler, P., & Stocks, N. P. (2017). Effect of cocoa on blood pressure. *Cochrane Database of Systematic Reviews, 2017*(4). Available from https://doi.org/10.1002/14651858.CD008893.pub3, http://as.wiley.com/WileyCDA/Brand/id-6.html.

Riso, P., Klimis-Zacas, D., Del Bo', C., Martini, D., Campolo, J., Vendrame, S., Møller, P., Loft, S., Maria, R. D., & Porrini, M. (2013). Effect of a wild blueberry (*Vaccinium angustifolium*) drink intervention on markers of oxidative stress, inflammation and endothelial function in humans with cardiovascular risk factors. *European Journal of Nutrition, 52*(3), 949–961. Available from https://doi.org/10.1007/s00394-012-0402-9.

Roleira, F. M. F., Tavares-Da-Silva, E. J., Varela, C. L., Costa, S. C., Silva, T., Garrido, J., & Borges, F. (2015). Plant derived and dietary phenolic antioxidants: Anticancer properties. *Food Chemistry, 183*, 235–258. Available from https://doi.org/10.1016/j.foodchem.2015.03.039, http://www.elsevier.com/locate/foodchem.

Rostom, B., Karaky, R., Kassab, I., & Sylla-Iyarreta Veitia, M. (2022). Coumarins derivatives and inflammation: Review of their effects on the inflammatory signaling pathways. *Eur J Pharmacol, 922*, 2022.

Saito, K., Yonekura-Sakakibara, K., Nakabayashi, R., Higashi, Y., Yamazaki, M., Tohge, T., & Fernie, A. R. (2013). The flavonoid biosynthetic pathway in Arabidopsis: Structural and genetic diversity. *Plant Physiology and Biochemistry, 72*, 21–34. Available from https://doi.org/10.1016/j.plaphy.2013.02.001.

Sarker, S. D., & Nahar, L. (2017). Progress in the chemistry of naturally occurring coumarins. *Progress in the chemistry of organic natural products, 106*, 241–304. Available from https://doi.org/10.1007/978-3-319-59542-9_3.

Sattarinezhad, A., Roozbeh, J., Shirazi Yeganeh, B., Omrani, G. R., & Shams, M. (2019). Resveratrol reduces albuminuria in diabetic nephropathy: A randomized double-blind placebo-controlled clinical trial. *Diabetes & Metabolism, 45*(1), 53–59. Available from https://doi.org/10.1016/j.diabet.2018.05.010.

Scallon, S., & Bell-Syer, A. Z. (2013). Flavonoids for treating venous leg ulcers. *Cochrane Database of Systematic Reviews, 5*.

Schindler, G., Patzak, U., Brinkhaus, B., von Nieciecki, A., Wittig, J., Krähmer, N., Glöckl, I., & Veit, M. (2002). Urinary excretion and metabolism of arbutin after oral administration of Arctostaphylos uvae ursi extract as film-coated tablets and aqueous solution in healthy humans. *The Journal of Clinical Pharmacology*, *42*(8), 920–927. Available from https://doi.org/10.1177/00912700240l102740.

Sgarbi, G., Liuzzi, F., Baracca, A., & Solaini, G. (2018). Resveratrol preserves mitochondrial function in a human post-mitotic cell model. *The Journal of Nutritional Biochemistry*, *62*, 9–17. Available from https://doi.org/10.1016/j.jnutbio.2018.07.017.

Shamsudin, Q.U., Ahmed, S., Mahmood, S.A. A., Shah, M.N., Sarian, M.K., et al. Flavonoids as antidiabetic and anti-inflammatory agents: A review on structural activity relationship-based studies and meta-analysis. International Journal of Molecular Sciences.

Sharifi-Rad, J., Quispe, C., Castillo, C. M. S., Caroca, R., Lazo-Vélez, M. A., Antonyak, H., Polishchuk, A., Lysiuk, R., Oliinyk, P., De Masi, L., Bontempo, P., Martorell, M., Daştan, S. D., Rigano, D., Wink, M., & Cho, W. C. (2022). Ellagic acid: A review on its natural sources, chemical stability, and therapeutic potential. *Oxidative Medicine and Cellular Longevity, 2022*. Available from https://doi.org/10.1155/2022/3848084, http://www.hindawi.com/journals/oximed/.

Sharma, C., Suhalka, P., & Bhatnagar, M. (2018). Curcumin and resveratrol rescue cortical–hippocampal system from chronic fluoride-induced neurodegeneration and enhance memory retrieval. *International Journal of Neuroscience*, *128*(11), 1007–1021. Available from https://doi.org/10.1080/00207454.2018.1458727.

Sheikh, P., Lohsiriwat, V., & Shelygin, Y. (2020). Micronized purified flavonoid fraction in hemorrhoid disease: A systematic review and meta-analysis. *Advances in Therapy*, *37* (6), 2792–2812. Available from https://doi.org/10.1007/s12325-020-01353-7, 0741-238X.

Shen, Y., Xu, Z., & Sheng, Z. (2017). Ability of resveratrol to inhibit advanced glycation end product formation and carbohydrate-hydrolyzing enzyme activity, and to conjugate methylglyoxal. *Food Chemistry*, *216*, 153–160. Available from https://doi.org/10.1016/j.foodchem.2016.08.034.

Shim, S. H., Kim, J. M., Choi, C. Y., Kim, C. Y., & Park, K. H. (2012). Ginkgo biloba extract and bilberry anthocyanins improve visual function in patients with normal tension glaucoma. *Journal of Medicinal Food*, *15*(9), 818–823. Available from https://doi.org/10.1089/jmf.2012.2241.

Sies, H. (2015). Oxidative stress: a concept in redox biology and medicine. *Redox Biology*, *4*, 180–183. Available from https://doi.org/10.1016/j.redox.2015.01.002.

Silva, L. B. A. R., Pinheiro-Castro, N., Novaes, G. M., De, G., Pascoal, F. L., & Ong, T. P. (2019). Bioactive food compounds, epigenetics and chronic disease prevention: Focus on early-life interventions with polyphenols. *Food Research International*, *125*, 108646. Available from https://doi.org/10.1016/j.foodres.2019.108646.

Simoben, C. V., Ibezim, A., Ntie-Kang, F., Nwodo, J. N., & Lifongo, L. L. (2015). Exploring cancer therapeutics with natural products from African medicinal plants, part I: Xanthones, quinones, steroids, coumarins, phenolics and other classes of compounds. *Anti-Cancer Agents in Medicinal Chemistry*, *15*(9), 1092–1111. Available from https://doi.org/10.2174/1871520615666150113110241, http://www.benthamdirect.org/pages/all_b_bypublication.php.

Singh, M., Kaur, M., & Silakari, O. (2014). Flavones: An important scaffold for medicinal chemistry. *European Journal of Medicinal Chemistry*, *84*, 206–239. Available from https://doi.org/10.1016/j.ejmech.2014.07.013.

Singh, R. P., Sharma, G., Dhanalakshmi, S., Agarwal, C., & Agarwal, R. (2003). Suppression of advanced human prostate tumor growth in athymic mice by silibinin feeding is associated with reduced cell proliferation, increased apoptosis, and inhibition of angiogenesis. *Cancer Epidemiology Biomarkers and Prevention*, *12*(9), 933–939.

Solhi, H., Ghahremani, R., Kazemifar, A. M., & Yazdi, Z. H. (2014). Silymarin in treatment of non-alcoholic steatohepatitis: A randomized clinical trial. *Caspian Journal of Internal Medicine*, *5*(1), 9–12. Available from http://www.caspjim.com/browse.php?a_id = 31&slc_lang = en&sid = 1&ftxt = 1.

Song, F., Zhu, Y., Shi, Z., Tian, J., Deng, X., Ren, J., Andrews, M. C., Ni, H., Ling, W., & Yang, Y. (2014). Plant food anthocyanins inhibit platelet granule secretion in hypercholesterolaemia: Involving the signalling pathway of PI3K–Akt. *Thrombosis and Haemostasis*, *112*(5), 981–991. Available from https://doi.org/10.1160/TH13-12-1002, http://th.schattauer.de/en/contents/archive/issue/2145/manuscript/21780/download.html.

Di Stasi, L. C. (2021). Coumarin derivatives in inflammatory bowel disease. *Molecules*, *26*(2), 2021.

St-Onge, M. P., & Zuraikat, F. M. (2019). Reciprocal roles of sleep and diet in cardiovascular health: A review of recent evidence and a potential mechanism. *Current Atherosclerosis Reports*, *21*(3). Available from https://doi.org/10.1007/s11883-019-0772-z, http://www.springer.com.

Summerlin, N., Soo, E., Thakur, S., Qu, Z., Jambhrunkar, S., & Popat, A. (2015). Resveratrol nanoformulations: Challenges and opportunities. *International Journal of Pharmaceutics*, *479*(2), 282–290. Available from https://doi.org/10.1016/j.ijpharm.2015.01.003.

Tan, J., Hu, J., He, Y., & Cui, F. (2015). Protective role of silymarin in a mouse model of renal Ischemia-Reperfusion injury. *Diagnostic Pathology*, *198*, 19.

Tang, N. P., Zhou, B., Wang, B., Yu, R. B., & Ma, J. (2009). Flavonoids intake and risk of lung cancer: A meta-analysis. *Japanese Journal of Clinical Oncology*, *39*(6), 352–359. Available from https://doi.org/10.1093/jjco/hyp028.

Tao, L., Ding, Q., Gao, C., & Sun, X. (2016). Resveratrol attenuates neuropathic pain through balancing pro-inflammatory and anti-inflammatory cytokines release in mice. *International Immunopharmacology*, *34*, 165–172. Available from https://doi.org/10.1016/j.intimp.2016.02.033.

Toledo, E., Salas-Salvado, J., Donat-Vargas, C., Buil-Cosiales, P., Estruch, R., Ros, E., Corella, D., Fito, M., Hu, F. B., Aros, F., Gomez-Gracia, E., Romaguera, D., Ortega-Calvo, M., Serra-Majem, L., Pinto, X., Schroder, H., Basora, J., Sorli, J. V., Bullo, M., Serra-Mir, M., & Martinez-Gonzalez, M. A. (2015). Mediterranean diet and invasive breast cancer risk among women at high cardiovascular risk in the predimed trial a randomized clinical trial. *JAMA Internal Medicine*, *175*(11), 1752–1760. Available from https://doi.org/10.1001/jamainternmed.2015.4838, http://archinte.jamanetwork.com/data/Journals/INTEMED/934633/ioi150080.pdf.

Tsao, R. (2010). Chemistry and biochemistry of dietary polyphenols. *Nutrients*, *2*(12), 1231–1246. Available from https://doi.org/10.3390/nu2121231.

Tse, G., & Eslick, G. D. (2016). Soy and isoflavone consumption and risk of gastrointestinal cancer: A systematic review and meta-analysis. *European Journal of Nutrition*, *55*(1), 63–73. Available from https://doi.org/10.1007/s00394-014-0824-7, http://www.springerlink.com/content/1436-6207.

Tung, X. Y., Yip, J. Q., & Gew, L. T. (2023). Searching for natural plants with antimelanogenesis and antityrosinase properties for cosmeceutical or nutricosmetics applications: A systematic review. *ACS Omega*, *8*(37), 33115–33201. Available from https://doi.org/10.1021/acsomega.3c02994, pubs.acs.org/journal/acsodf.

Tutino, V., De Nunzio, V., Milella, R. A., Gasparro, M., Cisternino, A. M., Gigante, I., Lanzilotta, E., Iacovazzi, P. A., Lippolis, A., Lippolis, T., Caruso, M. G., & Notarnicola, M. (2021). Impact of fresh table grape intake on circulating microRNAs levels in healthy subjects: a significant modulation of gastrointestinal cancer-related pathways. *Molecular Nutrition & Food Research*, *65*(21). Available from https://doi.org/10.1002/mnfr.202100428.

Vajdi, M., Karimi, A., Karimi, M., Abbasalizad Farhangi, M., & Askari, G. (2023). Effects of luteolin on sepsis: A comprehensive systematic review. *Phytomedicine*, *113*, 154734. Available from https://doi.org/10.1016/j.phymed.2023.154734.

Vauzour, D. (2012). Dietary polyphenols as modulators of brain functions: Biological actions and molecular mechanisms underpinning their beneficial effects. *Oxidative Medicine and Cellular Longevity*, *2012*, 1–16. Available from https://doi.org/10.1155/2012/914273.

Velderrain-Rodríguez, G. R., Quero, J., Osada, J., Martín-Belloso, O., & Rodríguez-Yoldi, M. J. (2021). Phenolic-rich extracts from avocado fruit residues as functional food ingredients with antioxidant and antiproliferative properties. *Biomolecules*, *11*(7). Available from https://doi.org/10.3390/biom11070977, https://www.mdpi.com/2218-273X/11/7/977/pdf.

Venkata Sairam, K., M. Gurupadayya, B., S. Chandan, R., K. Nagesha, D., & Vishwanathan, B. (2016). A review on chemical profile of coumarins and their therapeutic role in the treatment of cancer. *Current Drug Delivery*, *13*(2), 186–201. Available from https://doi.org/10.2174/1567201812666150702102800.

Wallace, T. C., & Monica Giusti, M. (2015). Anthocyanins. *Advances in Nutrition*, *6*(5), 620–622. Available from https://doi.org/10.3945/an.115.009233.

Wen, C., Dechsupa, N., Yu, Z., Zhang, X., Liang, S., Lei, X., Xu, T., Gao, X., Hu, Q., Innuan, P., Kantapan, J., & Lü, M. (2023). Pentagalloyl glucose: A review of anticancer properties, molecular targets, mechanisms of action, pharmacokinetics, and safety profile. *Molecules*, *28*(12), 4856. Available from https://doi.org/10.3390/molecules28124856.

Wick, J. Y. (2012). Aspirin: A history, a love story. *Consultant Pharmacist*, *27*(5), 322–329. Available from https://doi.org/10.4140/TCP.n.2012.322, http://ascp.metapress.com/content/r207k522419211m0/fulltext.pdf.

Woo, H. D., Kim, J., & Davis, K. R. (2013). Dietary flavonoid intake and smoking-related cancer risk: A meta-analysis. *PLoS ONE*, *8*(9), e75604. Available from https://doi.org/10.1371/journal.pone.0075604.

Wu, X., Xu, Y., Zhu, B., Liu, Q., Yao, Q., & Zhao, G. (2018). Resveratrol induces apoptosis in SGC-7901 gastric cancer cells. *Oncology Letters*, *16*(3), 2949–2956. Available from https://doi.org/10.3892/ol.2018.9045, http://www.spandidos-publications.com/ol/16/3/2949/download.

Xie, J., Liu, S., Wu, B., Li, G., Rao, S., Zou, L., Yi, Z., Zhang, C., Jia, T., Zhao, S., Schmalzing, G., Hausmann, R., Nie, H., Li, G., & Liang, S. (2017). The protective effect of resveratrol in the transmission of neuropathic pain mediated by the P2X7 receptor in the dorsal root ganglia. *Neurochemistry International*, *103*, 24–35. Available from https://doi.org/10.1016/j.neuint.2016.12.006.

Xie Y., Zhang D., Zhang J., Yuan, Metabolism, transport and drug-drug interactions of silymarin. *Molecules*.

Xue, H., Wei, M., & Ji, L. (2023). Chlorogenic acids: A pharmacological systematic review on their hepatoprotective effects. *Phytomedicine*, *118*, 154961. Available from https://doi.org/10.1016/j.phymed.2023.154961.

Yang, F., Oz, H. S., Barve, S., De Villiers, W. J. S., McClain, C. J., & Varilek, G. W. (2001). The green tea polyphenol (-)-epigallocatechin-3-gallate blocks nuclear factor-κB activation by inhibiting IκB kinase activity in the intestinal epithelial cell line IEC-6. *Molecular Pharmacology*, *60*(3), 528–533.

Yao, L. H., Jiang, Y. M., Shi, J., Tomás-Barberán, F. A., Datta, N., Singanusong, R., & Chen, S. S. (2004). Flavonoids in food and their health benefits. *Plant*

Foods for Human Nutrition, *59*(3), 113–122. Available from https://doi.org/10.1007/s11130-004-0049-7.

Yu, Y., Jing, X., Li, H., Zhao, X., & Wang, D. (2016). Soy isoflavone consumption and colorectal cancer risk: A systematic review and meta-analysis. *Scientific Reports*, *6*(1). Available from https://doi.org/10.1038/srep25939.

Yılmaz, Z., Kalaz, E. B., Aydın, A. F., Olgaç, V., Doğru-Abbasoğlu, S., Uysal, M., & Koçak-Toker, N. (2018). The effect of resveratrol on glycation and oxidation products in plasma and liver of chronic methylglyoxal-treated rats. *Pharmacological Reports*, *70*(3), 584–590. Available from https://doi.org/10.1016/j.pharep.2017.12.005.

Zeng, L., Xiang, R., Fu, C., Qu, Z., & Liu, C. (2022). The Regulatory effect of chlorogenic acid on gut-brain function and its mechanism: A systematic review. *Biomedicine & Pharmacotherapy*, *149*, 112831. Available from https://doi.org/10.1016/j.biopha.2022.112831.

Zeng, Y. X., Wang, S., Wei, L., Cui, Y. Y., & Chen, Y. H. (2020). Proanthocyanidins: Components, pharmacokinetics and biomedical properties. World Scientific Publishing Co. Pte Ltd, China. *American Journal of Chinese Medicine*, *48*(4), 813–869. Available from https://doi.org/10.1142/S0192415X2050041X, http://www.worldscinet.com/ajcm/ajcm.shtml.

Zhang, Q., Feng, H., Qluwakemi, B., Wang, J., Yao, S., Cheng, G., Xu, H., Qiu, H., Zhu, L., & Yuan, M. (2017). Phytoestrogens and risk of prostate cancer: an updated meta-analysis of epidemiologic studies. *International Journal of Food Sciences and Nutrition*, *68*(1), 28–42. Available from https://doi.org/10.1080/09637486.2016.1216525.

Zhang, S., Zhu, Q., Chen, J.-Y., OuYang, D., & Lu, J.-H. (2020). The pharmacological activity of epigallocatechin-3-gallate (EGCG) on Alzheimer's disease animal model: A systematic review. *Phytomedicine*, *79*, 153316. Available from https://doi.org/10.1016/j.phymed.2020.153316.

Zhang, Y., Ge, Y., Ping, X., Yu, M., Lou, D., & Shi, W. (2018). Synergistic apoptotic effects of silibinin in enhancing paclitaxel toxicity in human gastric cancer cell lines. *Molecular Medicine Reports*, *18*(2), 1835–1841. Available from https://doi.org/10.3892/mmr.2018.9129, https://www.spandidos-publications.com/mmr/18/2/1835/download.

Zhao, Y., Yuan, X., Li, X., & Zhang, Y. (2019). Resveratrol significantly inhibits the occurrence and development of cervical cancer by regulating phospholipid scramblase 1. *Journal of Cellular Biochemistry*, *120*(2), 1527–1531. Available from https://doi.org/10.1002/jcb.27335.

Zheng, X., Jia, B., Tian, X. T., Song, X., Wu, M. L., Kong, Q. Y., Li, H., & Liu, J. (2018). Correlation of reactive oxygen species levels with resveratrol sensitivities of anaplastic thyroid cancer cells. Hindawi Limited, China. *Oxidative Medicine and Cellular Longevity*, *2018*. Available from https://doi.org/10.1155/2018/6235417, http://www.hindawi.com/journals/oximed/.

Zhou, C., Qian, W., Ma, J., Cheng, L., Jiang, Z., Yan, B., Li, J., Duan, W., Sun, L., Cao, J., Wang, F., Wu, E., Wu, Z., Ma, Q., & Li, X. (2019). Resveratrol enhances the chemotherapeutic response and reverses the stemness induced by gemcitabine in pancreatic cancer cells via targeting SREBP1. *Cell Proliferation*, *52*(1). Available from https://doi.org/10.1111/cpr.12514, http://onlinelibrary.wiley.com/journal/10.1111/(ISSN)1365-2184.

Zhou, J., He, X., Sun, R., Yu, Z., Wang, C., Deng, S., Zhang, B., Huang, S., Han, C., & Li, D. (2022). Lignans from Bupleurum marginatum and their antioxidant activity. *Natural Product Research*, *36*(19), 5016–5021. Available from https://doi.org/10.1080/14786419.2021.1917570.

CHAPTER TWO

Metabolization pathways and bioavailability of polyphenols

Patricia Fernández-Moreno, María del Carmen Villegas-Aguilar, Felipe Huertas-Camarasa and Álvaro Fernández-Ochoa
Department of Analytical Chemistry, University of Granada, Granada, Granada, Spain

Introduction

Phenolic compounds are bioactive molecules derived from the secondary metabolism of plants. More than 10,000 phenolic structures have been identified naturally in different parts of plants including fruits, stems, roots, seeds, and leaves (Albuquerque et al., 2021; Silva et al., 2023; Tungmunnithum et al., 2018). These compounds are synthesized mainly through the shikimate and phenylpropanoid pathways, and to a lesser extent by the pentose phosphate pathway (Dias et al., 2021). These phytochemicals perform important functions in plants, being responsible for pigmentation to attract pollinating insects, providing an astringent effect that protects them from herbivorous animals, offering protection against ultraviolet radiation and protection against parasites and microorganisms (Albuquerque et al., 2021; Corso et al., 2020). Protection against ultraviolet B radiation is particularly important, as this is a stress factor for plants causing damage to DNA and proteins, as well as generating Reactive Oxygen Species (ROS) that cause oxidative stress (Yoon et al., 2021). Phenolic compounds can absorb this radiation and, due to their antioxidant capacity, prevent the formation of free radicals (Yoon et al., 2021). These compounds have been extensively researched and are currently of great interest because of their wide range of biological activities, particularly their high antioxidant capacity.

In general, phenolic compounds have shown a wide range of important and beneficial properties for plants and humans. These compounds are currently abundant constituents of the human diet. The primary sources of phenolic compounds are vegetables, fruits, plants, cereals, and

Bioactive Polyphenols for Health and Pathology Treatment.
DOI: https://doi.org/10.1016/B978-0-443-13821-8.00002-0

beverages, namely, wine, coffee, or tea (Bilal Hussain et al., 2019; Rahman et al., 2021). In the last few years, their dietary intake has been closely related to health-promoting effects, such as anti-inflammatory, antioxidant, anticarcinogenic, or antimicrobial, among others (Caponio et al., 2022; Karakaya, 2004). Consequently, they receive special interest for their presumed function in the prevention of various degenerative diseases such as cancer, Alzheimer's disease, or cardiovascular pathologies (Scalbert et al., 2002).

Their anti-inflammatory activity is mediated by their ability to neutralize the overproduction of free radicals, such as ROS, which induce the synthesis of pro-inflammatory cytokines that activate signaling cascades of the inflammatory response (Zhang & Tsao, 2016). The antiinflammatory activity of these compounds is attributed to their ability to inhibit other pro-inflammatory pathways, such as the degranulation of granulocytes or the activation of NF-κB (Sharma et al., 2018; Zhang & Tsao, 2016).

Free radicals cause multiple cellular damages due to oxidative stress, which is involved in the increased inflammatory response, leading to various pathologies associated with chronic inflammation such as cancer, cardiovascular and neurological diseases, and diabetes, with phenolic compounds being studied for the treatment and prevention of these conditions (Puangpraphant et al., 2022).

Phenolic compounds not only show activity against noncommunicable chronic diseases, but have also demonstrated therapeutic potential as antibacterial, antiviral, and antifungal agents (Laganà et al., 2019). For example, some of them, such as podophyllotoxin, are currently marketed as topical drugs for the treatment of anogenital warts caused by the human papillomavirus (Gilson et al., 2020).

According to the chemical structure, phenolic compounds contain one or more hydroxyl groups attached to one or more aromatic rings and are classified as flavonoids and nonflavonoids (Albuquerque et al., 2021; de la Rosa et al., 2019; Silva et al., 2023) . The classification of phenolic compounds is detailed in Fig. 2.1.

Flavonoids have a basic structure of three rings ($C_6-C_3-C_6$), where C_6 represents a benzene ring and C_n are chains containing one to four or seven carbon atoms, so their structure is more complex as they have two benzene rings connected by a three-carbon chain that forms a pyran ring (Shen et al., 2022; Silva et al., 2023). Therefore flavonoids have a phenylbenzopyran skeleton (Fig. 2.2), with a structure composed of three rings. A ring is the condensed benzene with furan, C, and the other benzene ring, B, is

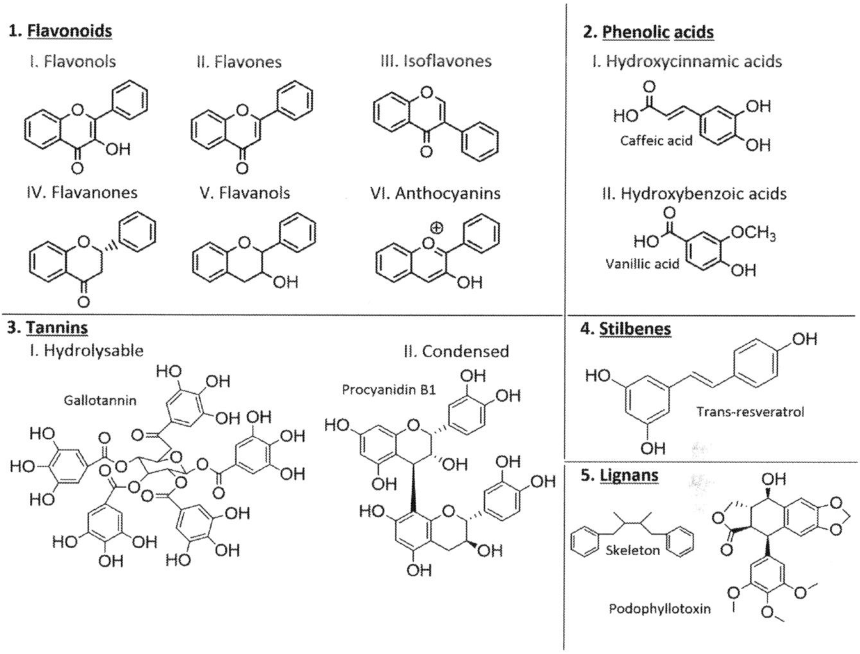

Figure 2.1 Classification of phenolic compounds and subfamilies.

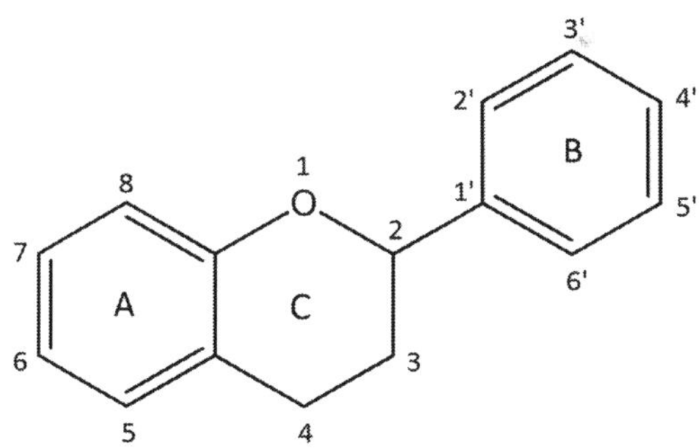

Figure 2.2 Phenylbenzopyran skeleton (basic structure of flavonoids) and nomenclature.

usually connected to C_2. Flavonoids are mostly found in conjugated forms with sugars, forming glucosides, esters, or polymers, and they can also be found as aglycones (Sasot et al., 2017; Tungmunnithum et al., 2018).

Flavonoids are subdivided into six subgroups: flavanols, anthocyanidins, flavanones, flavonols, isoflavones, and flavones. Flavonols, also known as 3-hydroxyflavones, have a double bond between C_2 and C_3, with a hydroxyl group at C_3 and a ketone group at C_4 (Silva et al., 2023). They exhibit a high antioxidant capacity due to the presence of the double bond next to the ketone group, which allows for delocalization of the charge of the phenoxy radical formed in B ring toward C ring and the presence of the hydroxyl group next to the double bond further enhances the stabilization of the delocalized electron (Fernandez-Panchon et al., 2008; Rolt & Cox, 2020). Some representative flavonols are quercetin, myricetin, and kaempferol.

Flavones have a double bond between C_2 and C_3 with a ketone group at C_4. This allows the stabilization of the phenoxy radical formed when interacting with free radicals by shifting the charge from the B ring to the C ring (Fernandez-Panchon et al., 2008). Isoflavones are structurally identical to flavones, except that the B ring is attached to C_3, which affects their antioxidant capacity as the phenoxy radical formed when interacting with free radicals cannot be stabilized (Bors et al., 1990). They have some structural similarity to estrogens, which gives them activity as agonists or antagonists depending on the concentrations of these hormones (Yu et al., 2016). Representative flavones include apigenin, luteolin, and diosmetin, which are mostly found in conjugated forms. The most common isoflavones are genistein, daidzein and glycitein.

Flavanones are structurally like flavones, except that they lack the double bond between C_2 and C_3, which reduces their antioxidant power (Albuquerque et al., 2021; Fernandez-Panchon et al., 2008). They have a stereogenic center at C_2, usually with (S) configuration, making them optically active (Dias et al., 2021). They are mainly found in citrus fruits, being naringenin and hesperetin the most abundant.

Flavanols, also known as flavan-3-ols, also lack the double bond between C_2 and C_3 and lose the ketone group at C_4, while maintaining the hydroxyl group at C_3 (de la Rosa et al., 2019; Dias et al., 2021). The two most representative compounds are catechin and epicatechin.

Anthocyanins have a unique structure as they lack the ketone group at C_4 but have two double bonds instead of one, making them cationic structures (Di Lorenzo et al., 2021). They have a hydroxyl group at C_3. The main anthocyanins are cyanidin, delphinidin, peonidin, petunidin, and pelargonidin.

Non-flavonoids include a variety of structures, which are generally smaller and simpler than those of flavonoids, such as phenolic acids, but they can also have highly complex structures, like condensed tannins. The most basic structure present within the nonflavonoids group are phenolic acids which consist of a phenolic ring attached to a carboxylic group and can contain one or more phenols. There are two groups of phenolic acids: hydroxycinnamic and hydroxybenzoic acids. Hydroxycinnamic acids are derived from cinnamic acid with a $C_6—C_3$ skeleton, which includes ferulic acid, p-coumaric acid, caffeic acid, and chlorogenic acid (Kumar & Goel, 2019). Hydroxybenzoic acids have a $C_6—C_1$ skeleton and derive from benzoic acid, with the most common ones being vanillic acid and p-hydroxybenzoic acid (Kumar & Goel, 2019).

Tannins constitute the subgroup of phenolic compounds with the largest size, as they are phenolic polymers. They are divided into hydrolyzable tannins, which are composed of simple phenols linked to a carbohydrate, and condensed tannins, which are formed by the condensation of flavan-3-ol monomers, also known as proanthocyanidins. Depending on the type of linkage, condensed tannins can be of type B when there are linkages C_4-C_6 or C_4-C_8, and of type A when a C_2-O_7 or C_2-O_5 double bond is formed (Hellström et al., 2009). The most common are the B-type proanthocyanidins such as procyanidin B1, and procyanidin B2.

Stilbenes are compounds that have a basic C_6-C_2-C_6 structure, joining two benzene rings with a double bond in the central chain, giving rise to Z/E isomers (De Filippis et al., 2017). (E)-resveratrol is the most representative compound belonging to this group because of its biological activity, unlike the Z isomer, which lacks biological activity. Currently, research is being conducted to modify (E)-resveratrol to obtain derivatives that enhance its anticancer activity (De Filippis et al., 2017; Yang et al., 2020).

The basic structure of lignans is phenylpropanoid chains, with the monomers that make them up being cinnamic acid, allylbenzene, propenylbenzene, and cinnamyl alcohol, which can be made up of two or more monomeric units (Cui et al., 2020). They are linked through the central carbons of the side chains by β-β' bonds, resulting in what is known as lignans, while when the link between the monomers occurs in any other position or involves other carbons, they are called as neolignans (Cui et al., 2020; MacRae & Towers, 1984; Sangiorgio et al., 2023). Lignans are subdivided into eight subgroups and neolignans into fifteen subgroups. One of the most important substances belonging to this group is podophyllotoxin, a lignan belonging to the aryltetralin subgroup.

Considering the great diversity of phenolic compounds and the general interest due to their bioactive properties, it is notable to note that these compounds generally do not reach the target tissues in their native form (Villegas-Aguilar et al., 2021). For example, glycosylated phenolic structures cannot be absorbed in their native form, so they need to be hydrolyzed to be absorbed, and thus the aglycone form become bioavailable (Dias et al., 2021; Fernandez-Panchon et al., 2008; Sasot et al., 2017; Shen et al., 2022; Wang et al., 2019). For that reason, it is of high importance to know the bioavailability and metabolic transformations of phenolic compounds since they must be considered to better understand the biological effects of them. Fortunately, recent technological advances are allowing greater knowledge about the metabolism and bioavailability of phenolic compounds. Numerous studies have recently been carried out in animal or human models using metabolomic methodologies for this purpose (Fernández-Ochoa et al., 2022). In this context, this chapter aims to describe the general metabolization pathways of phenolic compounds as well as report scientific evidence on the bioavailability of these compounds.

General common aspects of the bioavailability and metabolism of phenolic compounds

Bioavailability is defined as the fraction of compounds that have been absorbed, reaching the bloodstream, and are available at the potential targets of action where they can exhibit their bioactivity. It is closely related to the absorption, distribution, metabolism, and excretion processes (ADME). The bioavailability of phenolic compounds is generally difficult to predict since there are many factors that can influence this aspect. Some of these factors that influence bioavailability are the chemical structure, the concentration of the compound, interactions with other compounds, and the conditions of the individual (age, gender, lifestyle, health status, genetics, etc.), among others.

Metabolism is part of the ADME processes and is defined as a set of enzyme-mediated chemical reactions occurring in the body to modify the structure of chemical compounds, increasing their polarity to facilitate elimination. Metabolism can be divided into two types of reactions: phase I reactions, known as oxidation-reduction reactions, and phase II reactions, known as conjugation reactions, which involve the binding of compounds with various endogenous substances.

Most phenolic compounds are found in nature in their glucose-bound form (glycosylated form), which is not compatible with direct absorption by epithelial cells and needs to be transformed into simpler forms by

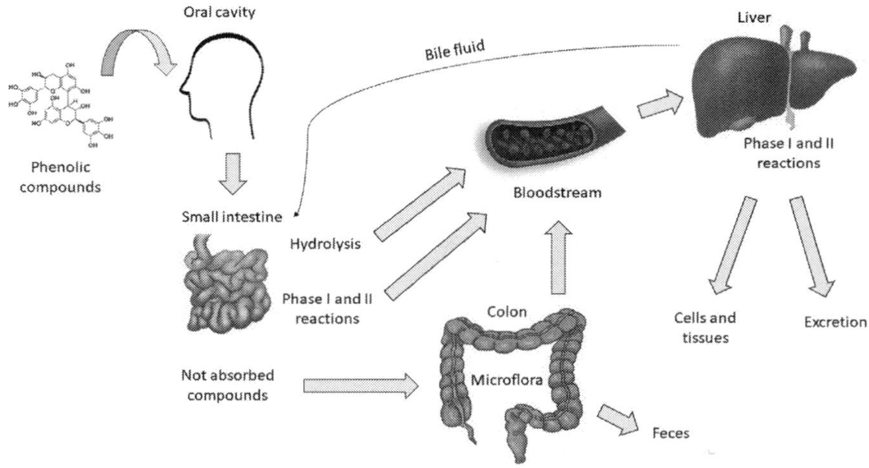

Figure 2.3 Schematic representation of the main metabolic reactions of phenolic compounds.

enzymes in the small intestine or colon (Velderrain-Rodríguez et al., 2014). Thus these compounds reaching the small intestine undergo enzymatic hydrolysis reactions which give rise to aglycone structures. After this first stage, aglycones and other simple phenols, before entering the bloodstream, may undergo further changes through phase I and phase II metabolic reactions. Some of the phenolic compounds that cannot be absorbed at the small intestinal level may be modified and transformed by the microbiota of the colon and subsequently absorbed in this part of the gastrointestinal tract or eliminated through the feces. In addition, metabolites that pass into the bloodstream can reach the liver, where they can undergo further transformation as part of phase I and phase II metabolism. After undergoing these transformations, the modified compounds can either reach the kidneys and be excreted in the urine or be transported from the intestine via bile fluid, where they may follow various pathways, be reabsorbed again, or be carried through the bloodstream to the target cells and tissues (Fig. 2.3) (Crozier et al., 2010; Velderrain-Rodríguez et al., 2014).

Phase I metabolism

During this phase I of metabolism, phenolic compounds can undergo hydrolysis, oxidation, and reduction reactions. Following these transformations, the bioactive effect of the resulting compounds may change. The main purpose of such transformations is to increase the polarity of the heterogeneous phenolic compounds to facilitate their excretion (Velderrain-Rodríguez et al., 2014). Hydrolysis reactions involve functional groups such as amides, lactones, and carboxylic esters (Bilal Hussain et al., 2019). Oxidation of phenolic compounds is the most important reaction occurring in phase I metabolism. These transformations are performed by enzymes that conduct oxidation processes through the microsomal cytochrome CYP450 (Basheer & Kerem, 2015). Most of the absorption of phenolic compounds occurs in the large intestine, where the colonic microbiota plays a significant role in the catabolic reactions that these compounds undergo, that is, reduction reactions (Aura, 2008).

Phase II metabolism

The main objective of phase II reactions of metabolism is to increase the polarity of compounds with the final aim of favoring the elimination of exogenous compounds (Kroon et al., 2004). In this sense, the main reactions that occur in phase II metabolism are glucuronidation, sulphation, and methylation.

The enzymes responsible for conducting the reactions that phenolic compounds undergo in this phase are uridine 5′-diphosphate glucuronide transferase, sulphonate transferase, and catechol *O*-methyltransferase (Crozier et al., 2010).

Glucuronidation is the main conjugation-based reaction in humans. To carry out this reaction, UDP-glucuronide transferase utilizes glucuronic acid diphosphate and causes it to eventually bind to the phenolic compound as glucuronic acid. This type of reaction is the main way in which they react during metabolism in humans (Bilal Hussain et al., 2019).

Methylation is another reaction that phenolic compounds undergo to increase their transportation ability across biological membranes. There are studies that indicate that when there is an increase in the number of methyl groups and a decrease in the number of hydroxyl groups that can be conjugated with glucuronic acid groups and sulfate groups, the stability and capacity to cross biological membranes increases (Gutiérrez-Grijalva et al., 2015).

Methodological approaches for the study of the bioavailability and metabolism of phenolic compounds

The continuous increase in knowledge of the bioavailability and metabolism of phenolic compounds is possible thanks to the technological development of analytical techniques (e.g., mass spectrometry, chromatography, nuclear magnetic resonance), that allow the detection of phenolic compounds or their derived metabolites in different types of biological samples. In general, analytical techniques, especially mass spectrometry, increasingly have more resolution, selectivity and sensitivity, which enable the detection of metabolites related to phenolic compounds present in lower concentrations or improve the characterization of isomeric chemical structures (Fernández-Ochoa et al., 2022). Likewise, advances in software, data processing tools, mass banks, and databases (e.g., FoodDB, HMBD, KEGG, PlantCyc, etc.), which increasingly contain more information and data on phenolic compounds, are also contributing to a better knowledge of the bioavailability and metabolism of phenolic compounds.

There is considerable diversity in relation to the experimental methodology of studies focused on studying the bioavailability and metabolism of phenolic compounds. In this context, these studies can be based on in vitro simulation of gastrointestinal digestion (e.g., INFOGEST (Brodkorb et al., 2019)), cell (e.g., caco-2 cell line (Lea, 2015)), or in vivo animals or human models (Fernández-Ochoa et al., 2022). Regarding in vivo models, animal models have the advantage that they allow a greater variety of tissues and organs (e.g., brain, liver, lung, etc.) to be collected where bioavailability can be studied. However, in both animal and human models, the biological samples most used for the study of bioavailability and metabolism are blood, urine, and feces, given their ease of collection and that several samples can be collected at different times. For bioavailability and metabolism studies, acute intervention studies are generally used. In these studies, the most typical biological samples, urine and blood, are collected at different times after the intake of the phenolic compounds or extracts of interest. These times can range from a few minutes to 24 or 48 hours after ingestion, depending on the type of samples. For example, blood samples are generally collected between 0 and 8 hours, and urine samples can be collected up to 24 or 48 hours after ingestion. Generally, bioavailable phenolic compounds or their derived metabolites reach the bloodstream quite rapidly after intake. However, depending on the family or compounds, these times may be longer, especially considering if the metabolites are absorbed after the action of the colon microbiota. More information about the recent advances and methodological information of bioavailability and metabolism studies have been recently reviewed (Fernández-Ochoa et al., 2022).

Bioavailability and metabolization studies of the different families of phenolic compounds

In recent years, much progress has been done in the investigation of the bioavailability of polyphenols by different in vitro or in vivo (human and animal models) approaches. Thus, this section describes studies that have demonstrated the bioavailability and metabolism of phenolic compounds according to the different families. Table 2.1 summarizes the most relevant studies on this topic.

Bioavailability and metabolism of flavonoids

Flavanols

Cocoa, tea leaves, wine, and apples are known dietary sources of flavan-3-ols, such as catechin, epicatechin (EC), or epigallocatechin (EGC) (Schramm et al., 2003). Tea is the main contributor of flavan-3-ols in several studies that have been carried out in different populations, especially in European countries and the United States, where people are used to drink tea very frequently (Zamora-Ros et al., 2010). Since they are found in so many different foods, the dietary source in the population is different demographically and, therefore, the type of flavan-3-ol consumed is also different (Campos et al., 2019). Although the inter-country dietary differences that can occur, numerous studies agree that the average consumption of flavan-3-ols is less than 1 g/day (Bernatova, 2018; Schramm et al., 2003). In an adult Spanish population, average intake of catechin and epicatechin is 19.06 ± 17.98 and 13.82 ± 10.78 mg/day, which corresponds to the 37.79% and 34.19% of the total daily flavan-3-ols consumption, respectively (Zamora-Ros et al., 2010). In accordance with this study, Somerset and Johannot estimated that the average daily epicatechin and catechin ingestion of adult Australians were 16.64 mg and 9.36 mg, respectively, which is also below 1 g (Somerset & Johannot, 2008; Thilakarathna & Rupasinghe, 2013).

Lambert et al. studied the bioavailability of epigallocatechin-3-gallate (EGCG) in mice after intravenous (*i.v.*) or intragastric (*i.g.*) administration. The bioavailability of EGCG after *i.g.* administration was 12.4%–25.6%. This low bioavailability is explained by the high rate of conjugation of the initial flavan-3-ol, which undergoes mainly phase II metabolism. Thus between 50% and 90% of EGCG appears in the plasma of mice in its conjugated forms. Interestingly, in the tissues of mice EGCG is found

Table 2.1 Outstanding studies on bioavailability and metabolism classified by the different families of phenolic compounds.

Phenolic compound and source	Model type	Dosage	Biological samples	Parameters	Bioavailability	References
Flavan-3-ols						
Pure epigallocatechin-3-gallate (EGCG)	Animal (mice)	21.8 μmol/kg b.w. intravenously and 163.8 μmol/kg b.w. intragastrically	Blood, urine, feces, lungs, liver, spleen, kidneys, colon, small intestine, prostate, and brain	Pharmacokinetic parameters and metabolite profiling	Potential bioavailability of phase II metabolites. Fecal excretion is the main route of elimination.	Lambert et al. (2003)
Epicatechin (EC) in 0.5% sodium carboxymethyl cellulose solution	Animal (rats)	350 mg/kg b.w.	Blood, urine, feces, heart, liver, spleen, lung, kidney, and brain	Metabolite profiling	67 metabolites were identified, mainly phase II metabolites, but also including the parent EC.	Shang et al. (2017)
Epicatechin (EC) from dark chocolate	Human	100 g of dark chocolate	Blood and urine	Pharmacokinetic parameters and metabolite profiling	Several phase II conjugates in plasma and urine were detected. 21 ± 2% of the total ingested EC was excreted in urine during the first 24 h.	Actis-Goretta et al. (2012)
Radiolabeled [2−14 C](−) epicatechin	Human	60 mg of [2−14 C] (−)EC	Blood, urine, and feces	Pharmacokinetic parameters, metabolite profiling and radioactivity measurement	EC is very well absorbed. >20 metabolites in plasma and urine (almost all of them sulfated or glucuronized)	Ottaviani et al. (2016)

(*Continued*)

Table 2.1 (Continued)

Phenolic compound and source	Model type	Dosage	Biological samples	Parameters	Bioavailability	References
Flavonols						
Quercetin from commercial quercetin-supplemented dog food	Animal (dogs)	10 mg/kg b.w.	Blood	Pharmacokinetic parameters and metabolite profiling	Absolute bioavailability of quercetin was only 4% in dogs. The major fraction of plasma flavonols consisted of glucuronidated and/or sulphated conjugates.	Reinboth et al. (2010)
Quercetin aglycone and quercetin glucoside	Animal (pigs)	30 μmol/kg b.w.	Blood	Pharmacokinetic parameters and metabolite profiling	Rapid absorption of quercetin. Increase dietary fat content enhanced the total bioavailability of both quercetin aglycone and quercetin glucoside.	Lesser et al. (2004)
Quercetin aglycone	Human	1095 mg	Blood	Pharmacokinetic parameters, metabolite profiling and biomarkers of antioxidants, oxidative stress, and inflammation	Dietary fat dose-dependently increases quercetin bioavailability.	Guo et al. (2013)
Quercetin glucosides from fried onions	Human	270 g of lightly fried onions	Blood and urine	Pharmacokinetic parameters and metabolite profiling	23 metabolites were identified in plasma and urine. Substantial phase II metabolism. Importance of urinary excretion of flavonol glucuronides.	Mullen et al. (2006)

Kaempferol from endive soup	Human	9 mg of kaempferol equivalent from 300 g of endive soup	Blood and urine	Pharmacokinetic parameters and metabolite profiling	Early absorption of kaempferol. 1.9% of the amount ingested was excreted in the urine. Kaempferol-3-glucuronide was the major compound detected in plasma and urine.	DuPont et al. (2004)
Anthocyanidins						
Anthocyanidin-3-glucosides from red wine and red grape juice	Human	400 mL of wine or juice	Blood and urine	Pharmacokinetic parameters and metabolite profiling	Higher bioavailability of dietary anthocyanidins contained in red grape juice. Low urinary excretion. The best bioavailable anthocyanin was petunidin-3-glucoside from red grape juice.	Frank et al. (2003)
Malvidin-3-glucoside (M-3-G) from red wine, dealcoholized red wine and red grape juice	Human	500 mL of wine or juice	Blood and urine	Pharmacokinetic parameters and metabolite profiling	Low urinary excretion of M-3-G. Dose-dependent bioavailability. Alcohol did not affect the portion of M-3-G absorbed.	Bub et al. (2001)
Cyanidin-3-glucoside (Cy-g) and Cyanidin-3,5-diglucoside (Cy-dg) from red fruit extract	Human and animal (rats)	160/320 mg of Cy-g and 20/40 mg of Cy-dg/kg b.w. (rats) 2.7 mg of Cy-g and 0.25 mg of Cy-dg/kg b.w. (humans)	Blood and liver (rats) Blood (human)	Pharmacokinetic parameters and metabolite profiling	Dose-dependent bioavailability. No aglycones or conjugates were found in rat and human plasma. Presence of methylated metabolites in rat liver.	Miyazawa et al. (1999)

(*Continued*)

Table 2.1 (Continued)

Phenolic compound and source	Model type	Dosage	Biological samples	Parameters	Bioavailability	References
Isoflavones						
Genistein and its glycoside genistin, either as pure compounds or from a soy protein isolate extract	Animal (rats)	56 μmol/kg b.w. of genistein or 35 μmol/kg b.w. of genistin	Blood, feces, and luminal contents	Pharmacokinetic parameters and metabolite profiling	Early absorption of isoflavones. Bioavailability of genistein is higher for the aglycon if compared to the glycosidic form.	Steensma et al. (2006)
Radiolabeled [^{14}C] genistein	Animal (rats)	4 mg/kg b.w.	Blood, urine, feces, heart, lungs, thymus, spleen, liver, kidney, abdominal adipose tissue, thigh, femur, brain, ovary or testis, uterus or prostate, vagina, stomach wall, jejunum, and cecum	Pharmacokinetic parameters and metabolite profiling	67% of the dose was excreted in urine. Conjugation of genistein in the intestinal wall and liver. Clear accumulation of genistein or its metabolites in reproductive tissues.	Coldham and Sauer (2000)
Isoflavones from *doenjang* paste	Animal (estrogen-deficient rats)	1 g/kg b.w. of *doenjang*	Blood	Pharmacokinetic parameters and metabolite profiling	Identification of 26 metabolites (parent isoflavones, phase I and II, and gut-mediated metabolites). Isoflavone aglycones were rapidly absorbed.	Lee et al. (2018)

Flavanones						
Naringenin from cooked tomato paste	Human	3.8 mg of naringenin aglycone	Blood	Pharmacokinetic parameters and metabolite profiling	Dose-dependent bioavailability. Naringenin is largely metabolized.	Bugianesi et al. (2002)
Naringenin and hesperetin glycosides from orange or grapefruit extract	Human	400–760 mL of orange juice and 400–720 mL of grapefruit juice	Blood and urine	Pharmacokinetic parameters and metabolite profiling	Naringenin and hesperetin are bioavailable from orange juice and grapefruit juice. After 4–8 hours of ingestion, 83 and 63% of naringenin was excreted in the urine from orange juice and grapefruit juice, respectively.	Erlund et al. (2001)
Naringenin from grapefruit extract	Human	1.5 L of pure grapefruit juice for a week (140 mg naringenin/L juice)	Urine	Pharmacokinetic parameters and metabolite profiling	14% of naringenin was recovered in the urine. It was excreted mainly as naringenin glucuronide.	Lee and Reidenberg (1998)
Naringenin and naringin from grapefruit extract	Human	20 mL of grapefruit juice/kg b.w. (621 μg naringenin/L juice)	Blood and urine	Pharmacokinetic parameters and metabolite profiling	Naringin or naringin glucuronides were not detected in urine. Naringenin and its glucuronides were mainly excreted after 2 hours.	Fuhr and Kummert (1995)

(*Continued*)

Table 2.1 (Continued)

Phenolic compound and source	Model type	Dosage	Biological samples	Parameters	Bioavailability	References
Flavones						
Luteolin and luteolin 7-O-β-glucoside	Human and animal (rats)	50 μmol/kg b.w. of luteolin and luteolin 7-O-β-glucoside (rats) 50 mg of luteolin (humans)	Blood and jejunum (rats) Blood (humans)	Pharmacokinetic parameters and metabolite profiling	Luteolin 7-O-β-glucoside was hardly absorbed but did not appear in rat plasma, so it may be first hydrolyzed to luteolin. Free luteolin and phase II conjugates were detected in rat and human plasma.	Shimoi et al. (1998)
Pure luteolin or luteolin from peanut hull extract (PHE)	Animal (rats)	14.3 mg/kg b.w.	Blood	Pharmacokinetic parameters, metabolite profiling and *in vitro* studies	Luteolin from PHE was absorbed more efficiently than pure luteolin. Absorption is more efficient in the duodenum and jejunum than in the colon and ileum.	Zhou et al. (2008)
Apigenin from parsley	Human	65.8 ± 15.5 μmol/kg b.w.	Blood and urine	Pharmacokinetic parameters and metabolite profiling	Apigenin is bioavailable after parsley consumption. However, plasma concentrations were low and only 0.22% of the ingested dose was found in the 24-hour urine samples. Plasma luteolin concentrations showed no response to parsley consumption.	Meyer et al. (2006)

Nonflavonoids						
Resveratrol from red wine and grape extract tablets	Human	375 mL of red wine (6.30 ± 0.09 mg of resveratrol) or 15 grape extract tablets (4.72 ± 0.07 mg of resveratrol)	Blood and urine	Pharmacokinetic parameters and metabolite profiling	17 resveratrol and piceid metabolites were detected after administration. Sulfate and glucuronide conjugated were found in urine, but only glucuronides were identified in plasma.	Rotches-Ribalta et al. (2012)
Resveratrol from red wine	Human	250 mL of red wine	LDL and urine	Pharmacokinetic parameters and metabolite profiling	Sulfated and glucuronized metabolites were identified in LDL and urine. Metabolites are principally excreted in urine. Resveratrol and its metabolites can be used as nutritional biomarkers of wine intake.	Urpi-Sarda et al. (2007)
Resveratrol	Animal (rats and mouse)	20 mg/kg b.w.	Blood and urine	Pharmacokinetic parameters and metabolite profiling	Resveratrol glucuronide and resveratrol sulfate were detected in serum samples. No phase I metabolites of resveratrol were detected.	Yu et al. (2002)

(*Continued*)

Table 2.1 (Continued)

Phenolic compound and source	Model type	Dosage	Biological samples	Parameters	Bioavailability	References
Chlorogenic acid, caffeic acid, p-coumaric acid, and ferulic acid	Animal (rats)	40 mg/kg b.w.	Urine	Pharmacokinetic parameters and metabolite profiling	Hydroxycinnamic acids were detected as free forms or phase II metabolites (sulfates/glucuronides). Chlorogenic acid was the less bioavailable although it was excreted more slowly.	Kishida and Matsumoto (2019)
Trans-cinnamic and p-coumaric acids	Animal (rats)	0.233 mmol/kg b.w.	Urine	Pharmacokinetic parameters, metabolite profiling and *in vitro* studies	Both acids were eliminated by renal excretion. Urinary excretion of coumaric acid was shorter, but its urinary recovery rate was 77 times higher than that of cinnamic acid (23% vs 0.3%) of the administered dose.	Garrait et al. (2006)

primarily in the free form. In addition, EGCG is essentially excreted through the feces (57.9−1572.1 nmol/g after *i.g.* administration), above the urine (0.11−9.39 μmol/g after *i.g.* administration). This result confirms the expected poor absorption of EGCG through the intestine because of its polar surface area (Lambert et al., 2003).

Another similar study was conducted by Shang et al., in which EC was suspended in 0.5% sodium carboxymethyl cellulose solution and orally administered to rats at a dose of 350 mg/kg *b.w.* Plasma, urine, feces, and six different tissues (heart, liver, spleen, lung, kidney, and brain) were collected and a total of 67 metabolites were identified, including the EC compound. Among them, 67 structures were found in urine, 40 in feces, and only 9 in plasma samples. Again, in agreement with other studies, phase II metabolites were the predominant ones. The discovery of these metabolites, which is one of the most challenging tasks, allows to increase the knowledge about the possible properties of these circulating metabolites (Shang et al., 2017).

Results described for human studies are in agreement with those obtained from animal studies (Bernatova, 2018). For instance, Ottaviani et al. performed a clinical trial with the aim of understanding the absorption, distribution, metabolization, and excretion of epicatechin in humans. The study consisted of the oral intake of 60 mg of radiolabeled [2−^{14}C] (−)epicatechin. More than 20 conjugated metabolites were detected in plasma and urine, being almost all of them sulfated or glucuronized derivates, in addition to very low levels of unmetabolized EC (Ottaviani et al., 2016). Other studies have shown similar trends reveling the importance of the phase II metabolism for flavan-3-ols compounds. For example, in a study analyzing plasma and urine from human subjects who ingested green tea containing 634 μmol of flavan-3-ols, metabolites were mainly found in sulfated, methylated, and glucuronidated forms of EC and EGC (Stalmach et al., 2010). Similar discoveries were reported by Actis-Goretta et al., who detected several phase II conjugates in human plasma and urine after oral consumption of 100 g of dark chocolate (Actis-Goretta et al., 2012).

Flavonols

Flavonols are widely distributed in foods, such as onions, lettuces, apples, grapes, kale, and berries, but also in beverages such as wine and tea (Thilakarathna & Rupasinghe, 2013; Zamora-Ros et al., 2010). In plants, flavonols mainly occur bound to sugars as lipophobic β-glycosides, but only aglycones can be absorbable (Hollman & Arts, 2000). According to a few epidemiological cohort studies in different populations, the average intake of flavonols is generally about 20–35 mg/day (Hollman & Arts, 2000; Zamora-Ros et al., 2010). Quercetin, myricetin, fisetin, or kaempferol are some of the most studied flavonols in recent decades (Crozier et al., 2000; M. Wang et al., 2023). In this group of compounds, it should be noted that most of the conjugated flavonols (over 80%) circulate bound to albumin. and not in their free form (Dufour & Dangles, 2005).

A study conducted by Reinboth et al. demonstrated that the absolute bioavailability of quercetin was only 4% after supplementing beagle dogs with this flavonol. Similar to other families of flavonoids, it was discovered that the sulfated and glucuronidated conjugates of quercetin were the principal circulating metabolites. However, in contrast to other studies in monogastric species, the bioavailability of quercetin after ingestion of rutin (quercetin rutinoside) was relatively higher when compared to the ingestion of quercetin aglycone itself (Berger et al., 2012; Reinboth et al., 2010). Several researchers reported that the absorption of flavonols is enhanced by conjugation with glucose (Dabeek & Marra, 2019; Hollman & Arts, 2000).

Exploring how to improve the bioavailability of flavonols, another study in pigs found that ingestion of fat-rich foods favored the absorption of quercetin aglycone and quercetin glucoside maybe due to the improved bioaccessibility of quercetin when it is incorporated into the mixed micelles made up after lipid digestion, but also because of the higher levels of bile production, what likewise increase the number of mixed micelles that can solubilize and transport quercetin (Kandemir et al., 2022; Lesser et al., 2004). This result is in accordance with the outcomes reported by Guo et al., who also compared the bioavailability of quercetin after consumption of low- and high-fat muffins containing quercetin aglycone by overweight subjects (Guo et al., 2013).

Human blood and urine samples were collected by Mullen et al. after feeding the subjects with 270 g of fried onions. In total, 23 metabolites were identified in that study. Among them, quercetin-3′-sulfate, quercetin-3-glucuronide, isorhamnetin-3-glucuronide, a quercetin diglucuronide and a quercetin glucuronide sulfate were the main plasma metabolites, and - quercetin-3′-glucuronide, two quercetinglucoside sulfates, and a methylquercetin diglucuronide were the main urinary metabolites. These results show clear evidence of the phase II metabolic reactions that occur after the release of the aglycone and suggest that urine is one of the main routes of excretion of flavonol glucuronides, consistent with the findings of other studies (Dabeek & Marra, 2019; Lehtonen et al., 2010; Mullen et al., 2006; Wang et al., 2003).

In the case of kaempferol, Dupont et al. revealed that it is more efficiently absorbed than quercetin after ingestion of kaempferol-3-glucuronide-rich endive soup. This outcome, which has also been proposed by other researchers, is suggested because no quercetin was detected in samples, which can also indicate that kaempferol is not converted to quercetin by phase I metabolism (DuPont et al., 2004; Nielsen et al., 1997). Almost 2% of the ingested kaempferol was excreted in the urine, which is a comparable result with the discoveries of other studies, suggesting that urinary kaempferol could be used as a biomarker of kaempferol consumption (De Vries et al., 1998; DuPont et al., 2004; Wang et al., 2003).

Finally, one study identified the metabolism of quercetin, luteolin, kaempferol-rutinoside, naringenin, and myricetin by rat colonic microflora. Quercetin-derived metabolites were phenylacetic acid; p-hydroxyphenylacetic acid; o-hydroxyphenylacetic acid; m-hydroxyphenylacetic acid; 3,4-dihydroxyphenylacetic acid; protocatechuic acid; homovainillic acid; 3-(3,4-dihydroxyphenyl) propionic acid; 3-(3,4-dihydroxyphenyl) propionic acid and p-hydroxybenzoic acid. Kaempferol-rutinoside-derived metabolites were phenylacetic acid; o-hydroxyphenylacetic acid and p-hydroxybenzoic acid. Naringenin-derived metabolites were phenylacetic acid; p-hydroxyphenylacetic acid; o-hydroxyphenylacetic acid; protocatechuic acid; 3-(4-hydroxyphenyl) propionic acid and 3-(2,4-dihydroxyphenyl) propionic acid. The metabolite that appeared derived from luteolin was 3-(2,4-dihydroxyphenyl) propionic acid and finally the myricetin-derived metabolites were phenylacetic acid; 3,4-dihydroxyphenylacetic acid and 3-(2,4-dihydroxyphenyl) propionic acid (Serra et al., 2012).

Anthocyanidins

Anthocyanins are water-soluble pigments belonging to the flavonoid group that are responsible for the color of flowers, fruits, vegetables, and some cereals. These compounds can be found in the form of glycosides, acylglucosides, or aglycones, with a C6−C3−C6 chemical skeleton. The main anthocyanins that can be found in plants are the following aglycones: delphinidin, pelargonidin, petunidin, cyanidin, peonidin, and malvidin, which differ in the number and degree of methylation of the hydroxyl groups on the B-ring. Typically, methylation increases redness and improves stability, while hydroxylation increases blueness and reduces stability (Luiza Koop et al., 2022).

Red wine and fruits such as apples, pears, cranberries, cherries, or grapes are the main contributors to anthocyanidins (Naeem et al., 2022). In a Spanish cohort, the average intake of anthocyanidins was 26.34 ± 26.29 mg/day in the case of male subjects, and 11.38 ± 10.93 mg/day for female participants (Zamora-Ros et al., 2010). Furthermore, according to the National Health and Nutrition Examination Survey (NHANES) of the United States, anthocyanidins are one of the least ingested polyphenols, with a value of about 5% of the total daily intake of flavonoids (Naeem et al., 2022).

There is a scarceness of studies on the bioavailability of anthocyanidins in comparison to other flavonoids. In 2003, the bioavailability of several anthocyanidin glucosides post-consumption of red wine or red grape juice in healthy volunteers was investigated. Results showed higher relative bioavailability of dietary anthocyanidins contained in red grape juice (76.3%) over those contained in red wine, with a low to high order of bioavailability that can be described as follows: cyanidin-, malvidin-, delphinidin-, peonidin-, and petunidin-3-glucoside from red grape juice (Frank et al., 2003). A similar investigation was conducted two years before, which was focused on studying the bioavailability of malvidin-3-glucoside (M-3-G) after ingestion of red wine, dealcoholized red wine, and red grape juice (Bub et al., 2001). Similar areas under the curve were observed in plasma samples after consumption of both wines, so that it can be considered that alcohol did not affect the portion of M-3-G absorbed. However, ethanol seemed to modify the time course of increase in M-3-G in plasma concentration. In addition, for all urinary samples that were analyzed, it was observed that 24 hours after ingestion of studied matrixes, M-3-G was not detected. This data may indicate a fast excretion of M-3-G, a result that has been reported for other anthocyanidins as well. Interestingly, a linear dependence was detected for the total excretion of M-3-G in plasma and urine and the dose consumed (Bub et al., 2001; Cao et al., 2001; Frank et al., 2005).

Seven human volunteers participated in a study aimed to determine the bioavailability of cyanidin-3-glucoside (Cy-g) and cyanidin-3,5-diglucoside (Cy-dg) after oral supplementation of 1.6 g of red fruit anthocyanin extract (Miyazawa et al., 1999). Moreover, they extended the study by introducing 0.8 g of the extract by stomach tube in rats. In both animal and human models, it was demonstrated that Cy-g and Cy-dg were absorbed from the digestive tract and incorporated into the plasma. In line with the results proposed by Bub et al., a dose-dependent incorporation of the anthocyanidins into the organisms was observed. Another interesting finding was that no aglycones or conjugates were found in rat and human plasma, so it may indicate that the glycosylated forms are responsible for the bioavailability of anthocyanidins (Bub et al., 2001; Miyazawa et al., 1999).

Several studies have shown the presence of unmetabolized (parent) anthocyanin glycosides in human blood and urine. However, although many researchers suggest that anthocyanins are not metabolized before release into the systemic circulation, recent evidence suggests otherwise. Recent studies have shown the presence of glucuronides, methylated forms, and sulphoconjugated forms of anthocyanins in urine as metabolized derivatives (Kay, 2006). Studies analyzing the metabolites resulting from the metabolic reactions of certain anthocyanins have shown that after ingestion of cyanidin and malvidin, the main metabolites that appear are their methylated and glucuronidated forms, while pelargonidin mainly undergoes glucuronidation reactions (Kalt, 2019).

Isoflavones

Isoflavones are natural constituents of plants that are structurally like mammalian oestrogens and have estrogenic effects. They have a diphenylpropane structure in which the B-ring is located at position 3. The presence of isoflavones is restricted to a few foodstuffs. Leguminous plants, and specifically soy, are the main sources of isoflavones, especially containing genistein and daidzein, which have been proposed to exert a function in the prevention of breast cancer and osteoporosis. An average intake of isoflavones of 30–40 mg/day was determined for Japanese population. However, in other countries, it could be lower due to reduced soy consumption (Scalbert & Williamson, 2000; Thilakarathna & Rupasinghe, 2013). Examples of isoflavones are daidzein and genistein, which according to studies can give rise to the metabolites dihydrodaidzein, equol, and *O*-desmethylangolensin in the case of daidzein and dihydrogenistein in the case of genistein (Uehara, 2013).

In the literature, it has been reviewed that the metabolism of isoflavones in humans is diverse, as the main reactions that can occur are deglycosylation, reduction, oxidation, methylation and demethylation, and conjugation with glucuronic and/or sulphonic acids before excretion. The most abundant phase I metabolites of the isoflavones, after deglycosylation, appear to be formed by reduction and, in the case of formononetin and biochanin A, which have methoxyl substituents at the 4-position, by demethylation (Heinonen et al., 2002).

Steensma et al. compared the bioavailability of genistein and its glycoside genistin when they were administered as pure compounds or as a soy protein isolate extract in a rat model (Steensma et al., 2006). The key reported outcome was that the bioavailability of genistein is higher for the aglycon than for its glycoside. It was also demonstrated that absorption of isoflavones is fast after oral intake, showing a maximum plasma concentration 15 minutes after consumption. Recovery of genistein and its metabolites occurred largely through feces, which is in accordance with other authors (Coldham & Sauer, 2000; Steensma et al., 2006).

[^{14}C]genistein was administered to rats by Coldham et al. at a dose of 4 mg/kg *b.w.* They observed efficient absorption in the gastrointestinal tract, since 67% of the administered dose was eliminated in the urine, although the bioavailability of the parent phytoestrogen was lower. Based on the results, they proposed that genistein follows a pathway of biotransformation into 4-hydroxyphenyl-2-propionic acid via reduction and heterocyclic ring fission, which is consistent with other studies (Coldham & Sauer, 2000; Coldham et al., 1999). Genistein glucuronides were also predominant plasma metabolites. Furthermore, a clear accumulation of genistein or its metabolites in reproductive tissues of male and female rats was remarkable, perhaps due to the higher concentrations of estrogen receptors or binding proteins (Coldham & Sauer, 2000).

Recently, Lee et al. investigated the bioavailability of isoflavone metabolites in estrogen-deficient rats after feeding them with *doenjang*, a well-known Korean fermented soybean paste, or boiled soybean (Lee et al., 2018). Twenty-six isoflavone metabolites were found in plasma samples, including three parent isoflavones (daidzein, genistein and glycitein), two-phase I, eight-phase II (e.g., daidzein-7-glucuronide, genistein 4′-sulfate), and 13 gut-mediated metabolites (e.g., equol 4′-sulfate, 2-(4-hydroxyphenyl) propionic acid, ferulic acid). Curiously, although fermentation of soybean seemed to reduce its total content of isoflavones, results suggested that ingestion of *doenjang* was more successful in enhancing the bioavailability of isoflavones in estrogen-deficient rats (Lee et al., 2018).

The biological relevance of sulfated and glucuronized metabolites is little knwon, and nowadays more researchers are focused on studying the bioactive properties of these circulating metabolites. Furthermore, it is a notable result that isoflavone aglycones are better and faster absorbed than the glucosides in human plasma, in contrast to the findings described for flavonols (Hsiao et al., 2020; Izumi et al., 2000). Notably, Naeem et al. published a classification of flavonoids based on their bioavailability, highlighting isoflavones as the most bioavailable. (Naeem et al., 2022).

Flavanones

Citrus fruits are the primary source of flavanones, namely, orange, grapefruit, or lemon, and to a lesser extent, tomatoes, and mint. In particular, hesperidin is the most consumed flavanone (Scalbert & Williamson, 2000). Other examples of flavanones could be naringenin, eriodyctiol, and hesperetin. As previously described for isoflavones, flavanone aglycones are quickly absorbed compared to glycosides (Thilakarathna & Rupasinghe, 2013). For instance, Bugianesi et al. designed a study for determining the human plasma levels of naringenin after consumption of cooked tomato paste that contained 3.8 mg of naringenin aglycone. Results showed a maximum concentration of naringenin aglycone in plasma two hours after the intake. This is consistent with the early absorption of aglycones in the digestive tract described by various authors (Bugianesi et al., 2002; Hollman & Arts, 2000). On the other hand, naringenin and hesperetin glycosides were administered to volunteers through the consumption of orange or grapefruit juices in a study conducted by Erlund et al. (2001). In this case, they found the maximum peaks of naringenin and hesperetin glycosides at approximately 4.8−5.5 hours post-ingestion of the juices. This would probably be consistent with the absorption of glycosides in the distal part of the small intestine (Bugianesi et al., 2002; Erlund et al., 2001).

Lee and Reidenberg studied the bioavailability of naringenin after consumption of 1500 mL of pure grapefruit juice for a whole week. About 14% of naringenin was recovered in the urine and it was excreted as naringenin glucuronide (Lee & Reidenberg, 1998). This result aling with the finding previously published by Fuhr and Kummert, who conducted a similar study based on tgrapefruit juice consumption. That study showed that naringin is metabolized to naringenin, which undergoes rapid and nearly complete glucuronidation, and then is excreted in the urine iat a rate similar to what other researchers have described (Fuhr & Kummert, 1995; Manach et al., 2005). Finally, the excretion of flavanones seems to be slower than that of catechins, but it is somewhat faster than that of quercetin (Manach et al., 2003).

Flavones

Flavones are the less abundant flavonoids. They exist mainly in celery, sweet red pepper, and green leafy herbs (chamomile, parsley). Apigenin and luteolin are the most studied flavones (Thilakarathna & Rupasinghe, 2013). In a Spanish cohort it was estimated that the contribution of apigenin and luteolin was only 0.40% and 0.66% of the total daily flavonoid ingestion, respectively (Zamora-Ros et al., 2010). Comparatively, daily average apigenin intake was greater than luteolin intake in Chinese and Australian adults (Thilakarathna & Rupasinghe, 2013). Hence, food habits play an important role in the flavone intake.

Absorption of luteolin and luteolin 7-O-β-glucoside in rats was investigated after oral administration. 7-O-β-glucoside was not detected in plasma, so it may be hydrolyzed to luteolin by bacteria of the intestinal mucosa. Plasma of rats that were fed with luteolin contained free luteolin, in addition to phase II conjugates, and reached their maximum levels 15–30 min after dosing with luteolin in propyleneglycol (Shimoi et al., 1998). Interestingly, according to Zhou et al., the oral bioavailability of lutein from plant extracts (e.g., peanut hull extract) in rats is significantly higher than that of the pure luteolin, which is a remarkable finding (Zhou et al., 2008).

Direct evidence of a human intervention study demonstrated that apigenin is enriched in plasma and urine after consumption of parsley (Meyer et al., 2006). Approximately, 0.22 ± 0.16% of ingested apigenin was found in the urinary excretion, a result that is in good agreement with the findings of Nielsen et al. and Borges et al. (Borges et al., 2022; Meyer et al., 2006; Shimoi et al., 1998).

Bioavailability and metabolism of nonflavonoids

One of the most important groups within non-flavonoid compounds is the phenolic acids. Two classes of phenolic acids can be distinguished depending on their structure: derivatives of benzoic acid and derivatives of cinnamic acid. The presence of hydroxybenzoic acids is linked to certain red fruits and onions, but they are not abundant in edible plants. Some of the most well-known hydroxybenzoic acids could be gallic, protocatechuic, or vanillic acids. In contrast, hydroxycinnamic acids are present in citrus fruits and others, such as apples, cherries, or peaches. Coumaric, caffeic, and ferulic acids are some of the most common hydroxycinnamic acids (Cosme et al., 2020; Scalbert & Williamson, 2000).

Ferulic acid is the most common hydroxycinnamic acid in vegetables and is ester-linked to arabinose in various plant polysaccharides such as arabinoxylans and pectins. One of the most important metabolic transformations this compound undergoes is its conversion to vanillin (3-methoxy-4-hydroxy-benzaldehyde). Other transformations that ferulic acid can undergo in humans include its conversion to caffeic acid, a process observed by *Penicillium rubrum* and *Pseudomonas* sp. under anaerobic conditions. Ferulic acid is also converted to 3-methoxy-4-hydroxy phenyl propionic acid by anaerobic bacteria such as *Pseudomonas cepacia* and *Wolinella succinogenes* and to coniferyl alcohol. In addition, the rat intestinal microflora could convert ferulic acid to m-hydroxyphenyl propionic acid, dihydrocaffeic acid and 4-ethyl catechol (Mathew & Abraham, 2006). In another study in which different types of bread were analyzed, it was observed that in humans ferulic acid gave rise to metabolites such as ferulic acid-4′-O sulfate, dihydroferulic acid-4′-O-sulfate, and dihydroferulic acid-O-glucuronide (Bresciani et al., 2016).

Kishida and Matsumoto aimed to examine the bioavailability of some representative hydroxycinnamic acids (chlorogenic acid, caffeic acid, p-coumaric acid, and ferulic acid) in rats. Results showed that caffeic, p-coumaric and ferulic acids were primarily excreted through the urine within 0–6 hours as free forms or sulfates/glucuronides, while the majority of chlorogenic acid was detected at 6–24 h or 24–48 h. In addition, the urinary excretion rate of caffeic, p-coumaric and ferulic acids were between 54.1% and 73.2% of the dose, whereas for chlorogenic acid it was 4.9% (Kishida & Matsumoto, 2019). In conclusion, chlorogenic acid was less bioavailable although it was excreted more slowly, and these results are consistent with the findings reported by other authors who worked similarly on hydroxycinnamic acids bioavailability in animals (Adam et al., 2002; Garrait et al., 2006; Gonthier et al., 2003; Kishida & Matsumoto, 2019).

Another important group of compounds belonging to this group are stilbenes, to which resveratrol belongs, an antioxidant compound present in grapes and wine. Stilbenes are not abundant in food plants but are present in grapes and wines. Resveratrol has received considerable attention during the last few years due to its anticarcinogenic properties (Cosme et al., 2020; Scalbert & Williamson, 2000). Among the metabolization reactions that resveratrol can undergo are glucuronidation and sulphation, and thus the appearance of the metabolites resveratrol-3-O-glucuronide, resveratrol-4′-O-glucuronide, resveratrol-3-O-sulfate, resveratrol-4′-O-sulfate and resveratrol-3, 4′-O-disulfate (Wang & Sang, 2018). Rotches-Ribalta et al. investigated the bioavailability of resveratrol and piceid from red wine and grape extract when they were administered to human volunteers. Seventeen metabolites were found, including conjugates of resveratrol, dihydroresveratrol, and piceid (Rotches-Ribalta et al., 2012). As other published studies have shown, glucuronides and sulfates are the main metabolites of resveratrol (Urpi-Sarda et al., 2007). Therefore they were identified in urine samples. However, only glucuronide metabolites were found in plasma samples. The low bioavailability of unchanged resveratrol is also described by other authors who detailed its intense metabolism involving both glucuronidation and sulfation in human and animal models (Meng et al., 2004; Walle et al., 2004; Yu et al., 2002).

A group also belonging to the non-flavonoid phenolic compounds is the tannins. These compounds can be divided into four major groups: gallotannins, ellagitannins, complex tannins, and condensed tannins. The oral intake of tannins has two potential mechanisms of action; they may exert their biological effects as nonabsorbable complexes or as absorbable simpler units. Studies in which the bioavailability of tannins has been studied show that ellagitannins, which are hydrolyzable tannins, gradually fragmented into ellagic acid glucuronides and various urolithins. These urolithins were detected in bile and plasma, indicating an active circulation. Urolithin A is also considered a biomarker of ellagitannin intake, as it is also detected in human plasma and urine (Fraga-Corral et al., 2021).

Finally, there is the group of lignans. Most of the lignans present in plant foods in humans are converted by the intestinal microflora in the upper part of the large intestine. A consortium of intestinal bacteria can convert plant lignans into enterolignans, enterodiol, and enterolactone. Differences in the bioavailability of various plant lignans from foods and variations in enterolignan production among gut microbial communities contribute to large inter-individual variation in enterolignan exposure (Chen et al., 2020). Thus in a study analyzing enterolignans in urine, enterodiol, and enterolactone were detected in the blood and urine of humans and animals, but only small amounts of the plant lignans cyclolariciresinol, lariciresinol, pinoresinol, and syringaresinol were found in human urine (Nurmi et al., 2003).

Conclusion

Scientific progress has shown that phenolic compounds may not reach target tissues in their native form since they undergo metabolization reactions. Among the different metabolization reactions, the hydrolysis reactions of glycosylated structures stand out since these structures cannot be absorbed in their native form. Furthermore, phase I and phase II metabolization reactions are shown as general metabolization reactions for most phenolic compounds, both flavonoids and non-flavonoid phenolics. The advance in knowledge related to the bioavailability and metabolism of phenolic compounds is being possible thanks to the increase in studies, both in vitro and in vivo, focused on this topic; together with the progress of analytical techniques. However, there is still a lack of studies focused on the mechanisms of biological action of metabolized compounds, which may be different from that of their native compounds. Therefore further study of the bioactive properties of these bioavailable circulating metabolites is needed to fully understand the beneficial effects of phenolic compounds.

Acknowledgments

M.d.C.V.-A. and P.F-M thanks the Spanish Ministry of Science, Innovation, and Universities for the grants FPU19/01146 and FPU21/02410, respectively.

Declaration of competing interest

The authors declare that they have no known competing financial interests or personal relationships that could have appeared to influence the work reported in this chapter.

References

Actis-Goretta, L., Lévèques, A., Giuffrida, F., Romanov-Michailidis, F., Viton, F., Barron, D., Duenas-Paton, M., Gonzalez-Manzano, S., Santos-Buelga, C., Williamson, G., & Dionisi, F. (2012). Elucidation of (−)-epicatechin metabolites after ingestion of chocolate by healthy humans. *Free Radical Biology and Medicine*, *53*(4), 787−795. Available from https://doi.org/10.1016/j.freeradbiomed.2012.05.023.

Adam, A., Crespy, V., Levrat-Verny, M. A., Leenhardt, F., Leuillet, M., Demigné, C., & Rémésy, C. (2002). The bioavailability of ferulic acid is governed primarily by the food matrix rather than its metabolism in intestine and liver in rats. *Journal of Nutrition*, *132*(7), 1962−1968. Available from https://doi.org/10.1093/jn/132.7.1962.

Albuquerque, B. R., Heleno, S. A. 8, Oliveira, M. B. P. P., Barros, L., & Ferreira, I. C. F. R. (2021). Phenolic compounds: Current industrial applications, limitations and future challenges. *Food and Function*, *12*(1), 14−29. Available from https://doi.org/10.1039/d0fo02324h.

Aura, A. M. (2008). Microbial metabolism of dietary phenolic compounds in the colon. *Phytochemistry Reviews*, 7(3), 407−429. Available from https://doi.org/10.1007/s11101-008-9095-3.

Basheer, L., & Kerem, Z. (2015). Interactions between CYP3A4 and dietary polyphenols. *Oxidative Medicine and Cellular Longevity*, *2015*, 1–15. Available from https://doi.org/10.1155/2015/854015.

Berger, L. M., Wein, S., Blank, R., Metges, C. C., & Wolffram, S. (2012). Bioavailability of the flavonol quercetin in cows after intraruminal application of quercetin aglycone and rutin. *Journal of Dairy Science*, *95*(9), 5047–5055. Available from https://doi.org/10.3168/jds.2012-5439.

Bernatova, I. (2018). Biological activities of (−)-epicatechin and (−)-epicatechin-containing foods: Focus on cardiovascular and neuropsychological health. *Biotechnology Advances*, *36*(3), 666–681. Available from https://doi.org/10.1016/j.biotechadv.2018.01.009.

Bilal Hussain, M., Hassan, S., Waheed, M., Javed, A., Adil Farooq, M., & Tahir, A. (2019). *Bioavailability and metabolic pathway of phenolic compounds*. IntechOpen. Available from 10.5772/intechopen.84745.

Borges, G., Fong, R. Y., Ensunsa, J. L., Kimball, J., Medici, V., Ottaviani, J. I., & Crozier, A. (2022). Absorption, distribution, metabolism and excretion of apigenin and its glycosides in healthy male adults. *Free Radical Biology and Medicine*, *185*, 90–96. Available from https://doi.org/10.1016/j.freeradbiomed.2022.04.007.

Bors, W., Heller, W., Michel, C., & Saran, M. (1990). [36] Flavonoids as antioxidants: Determination of radical-scavenging efficiencies. *Methods in Enzymology*, *186*, 343–355. Available from https://doi.org/10.1016/0076-6879(90)86128-i.

Bresciani, L., Scazzina, F., Leonardi, R., Dall'Aglio, E., Newell, M., Dall'Asta, M., Melegari, C., Ray, S., Brighenti, F., & Daniele, D. R. (2016). Bioavailability and metabolism of phenolic compounds from wholegrain wheat and aleurone-rich wheat bread. *Molecular Nutrition & Food Research*, *60*(11), 2343–2354. Available from https://doi.org/10.1002/mnfr.201600238.

Brodkorb, A., Egger, L., Alminger, M., Alvito, P., Assunção, R., Ballance, S., Bohn, T., Bourlieu-Lacanal, C., Boutrou, R., Carrière, F., Clemente, A., Corredig, M., Dupont, D., Dufour, C., Edwards, C., Golding, M., Karakaya, S., Kirkhus, B., Le Feunteun, S., … Recio, I. (2019). INFOGEST static in vitro simulation of gastrointestinal food digestion. *Nature Protocols*, *14*(4), 991–1014. Available from https://doi.org/10.1038/s41596-018-0119-1.

Bub, A., Watzl, B., Heeb, D., Rechkemmer, G., & Briviba, K. (2001). Malvidin-3-glucoside bioavailability in humans after ingestion of red wine, dealcoholized red wine and red grape juice. *European Journal of Nutrition*, *40*(3), 113–120. Available from https://doi.org/10.1007/s003940170011.

Bugianesi, R., Catasta, G., Spigno, P., D'Uva, A., & Maiani, G. (2002). Naringenin from cooked tomato paste is bioavailable in men. *Journal of Nutrition*, *132*(11), 3349–3352. Available from https://doi.org/10.1093/jn/132.11.3349.

Campos., Stehle., & Simon. (2019). Microbial metabolites of flavan-3-ols and their biological activity. *Nutrients*, *11*(10), 2260. Available from https://doi.org/10.3390/nu11102260.

Cao, G., Muccitelli, H. U., Sánchez-Moreno, C., & Prior, R. L. (2001). Anthocyanins are absorbed in glycated forms in elderly women: A pharmacokinetic study. *American Journal of Clinical Nutrition*, *73*(5), 920–926. Available from https://doi.org/10.1093/ajcn/73.5.920.

Caponio, G., Lippolis, T., Tutino, V., Gigante, I., De Nunzio, V., Milella, R., Gasparro, M., & Notarnicola, M. (2022). Nutraceuticals: Focus on anti-inflammatory, anti-cancer, antioxidant properties in gastrointestinal tract. *Antioxidants*, *11*(7), 1274. Available from https://doi.org/10.3390/antiox11071274.

Chen, J., Chen, Y., & Ye, X. (2020). *Lignans in Diets. Handbook of Dietary Phytochemicals* (pp. 1–22). Singapore: Springer. Available from https://doi.org/10.1007/978-981-13-1745-3_38-1.

Coldham, N. G., Howells, L. C., Santi, A., Montesissa, C., Langlais, C., King, L. J., Macpherson, D. D., & Sauer, M. J. (1999). Biotransformation of genistein in the rat: Elucidation of metabolite structure by product ion mass fragmentology. *Journal of Steroid Biochemistry and Molecular Biology*, *70*(4-6), 169—184. Available from https://doi.org/10.1016/S0960-0760(99)00104-1.

Coldham, N. G., & Sauer, M. J. (2000). Pharmacokinetics of [14C]genistein in the rat: Gender-related differences, potential mechanisms of biological action, and implications for human health. *Toxicology and Applied Pharmacology*, *164*(2), 206—215. Available from https://doi.org/10.1006/taap.2000.8902, http://www.elsevier.com/inca/publications/store/6/2/2/9/5/1/index.htt.

Corso, M., Perreau, F., Mouille, G., & Lepiniec, L. (2020). Specialized phenolic compounds in seeds: structures, functions, and regulations. *Plant Science*, *296*110471. Available from https://doi.org/10.1016/j.plantsci.2020.110471.

Cosme, P., Rodríguez, A. B., Espino, J., & Garrido, M. (2020). Plant phenolics: Bioavailability as a key determinant of their potential health-promoting applications. *Antioxidants*, *9*(12), 1263. Available from https://doi.org/10.3390/antiox9121263.

Crozier, A., Burns, J., Aziz, A. A., Stewart, A. J., Rabiasz, H. S., Jenkins, G. I., Edwards, C. A., & Lean, M. E. J. (2000). Antioxidant flavonols from fruits, vegetables and beverages: Measurements and bioavailability. *Biological Research*, *33*(2), 79—88. Available from https://doi.org/10.4067/S0716-97602000000200007.

Crozier, A., Del Rio, D., & Clifford, M. N. (2010). Bioavailability of dietary flavonoids and phenolic compounds. *Molecular Aspects of Medicine*, *31*(6), 446—467. Available from https://doi.org/10.1016/j.mam.2010.09.007.

Cui, Q., Du, R., Liu, M., & Rong, L. (2020). Lignans and their derivatives from plants as antivirals. *Molecules*, *25*(1), 183. Available from https://doi.org/10.3390/molecules25010183.

Dabeek, W. M., & Marra, M. V. (2019). Dietary quercetin and kaempferol: Bioavailability and potential cardiovascular-related bioactivity in humans. *Nutrients*, *11*(10). Available from https://doi.org/10.3390/nu11102288, https://www.mdpi.com/2072-6643/11/10/2288/pdf.

De Filippis, B., Ammazzalorso, A., Fantacuzzi, M., Giampietro, L., Maccallini, C., & Amoroso, R. (2017). Anticancer Activity of Stilbene-Based Derivatives. *ChemMedChem*, *12*, 558—570. Available from https://doi.org/10.1002/cmdc.201700045.

de la Rosa, L. A., Moreno-Escamilla, J. O., Rodrigo-García, J., & Alvarez-Parrilla, E. (2019). *Chapter 12 - Phenolic Compounds. Postharvest Physiology and Biochemistry of Fruits and Vegetables* (pp. 253—271). Woodhead Publishing. Available from https://doi.org/10.1016/B978-0-12-813278-4.00012-9.

De Vries, J. H. M., Hollman, P. C. H., Meyboom, S., Buysman, M. N. C. P., Zock, P. L., Van Staveren, W. A., & Katan, M. B. (1998). Plasma concentrations and urinary excretion of the antioxidant flavonols quercetin and kaempferol as biomarkers for dietary intake. *American Journal of Clinical Nutrition*, *68*(1), 60—65. Available from https://doi.org/10.1093/ajcn/68.1.60.

Di Lorenzo, C., Colombo, F., Biella, S., Stockley, C., & Restani, P. (2021). Polyphenols and human health: The role of bioavailability. *Nutrients*, *13*(1), 273. Available from https://doi.org/10.3390/nu13010273.

Dias, M. C., Pinto, D. C. G. A., & Silva, A. M. S. (2021). Plant flavonoids: Chemical characteristics and biological activity. *Molecules*, *26*(17). Available from https://doi.org/10.3390/molecules26175377, https://www.mdpi.com/1420-3049/26/17/5377/pdf.

Dufour, C., & Dangles, O. (2005). Flavonoid—serum albumin complexation: determination of binding constants and binding sites by fluorescence spectroscopy. *Biochimica et Biophysica Acta (BBA) - General Subjects*, *1721*(1-3), 164—173. Available from https://doi.org/10.1016/j.bbagen.2004.10.013.

DuPont, M. S., Day, A. J., Bennett, R. N., Mellon, F. A., & Kroon, P. A. (2004). Absorption of kaempferol from endive, a source of kaempferol-3-glucuronide, in humans. *European Journal of Clinical Nutrition, 58*(6), 947–954. Available from https://doi.org/10.1038/sj.ejcn.1601916.

Erlund, I., Meririnne, E., Alfthan, G., & Aro, A. (2001). Plasma kinetics and urinary excretion of the flavanones naringenin and hesperetin in humans after ingestion of orange juice and grapefruit juice. *The Journal of Nutrition, 131*(2), 235–241. Available from https://doi.org/10.1093/jn/131.2.235.

Fernández-Ochoa, Á., Cádiz-Gurrea, M. de la L., Fernández-Moreno, P., Rojas-García, A., Arráez-Román, D., & Segura-Carretero, A. (2022). Recent analytical approaches for the study of bioavailability and metabolism of bioactive phenolic compounds. *Molecules, 27*(3). Available from https://doi.org/10.3390/molecules27030777, https://www.mdpi.com/1420-3049/27/3/777/pdf.

Fernandez-Panchon, M. S., Villano, D., Troncoso, A. M., & Garcia-Parrilla, M. C. (2008). Antioxidant activity of phenolic compounds: From in vitro results to in vivo evidence. *Critical Reviews in Food Science and Nutrition, 48*(7), 649–671. Available from https://doi.org/10.1080/10408390701761845.

Fraga-Corral, M., Otero, P., Cassani, L., Echave, J., Garcia-Oliveira, P., Carpena, M., Chamorro, F., Lourenço-Lopes, C., Prieto, M. A., & Simal-Gandara, J. (2021). Traditional applications of tannin rich extracts supported by scientific data: Chemical composition, bioavailability and bioaccessibility. *Foods, 10*(2). Available from https://doi.org/10.3390/foods10020251, https://www.mdpi.com/2304-8158/10/2/251/pdf.

Frank, T., Janßen, M., Netzel, M., Straß, G., Kler, A., Kriesl, E., & Bitsch, I. (2005). Pharmacokinetics of anthocyanidin-3-glycosides following consumption of *Hibiscus sabdariffa* L. extract. *The Journal of Clinical Pharmacology, 45*(2), 203–210. Available from https://doi.org/10.1177/0091270004270561.

Frank, T., Netzel, M., Strass, G., Bitsch, R., & Bitsch, I. (2003). Bioavailability of anthocyanidin-3-glucosides following consumption of red wine and red grape juice. *Canadian Journal of Physiology and Pharmacology, 81*(5), 423–435. Available from https://doi.org/10.1139/y03-038.

Fuhr, U., & Kummert, A. L. (1995). The fate of naringin in humans: A key to grapefruit juice-drug interactions? *Clinical Pharmacology & Therapeutics, 58*(4), 365–373. Available from https://doi.org/10.1016/0009-9236(95)90048-9.

Garrait, G., Jarrige, J. F., Blanquet, S., Beyssac, E., Cardot, J. M., & Alric, M. (2006). Gastrointestinal absorption and urinary excretion of trans-cinnamic and p-coumaric acids in rats. *Journal of Agricultural and Food Chemistry, 54*(8), 2944–2950. Available from https://doi.org/10.1021/jf053169a.

Gilson, R., Nugent, D., Bennett, K., Doré, C. J., Murray, M. L., Meadows, J., Haddow, L. J., Lacey, C., Sandmann, F., Jit, M., Soldan, K., Tetlow, M., Caverly, E., Nathan, M., & Copas, A. J. (2020). Imiquimod versus podophyllotoxin, with and without human papillomavirus vaccine, for anogenital warts: the HIPvac factorial RCT. *Health Technology Assessment, 24*(47), 1–86. Available from https://doi.org/10.3310/hta24470.

Gonthier, M. P., Verny, M. A., Besson, C., Rémésy, C., & Scalbert, A. (2003). Chlorogenic acid bioavailability largely depends on its metabolism by the gut microflora in rats. *Journal of Nutrition, 133*(6), 1853–1859. Available from https://doi.org/10.1093/jn/133.6.1853.

Guo, Y., Mah, E., Davis, C. G., Jalili, T., Ferruzzi, M. G., Chun, O. K., & Bruno, R. S. (2013). Dietary fat increases quercetin bioavailability in overweight adults. *Molecular Nutrition and Food Research, 57*(5), 896–905. Available from https://doi.org/10.1002/mnfr.201200619.

Gutiérrez-Grijalva, E. G.-G. P., Ambriz-Pérez, D. L., Leyva-López, N., Castillo-López, R. I., & Heredia, J. B. (2015). Biodisponibilidad de compuestos fenólicos dietéticos. *Revisión. Revista Española de Nutrición Humana y Dietética, 20*(2), 140–147. Available from https://doi.org/10.14306/renhyd.20.2.184.

Heinonen, S. M., Wähälä, K., & Adlercreutz, H. (2002). Metabolism of isoflavones in human subjects. *Phytochemistry Reviews, 1*(2), 175–182. Available from https://doi.org/10.1023/A:1022555216303, Finland.

Hellström, J. K., Törrönen, A. R., & Mattila, P. H. (2009). Proanthocyanidins in Common Food Products of Plant Origin. *Journal of Agricultural and Food Chemistry, 57* (17), 7899–7906. Available from https://doi.org/10.1021/jf901434d.

Hollman, P. C. H., & Arts, I. C. W. (2000). Flavonols, flavones and flavanols – Nature, occurrence and dietary burden. *Journal of the Science of Food and Agriculture, 80*(7), 1081–1093. Available from https://doi.org/10.1002/(SICI)1097-0010(20000515)80:7 < 1081::AID-JSFA566 > 3.0.CO;2-G.

Hsiao, Y. H., Ho, C. T., & Pan, M. H. (2020). Bioavailability and health benefits of major isoflavone aglycones and their metabolites. *Journal of Functional Foods, 74*. Available from https://doi.org/10.1016/j.jff.2020.104164, http://www.elsevier.com/wps/find/journaldescription.cws_home/717426/description#description.

Izumi, T., Osawa, S., Obata, A., Tobe, K., Saito, M., Kataoka, S., Kubota, Y., Kikuchi, M., Piskula, M. K., & Kubota, Y. (2000). Soy isoflavone aglycones are absorbed faster and in higher amounts than their glucosides in humans. *Journal of Nutrition, 130*(7), 1695–1699. Available from https://doi.org/10.1093/jn/130.7.1695.

Kalt, W. (2019). Anthocyanins and their C6-C3-C6 metabolites in humans and animals. *Molecules, 24*(22), 4024. Available from https://doi.org/10.3390/molecules24224024.

Kandemir, K., Tomas, M., McClements, D. J., & Capanoglu, E. (2022). Recent advances on the improvement of quercetin bioavailability. *Trends in Food Science and Technology, 119*, 192–200. Available from https://doi.org/10.1016/j.tifs.2021.11.032.

Karakaya, S. (2004). Bioavailability of phenolic compounds. *Critical Reviews in Food Science and Nutrition, 44*(6), 453–464. Available from https://doi.org/10.1080/10408690490886683.

Kay, C. D. (2006). Aspects of anthocyanin absorption, metabolism and pharmacokinetics in humans. *Nutrition Research Reviews, 19*(1), 137–146. Available from https://doi.org/10.1079/NRR2005116.

Kishida, K., & Matsumoto, H. (2019). Urinary excretion rate and bioavailability of chlorogenic acid, caffeic acid, p-coumaric acid, and ferulic acid in non-fasted rats maintained under physiological conditions. *Heliyon, 5*(10), e02708. Available from https://doi.org/10.1016/j.heliyon.2019.e02708.

Kroon, P. A., Clifford, M. N., Crozier, A., Day, A. J., Donovan, J. L., Manach, C., & Williamson, G. (2004). How should we assess the effects of exposure to dietary polyphenols in vitro? *American Journal of Clinical Nutrition, 80*(1), 15–21. Available from https://doi.org/10.1093/ajcn/80.1.15.

Kumar, N., & Goel, N. (2019). Phenolic acids: Natural versatile molecules with promising therapeutic applications. *Biotechnology Reports, 24*, e00370. Available from https://doi.org/10.1016/j.btre.2019.e00370.

Laganà, P., Anastasi, G., Marano, F., Piccione, S., Singla, R. K., Dubey, A. K., Delia, S., Coniglio, M. A., Facciolà, A., Di Pietro, A., Haddad, M. A., Al-Hiary, M., & Caruso, G. (2019). Phenolic substances in foods: health effects as anti-inflammatory and antimicrobial agents. *Journal of AOAC International, 102*(5), 1378–1387. Available from https://doi.org/10.1093/jaoac/102.5.1378.

Lambert, J. D., Lee, M. J., Lu, H., Meng, X., Hong, J. J. J., Seril, D. N., Yang, C. S., & Sturgill, M. G. (2003). Epigallocatechin-3-gallate is absorbed but extensively glucuronidated following oral administration to mice. *Journal of Nutrition, 133*(12), 4172–4177. Available from https://doi.org/10.1093/jn/133.12.4172.

Lea, T. (2015). Caco-2 cell line. In K. Verhoeckx, et al. (Eds.), *The Impact of Food Bioactives on Health* (pp. 103–111). Cham: Springer. Available from https://doi.org/10.1007/978-3-319-16104-4_10.

Lee, D. H., Kim, M. J., Park, S. H., Song, E. J., Nam, Y. D., Ahn, J., Jang, Y. J., Ha, T. Y., & Jung, C. H. (2018). Bioavailability of isoflavone metabolites after Korean Fermented Soybean Paste (Doenjang) ingestion in estrogen-deficient rats. *Journal of Food Science, 83*(8), 2212–2221. Available from https://doi.org/10.1111/1750-3841.14214.

Lee, Y. S., & Reidenberg, M. M. (1998). A method for measuring naringenin in biological fluids and its disposition from grapefruit juice by man. *Pharmacology, 56*(6), 314–317. Available from https://doi.org/10.1159/000028215.

Lehtonen, H. M., Lehtinen, O., Suomela, J. P., Viitanen, M., & Kallio, H. (2010). Flavonol glycosides of sea buckthorn (hippophaë rhamnoides ssp. sinensis) and lingonberry (vaccinium vitis-idaea) are bioavailable in humans and monoglucuronidated for excretion. *Journal of Agricultural and Food Chemistry, 58*(1), 620–627. Available from https://doi.org/10.1021/jf9029942, http://pubs.acs.org/doi/pdfplus/10.1021/jf9029942.

Lesser, S., Cermak, R., & Wolffram, S. (2004). Bioavailability of quercetin in pigs is influenced by the dietary fat content. *The Journal of Nutrition, 134*(6), 1508–1511. Available from https://doi.org/10.1093/jn/134.6.1508.

Luiza Koop, B., Nascimento da Silva, M., Diniz da Silva, F., Thayres dos Santos Lima, K., Santos Soares, L., José de Andrade, C., Ayala Valencia, G., & Rodrigues Monteiro, A. (2022). Flavonoids, anthocyanins, betalains, curcumin, and carotenoids: Sources, classification and enhanced stabilization by encapsulation and adsorption. *Food Research International, 153*110929. Available from https://doi.org/10.1016/j.foodres.2021.110929.

MacRae, W. D., & Towers, G. H. N. (1984). Biological activities of lignans. *Phytochemistry, 23*(6), 1207–1220. Available from https://doi.org/10.1016/S0031-9422(00)80428-8.

Manach, C., Morand, C., Gil-Izquierdo, A., Bouteloup-Demange, C., & Rémésy, C. (2003). Bioavailability in humans of the flavanones hesperidin and narirutin after the ingestion of two doses of orange juice. *European Journal of Clinical Nutrition, 57*(2), 235–242. Available from https://doi.org/10.1038/sj.ejcn.1601547.

Manach, C., Williamson, G., Morand, C., Scalbert, A., & Rémésy, C. (2005). Bioavailability and bioefficacy of polyphenols in humans. I. Review of 97 bioavailability studies. *The American Journal of Clinical Nutrition, 81*(1), 230S–242S. Available from https://doi.org/10.1093/ajcn/81.1.230S.

Mathew, S., & Abraham, T. E. (2006). Bioconversions of ferulic acid, an hydroxycinnamic acid. *Critical Reviews in Microbiology, 32*(3), 115–125. Available from https://doi.org/10.1080/10408410600709628.

Meng, X., Maliakal, P., Lu, H., Lee, M. J., & Yang, C. S. (2004). Urinary and plasma levels of resveratrol and quercetin in humans, mice, and rats after ingestion of pure compounds and grape juice. *Journal of Agricultural and Food Chemistry, 52*(4), 935–942. Available from https://doi.org/10.1021/jf030582e.

Meyer, H., Bolarinwa, A., Wolfram, G., & Linseisen, J. (2006). Bioavailability of apigenin from apiin-rich parsley in humans. *Annals of Nutrition and Metabolism, 50*(3), 167–172. Available from https://doi.org/10.1159/000090736.

Miyazawa, T., Nakagawa, K., Kudo, M., Muraishi, K., & Someya, K. (1999). Direct intestinal absorption of red fruit anthocyanins, cyanidin-3-glucoside and cyanidin-3,5-

diglucoside, into rats and humans. *Journal of Agricultural and Food Chemistry*, *47*(3), 1083–1091. Available from https://doi.org/10.1021/jf9809582.

Mullen, W., Edwards, C. A., & Crozier, A. (2006). Absorption, excretion and metabolite profiling of methyl-, glucuronyl-, glucosyl- and sulpho-conjugates of quercetin in human plasma and urine after ingestion of onions. *British Journal of Nutrition*, *96*(01), 107. Available from https://doi.org/10.1079/bjn20061809.

Naeem, A., Ming, Y., Pengyi, H., Jie, K. Y., Yali, L., Haiyan, Z., Shuai, X., Wenjing, L., Ling, W., Xia, Z. M., Shan, L. S., & Qin, Z. (2022). The fate of flavonoids after oral administration: a comprehensive overview of its bioavailability. *Critical Reviews in Food Science and Nutrition*, *62*(22), 6169–6186. Available from https://doi.org/10.1080/10408398.2021.1898333, https://www.tandfonline.com/loi/bfsn20.

Nielsen, S. E., Kall, M., Justesen, U., Schou, A., & Dragsted, L. O. (1997). Denmark human absorption and excretion of flavonoids after broccoli consumption. *Cancer Letters*, *114*(1-2), 173–174. Available from https://doi.org/10.1016/S0304-3835(97)04654-5.

Nurmi, T., Voutilainen, S., Nyyssonen, K., Adlercreutz, H., & Salonen, J. (2003). Liquid chromatography method for plant and mammalian lignans in human urine. *Journal of Chromatography B*, *798*(1), 101–110. Available from https://doi.org/10.1016/j.jchromb.2003.09.018.

Ottaviani, J. I., Borges, G., Momma, T. Y., Spencer, J. P. E., Keen, C. L., Crozier, A., & Schroeter, H. (2016). The metabolome of [2-14C](-)-epicatechin in humans: Implications for the assessment of efficacy, safety, and mechanisms of action of polyphenolic bioactives. *Scientific Reports*, *6*. Available from https://doi.org/10.1038/srep29034.

Puangpraphant, S., Cuevas-Rodríguez, E. O., & Oseguera-Toledo, M. (2022). *Anti-inflammatory and antioxidant phenolic compounds. Current advances for development of functional foods modulating inflammation and oxidative stress* (pp. 165–180). Thailand: Elsevier. Available from https://www.sciencedirect.com/book/9780128234822, https://doi.org/10.1016/B978-0-12-823482-2.00018-2.

Rahman, M.M., Rahaman, M.S., Islam, M.R., Rahman, F., Mithi, F.M., Alqahtani, T., Almikhlafi, M.A., Alghamdi, S.Q., Alruwaili, A.S., Hossain, M.S., Ahmed, M., Das, R., Emran, T. Bin, & Uddin, M.S. (2021). Role of phenolic compounds in human disease: Current knowledge and future prospects. *Molecules*, *27*(1), 233. https://doi.org/10.3390/molecules27010233.

Reinboth, M., Wolffram, S., Abraham, G., Ungemach, F. R., & Cermak, R. (2010). Oral bioavailability of quercetin from different quercetin glycosides in dogs. *British Journal of Nutrition*, *104*(2), 198–203. Available from https://doi.org/10.1017/S000711451000053X.

Rolt, A., & Cox, L. S. (2020). Structural basis of the anti-ageing effects of polyphenolics: Mitigation of oxidative stress. *BMC Chemistry*, *14*(1). Available from https://doi.org/10.1186/s13065-020-00696-0.

Rotches-Ribalta, M., Andres-Lacueva, C., Estruch, R., Escribano, E., & Urpi-Sarda, M. (2012). Pharmacokinetics of resveratrol metabolic profile in healthy humans after moderate consumption of red wine and grape extract tablets. *Pharmacological Research*, *66*(5), 375–382. Available from https://doi.org/10.1016/j.phrs.2012.08.001.

Sangiorgio, P., Errico, S., Verardi, A., Moliterni, S., Tamasi, G., Rossi, C., & Balducchi, R. (2023). Bioactive lignans from flaxseed: Biological properties and patented recovery technologies. *Nutraceuticals*, *3*(1), 58–74. Available from https://doi.org/10.3390/nutraceuticals3010005.

Sasot, G., Martínez-Huélamo, M., Vallverdú-Queralt, A., Mercader-Martí, M., Estruch, R., & Lamuela-Raventós, R. M. (2017). Identification of phenolic metabolites in human urine after the intake of a functional food made from grape extract by a high resolution LTQ-Orbitrap-MS approach. *Food Research International, 100*, 435–444. Available from https://doi.org/10.1016/j.foodres.2017.01.020.

Scalbert, A., Morand, C., Manach, C., & Rémésy, C. (2002). Absorption and metabolism of polyphenols in the gut and impact on health. *Biomedicine & Pharmacotherapy, 56*(6), 276–282. Available from https://doi.org/10.1016/s0753-3322(02)00205-6.

Scalbert, A., & Williamson, G. (2000). Dietary intake and bioavailability of polyphenols. *The Journal of Nutrition, 130*(8), 2073S. Available from https://doi.org/10.1093/jn/130.8.2073s.

Schramm, D. D., Karim, M., Schrader, H. R., Holt, R. R., Kirkpatrick, N. J., Polagruto, J. A., Ensunsa, J. L., Schmitz, H. H., & Keen, C. L. (2003). Food effects on the absorption and pharmacokinetics of cocoa flavanols. *Life Sciences, 73*(7), 857–869. Available from https://doi.org/10.1016/S0024-3205(03)00373-4.

Serra, A., Maclà, A., Romero, M. P., Reguant, J., Ortega, N., & Motilva, M. J. (2012). Metabolic pathways of the colonic metabolism of flavonoids (flavonols, flavones and flavanones) and phenolic acids. *Food Chemistry, 130*(2), 383–393. Available from https://doi.org/10.1016/j.foodchem.2011.07.055.

Shang, Z., Wang, F., Dai, S., Lu, J., Wu, X., & Zhang, J. (2017). Profiling and identification of (−)-epicatechin metabolites in rats using ultra-high performance liquid chromatography coupled with linear trap-Orbitrap mass spectrometer. *Drug Testing and Analysis, 9*(8), 1224–1235. Available from https://doi.org/10.1002/dta.2155.

Sharma, A., Sharma, P., Singh Tuli, H., & Sharma, A. K. (2018). *Phytochemical and pharmacological properties of flavonols. Encyclopedia of life sciences* (pp. 1–12). Wiley. Available from 10.1002/9780470015902.a0027666.

Shen, N., Wang, T., Gan, Q., Liu, S., Wang, L., & Jin, B. (2022). Plant flavonoids: Classification, distribution, biosynthesis, and antioxidant activity. *Food Chemistry, 383*132531. Available from https://doi.org/10.1016/j.foodchem.2022.132531.

Shimoi, K., Okada, H., Furugori, M., Goda, T., Takase, S., Suzuki, M., Hara, Y., Yamamoto, H., & Kinae, N. (1998). Intestinal absorption of luteolin and luteolin 7-O-β-glucoside in rats and humans. *FEBS Letters, 438*(3), 220–224. Available from https://doi.org/10.1016/s0014-5793(98)01304-0.

Silva, A., Silva, V., Igrejas, G., Aires, A., Falco, V., Valentão, P., & Poeta, P. (2023). Phenolic compounds classification and their distribution in winemaking by-products. *European Food Research and Technology, 249*(2), 207–239. Available from https://doi.org/10.1007/s00217-022-04163-z.

Somerset, S. M., & Johannot, L. (2008). Dietary flavonoid sources in Australian adults. *Nutrition and Cancer, 60*(4), 442–449. Available from https://doi.org/10.1080/01635580802143836.

Stalmach, A., Mullen, W., Steiling, H., Williamson, G., Lean, M. E. J., & Crozier, A. (2010). Absorption, metabolism, and excretion of green tea flavan-3-ols in humans with an ileostomy. *Molecular Nutrition and Food Research, 54*(3), 323–334. Available from https://doi.org/10.1002/mnfr.200900194.

Steensma, A., Faassen-Peters, M. A. W., Noteborn, H. P. J. M., & Rietjens, I. M. C. M. (2006). Bioavailability of genistein and its glycoside genistin as measured in the portal vein of freely moving unanesthetized rats. *Journal of Agricultural and Food Chemistry, 54*(21), 8006–8012. Available from https://doi.org/10.1021/jf060783t.

Thilakarathna, S., & Rupasinghe, H. (2013). Flavonoid bioavailability and attempts for bioavailability enhancement. *Nutrients, 5*(9), 3367–3387. Available from https://doi.org/10.3390/nu5093367.

Tungmunnithum, D., Thongboonyou, A., Pholboon, A., & Yangsabai, A. (2018). Flavonoids and other phenolic compounds from medicinal plants for pharmaceutical and medical aspects: an overview. *Medicines, 5*(3), 93. Available from https://doi.org/10.3390/medicines5030093.

Uehara, M. (2013). Isoflavone metabolism and bone-sparing effects of daidzein-metabolites. *Journal of Clinical Biochemistry and Nutrition*, *52*(3), 193–201. Available from https://doi.org/10.3164/jcbn.13-2.

Urpi-Sarda, M., Zamora-Ros, R., Lamuela-Raventos, R., Cherubini, A., Jauregui, O., De La Torre, R., Covas, M. I., Estruch, R., Jaeger, W., & Andres-Lacueva, C. (2007). HPLC-tandem mass spectrometric method to characterize resveratrol metabolism in humans. *Clinical Chemistry*, *53*(2), 292–299. Available from https://doi.org/10.1373/clinchem.2006.071936.

Velderrain-Rodríguez, G. R., Palafox-Carlos, H., Wall-Medrano, A., Ayala-Zavala, J. F., Chen, C.-Y. O., Robles-Sánchez, M., Astiazaran-García, H., Alvarez-Parrilla, E., & González-Aguilar, G. A. (2014). Phenolic compounds: Their journey after intake. *Food Funct*, *5*(2), 189–197. Available from https://doi.org/10.1039/c3fo60361j.

Villegas-Aguilar, Md. C., Fernández-Ochoa, Á., Leyva-Jiménez, F. J., Miranda-Segura, Á., Cádiz-Gurrea, M. de la L, & Segura-Carretero, A. (2021). *Phenolic compounds bioactive food components activity in mechanistic approach* (pp. 27–53). Spain: Elsevier. Available from https://doi.org/10.1016/B978-0-12-823569-0.00001-1.

Walle, T., Hsieh, F., DeLegge, M. H., Oatis, J. E., & Walle, U. K. (2004). High absorption but very low bioavailability of oral resveratrol in humans. *Drug Metabolism and Disposition*, *32*(12), 1377–1382. Available from https://doi.org/10.1124/dmd.104.000885.

Wang, P., & Sang, S. (2018). Metabolism and pharmacokinetics of resveratrol and pterostilbene. *BioFactors*, *44*(1), 16–25. Available from https://doi.org/10.1002/biof.1410.

Wang, F. M., Yao, T. W., & Zeng, S. (2003). Determination of quercetin and kaempferol in human urine after orally administrated tablet of ginkgo biloba extract by HPLC. *Journal of Pharmaceutical and Biomedical Analysis*, *33*(2), 317–321. Available from https://doi.org/10.1016/s0731-7085(03)00255-3.

Wang, S. H., Hu, Y. L., & Liu, T. X. (2019). Plant distribution and pharmacological activity of flavonoids. *Traditional Medicine Research*, *4*(5), 269–287. Available from https://doi.org/10.53388/TMR20190824131.

Wang, M., Lu, Y., Wu, Q., Chen, G., Zhao, H., Ho, C. T., & Li, S. (2023). Biotransformation and gut microbiota-mediated bioactivity of flavonols. *Journal of Agricultural and Food Chemistry*, *71*(22), 8317–8331. Available from https://doi.org/10.1021/acs.jafc.3c01087.

Yang, M. -F., Yao, X., Chen, L. -M., Gu, J. -Y., Yang, Z. -H., Chen, H. -F., Zheng, X., & Zheng, Z. -T. (2020). Synthesis and biological evaluation of resveratrol derivatives with anti-breast cancer activity. *Archiv der Pharmazie*, *353*(7). Available from https://doi.org/10.1002/ardp.202000044.

Yoon, H. I., Kim, H. Y., Kim, J., Oh, M.-M., & Son, J. E. (2021). Quantitative analysis of UV-B radiation interception in 3d plant structures and intraindividual distribution of phenolic contents. *International Journal of Molecular Sciences*, *22*(5), 1–18. Available from https://doi.org/10.3390/ijms22052701, https://www.mdpi.com/1422-0067/22/5/2701/pdf.

Yu, J., Bi, X., Yu, B., & Chen, D. (2016). Isoflavones: Anti-Inflammatory Benefit and Possible Caveats. *Nutrients*, *8*(6), 1–16, 361. Available from https://doi.org/10.3390/nu8060361.

Yu, C., Shin, Y. G., Chow, A., Li, Y., Kosmeder, J. W., Lee, Y. S., Hirschelman, W. H., Pezzuto, J. M., Mehta, R. G., & Van Breemen, R. B. (2002). Human, rat, and mouse metabolism of resveratrol. *Pharmaceutical Research*, *19*(12), 1907–1914. Available from https://doi.org/10.1023/A:1021414129280.

Zamora-Ros, R., Andres-Lacueva, C., Lamuela-Raventós, R. M., Berenguer, T., Jakszyn, P., Barricarte, A., Ardanaz, E., Amiano, P., Dorronsoro, M., Larrañaga, N., Martínez, C., Sánchez, M. J., Navarro, C., Chirlaque, M. D., Tormo, M. J., Quirós, J. R., &

González, C. A. (2010). Estimation of dietary sources and flavonoid intake in a Spanish Adult Population (EPIC-Spain). *Journal of the American Dietetic Association, 110* (3), 390–398. Available from https://doi.org/10.1016/j.jada.2009.11.024.

Zhang, H., & Tsao, R. (2016). Dietary polyphenols, oxidative stress and antioxidant and anti-inflammatory effects. *Current Opinion in Food Science, 8*, 33–42. Available from https://doi.org/10.1016/j.cofs.2016.02.002.

Zhou, P., Li, L. P., Luo, S. Q., Jiang, H. D., & Zeng, S. (2008). Intestinal absorption of luteolin from peanut hull extract is more efficient than that from individual pure luteolin. *Journal of Agricultural and Food Chemistry, 56*(1), 296–300. Available from https://doi.org/10.1021/jf072612.

SECTION 2

Polyphenols and pathologies related with metabolic syndrome

CHAPTER THREE

Polyphenolic modulation of intestinal inflammatory response

Alba Rodriguez-Nogales[1,2] and Julio Galvez[1,2,3]
[1]Department of Pharmacology, Center for Biomedical Research (CIBM), University of Granada, Granada, Spain
[2]Instituto de Investigación Biosanitaria de Granada (ibs.GRANADA), Granada, Spain
[3]Centro de Investigación Biomédica En Red para Enfermedades Hepáticas y Digestivas (CIBER-EHD), School of Pharmacy, University of Granada, Granada, Spain

Abbreviations

CRP	C reaction protein
CD	Crohn's disease
COX-2	cyclooxygenase-2
GPx	glutathione peroxidase
HO-1	heme oxygenase-1
IDO	indoleamine 2,3-dioxygenase
IBD	Inflammatory bowel disease
IEC	intestinal epithelial cell
ICAM-1	intercellular adhesion molecule-1
LPS	lipopolysaccharide
MMPs	matrix metalloproteinases
MAPKs	mitogen-activated protein kinases
MCP-1	monocyte chemotactic protein 1
MPO	myeloperoxidase
NO	nitric oxide
iNOS	nitric oxide synthase
Nrf2	nuclear factor (erythroid-derived 2)-related factor
SOD	superoxide dismutase
TJ	tight junctions
TLR	toll-like receptor
TNBS	trinitrobenzene sulfonic acid
TNF-α	tumor necrosis factor-alpha
UC	ulcerative colitis
ZO	zonula occludens

Bioactive Polyphenols for Health and Pathology Treatment.
DOI: https://doi.org/10.1016/B978-0-443-13821-8.00003-2

Introduction

The term inflammatory bowel disease (IBD) usually refers to two different conditions: Crohn's disease (CD) and ulcerative colitis (UC), which are characterized by chronic intestinal inflammation with alternating periods of relapses and remissions of varying duration and severity (Baumgart & Sandborn, 2007). Both CD and UC display similar symptoms, including diarrhea, abdominal pain, fever, clinical signs of bowel obstruction, as well as passage of blood or mucus or both, being their intensity mainly dependent on disease location (Baumgart & Sandborn, 2007). This situation usually implies a deterioration in the quality of life of patients, who require prolonged pharmacological treatments, and some of them surgical interventions. In CD patients, the intestinal inflammation can affect any part of the digestive tract, from the mouth to the anus, and inflamed areas can alternate with others free of damage. In these patients, the histological analysis of the inflamed segments reveals transmural inflammation associated with the severe infiltration of acute and chronic inflammatory cells within the mucosa, submucosa, and muscularis propria. On the contrary, UC patients show nontransmural inflammation that is typically restricted to the large intestine and rectum, being the inflammatory changes limited to the mucosa and submucosa, with cryptitis and crypt abscesses, although the composition of the inflammatory cell infiltrate is similar to that observed in CD (Vucelic, 2009).

Nowadays, IBD is considered a global public health problem. The prevalence of IBD is highest in the Western world, being affected approximately 0.5% of the whole population, with an incidence ranging from 10 to 30 per 100,000. In fact, it has been estimated that over 1.5 million and 2 million people suffer from the disease in North America and Europe, respectively, with an estimation of 31% increase in the last three decades (Piovani, 2020; Wang, 2024). Of note, although it has been classically considered a disease characteristic of developed countries, newer epidemiological studies have revealed that this trend has been also evidenced in underdeveloped countries in South America, Africa, and Asia (Ramos & Papadakis, 2019).

At present, IBD etiology has not been fully elucidated, although it is generally accepted that both genetic and environmental factors interact in the subject, which would promote an altered epithelial barrier function. This results in the subsequent access of luminal components into the bowel wall that could trigger an exacerbated and uncontrolled immune response, thus contributing to the initiation and progression of the intestinal inflammatory process. The altered immune response is associated with the activation of different immune cells, including monocytes, neutrophils, T lymphocytes or macrophages, and the increased expression and release of pro-inflammatory mediators with a critical role in the pathogenesis of IBD, including cytokines, like interleukin (IL)-6, IL-1 β, and tumor necrosis factor-alpha (TNF-α), as well as activated reactive oxygen and nitrogen species (Zhang & Li, 2014).

Since the exact etiology of IBD is unknown nowadays, no specific curative treatment is currently available against human IBD. Alternatively, different drugs are being used to downregulate the inflammatory process that occurs in these intestinal conditions, including glucocorticoids, aminosalicylates, immunosuppressants, and, more recently, biologicals (Elhag, 2022; Juillerat, 2022). Unfortunately, these drugs do not always exhibit efficacy for long-term remission in many IBD patients; moreover, their administration can be associated with significant and potentially serious side effects (Quezada et al., 2018). In fact, the worsening quality of life that some IBD patients may experience with conventional drug treatments could lead to the use of complementary and/or alternative therapies to manage their disease. Different factors can facilitate this action, such as the general belief about the safety of traditional medicine along with the thought that their disease would be better controlled (Lin & Cheifetz, 2018). Of note, when considering alternative and/or complementary therapies, plant-based products have a significant position in the management of these intestinal conditions, most probably due to their well-accepted safety profile and their reputed empirical efficacy from ancient times. In these products, it has been proposed that the presence of polyphenols in their composition can play a prominent role in the possible effects for the management of these intestinal conditions.

This chapter will focus on the beneficial effects exerted by natural phenolic compounds in the treatment and/or prevention of IBD, describing the studies performed in humans, and characterizing the possible mechanisms involved in these effects, reported in both in vivo and in vitro preclinical studies.

Polyphenols

Polyphenols, also named as phenolic compounds, are secondary metabolites frequently found in plants, and typically characterized by the presence in their chemical structure of one or more phenyl rings and one or more hydroxyl groups linked directly to the aromatic ring. Polyphenols are usually classified into two main groups: flavonoids (mainly including flavonols, flavones, flavan-3-ols, isoflavones, flavanones, and anthocyanidins) and non-flavonoids (like phenolic acids, hydrolyzable tannins, and stilbenes, among others) (Del Rio et al., 2013).

It is well reported that most of the beneficial effects exerted by polyphenols against different inflammatory conditions are derived from the main biological activities ascribed to these compounds, including antioxidant properties, due to their ability to act as free radical scavengers, and the modulation of enzymatic activities that are closely related with the inflammatory process. When considering the latter in the context of IBD, polyphenols can inhibit both NF-κB and mitogen-activated protein kinases (MAPKs) signaling pathways, associated with the reduced transcription of inducible enzymes, like cyclooxygenase-2 (COX-2) or nitric oxide synthase (iNOS), and the downregulated production of pro-inflammatory cytokines, such as TNF-α, IL-6, and IL-1β, which are typically increased in intestinal inflammation (Liu, 2021) (Fig. 3.1). Moreover,

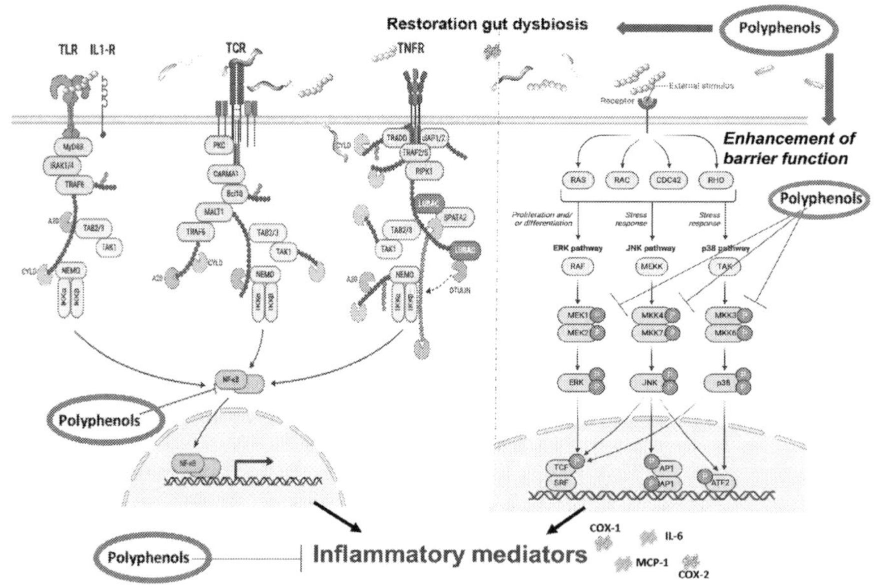

Figure 3.1 Major mechanisms of action of polyphenols. Polyphenols reduce inflammation by acting as antioxidants and inhibiting key signaling pathways, lowering pro-inflammatory cytokines, and boosting antioxidant enzymes.

these compounds have been reported to upregulate antioxidant enzymes, like superoxide dismutase (SOD) and glutathione peroxidase (GPx), thus potentiating the above-mentioned antioxidant properties, since these enzymes are able to eliminate oxygen and nitrogen reactive species, including superoxide, hydroxyl radical and nitric oxide (NO), which are overproduced in the inflamed gut tissue (Liu, 2021).

Curcumin

Curcumin is an orange-yellow polyphenolic chemical substance extracted from the rhizomes of *Curcuma longa* L. It is the main active ingredient of Chinese medicine Curcuma Longa and is now widely used in food processing, cosmetics, dyes, etc., and it has been used as a traditional medicine in India and Southeast Asia for thousands of years. In this sense, curcumin has been associated with a wide range of therapeutic effects derived from its anti-inflammatory, hypoglycemic, antioxidant, wound-healing, and antimicrobial activities, among others (Burge, 2019). In fact, and confirming its health benefits, numerous clinical studies have been performed with this compound in the last ten years, thus highlighting the growing interest and exploration of the therapeutic potential of curcumin in clinical settings. Many of these studies have primarily focused on evaluating the effect of curcumin on different types of cancer such as breast, colon, and pancreatic cancer. However, there are other studies that have highlighted the role of curcumin in intestinal inflammation. Thus one open-label study evaluated the efficacy of curcumin (550 mg twice daily for 1 month and then 550 mg three times daily for another month) in patients with ulcerative proctitis or CD (Holt et al., 2005). The results revealed a significant decrease in symptoms as well as in inflammatory indices (erythrocyte sedimentation rate and C reaction protein (CRP)) in all treated patients and a mean reduction of 55 points in the CD activity index. In another study, the efficacy of curcumin as a maintenance treatment was evaluated in UC patients (Hanai, 2006). For this randomized, double-blind, multicenter trial, one group of patients received curcumin, 1 g after breakfast and 1 g after the evening meal, plus sulfasalazine or mesalamine, and the other group received placebo plus sulfasalazine or mesalamine for 6 months. The relapse rates were around 5% in the curcumin-treated group and 21% in the placebo group. Similarly, recent studies have demonstrated that the intake of oral curcumin at a dosage of 500 mg/day in association with prednisone led to clinical and endoscopic remission in UC(Gupta et al., 2013). Following 1 year of curcumin and prednisone therapy, the results showed a reduction in bowel movements and reported improved well-being. Subsequent clinical evaluations confirmed the sustained clinical remission in these patients (Gupta et al., 2013). Moreover, all these clinical trials have revealed that curcumin has nontoxicity or important side effects, showing a high tolerability and great safety and a great potential for clinical application. Consequently, curcumin can be considered a promising treatment option or adjunct therapy in the management of IBD.

The beneficial clinical effects of curcumin in IBD have been attributed to a variety of mechanisms. In fact, several experimental models of IBD in rodents have confirmed the ability of curcumin to exert immunomodulatory and anti-inflammatory effects, being achieved by inhibiting the production and release of inflammatory mediators, like cytokines, through the regulation of gene expression, scavenging oxygen free radicals, and modulating the intestinal microbiome altered composition. It has been well described that IBD is largely characterized by upregulated toll-like receptor (TLR), specifically TLR4/NF-κB/AP-1 signaling. Since the activation of TLR4 initiates an innate immune response and subsequent inflammation (Boone & Ma, 2003), those treatments abrogating TLR4-dependent signal transduction have been reported to ameliorate intestinal inflammation (Hwang, 2018). In vitro experiments with different immune cells have revealed that curcumin supplementation to cell culture inhibited the TLR4 signaling at a number of downstream steps, including TRAF6 and IRAK1, as well as through immune-modulating (e.g., MCP-1, MIP-2) and signaling-associated cytokine blockades (Boozari et al., 2019; Gradisar, 2007). Supporting these results, in a Caco-2 model of the intestinal epithelium, treatment with curcumin resulted in diminished lipopolysaccharide (LPS)-induced pro-inflammatory cytokine release and tight junction (TJ) protein disruption (Wang et al., 2017), likely through a reduction of TLR4-dependent signaling. In in vivo experimental models of intestinal inflammation, like the trinitrobenzene sulfonic acid (TNBS)-induced model of colitis in rodent, curcumin has been shown to ameliorate the disease through a reduction in TLR4 signal transduction (Baliga, 2012; Lubbad et al., 2009). Following TLR4 activation, signaling proceeds through either NF-κB or AP-1 pathways, both of which play crucial roles in IBD; in fact, the block of NF-κB signaling has demonstrated a reduction in bowel injury (Atreya et al., 2008; Murano et al., 2000). Furthermore, inhibitors of p38 MAPK have shown efficacy in treating colitis (Waetzig, 2002). Curcumin's inhibition of NF-κB activation has been proposed to occur via IKKβ inhibition, thus leading to decreased IκB kinase activity, and hindering NF-κB subunit translocation into the nucleus (Pan et al., 2000). The proposed mechanism for the beneficial effects of curcumin in treating intestinal inflammatory diseases parallels that of steroids, by preventing IκBα degradation in the cytoplasm and specifically inhibiting the nuclear translocation of the p65 subunit (Schreiber et al., 1998). In AP-1 signaling, curcumin can inhibit MAPK, ERK1/2, JNK, and p38, directly and indirectly, thereby curbing the transcription of inflammatory target genes (Boozari et al., 2019). Likewise, other investigations have supported the ability of curcumin to inhibit NF-κB and/or AP-1 signaling pathways in epithelial cells (Holt et al., 2005; Jian et al., 2005). For instance, in a variety of intestinal epithelial cell (IEC) lines, curcumin treatment inhibited NF-κB-binding to DNA, degradation of IκBα, translocation to the nucleus of RelA, serine phosphorylation of IκB, and activity of IKK. In HT29 IECs, in vitro curcumin treatment suppressed TNF-α- and IL-1β-induced activation of p38 and JNK, while also preventing IκB degradation(Moon, 2006). These observations have been confirmed in vivo, in the experimental model of rat colitis induced after colonic instillation of TNBS, since curcumin administration significantly lowered protein expression of MyD88 and NF-Kb (Lubbad et al., 2009), inhibited IκB degradation (Jian et al., 2005), and mitigated colitis symptoms by reducing p38 MAPK activity (Camacho-Barquero et al., 2007). Similarly, curcumin exerted beneficial effects in experimental TNBS-colitis in mice through a reduction in NF-κB activity. Furthermore, pretreatment of mouse dendritic cells with curcumin suppressed NF-κB translocation to the nucleus and decreased phosphorylation of ERK, p38, and JNK(Cong, 2009).

Moreover, the immunomodulatory and anti-inflammatory properties exerted by curcumin have been also related with its ability to influence the operation, differentiation, and development of various immune cells actively involved in the onset and/or progression of IBD. Thus when dendritic cells were exposed to curcumin, a tend to foster the generation of intestinal T cells exhibiting a less responsive phenotype was observed, also displaying these dendritic cells restrained antigen-presenting capability, which in turn resulted in diminished activation of the adaptive immune system (Cong, 2009). It has been also reported that the curcumin-treated dendritic cells promote the differentiation of intestinal Tregs, which effectively impeded the progression of the disease in a mouse model of colitis (Cong, 2009). Other investigations have suggested that pretreatment with curcumin suppresses LPS-induced translocation of NF-κB p65 and phosphorylation of MAPK in dendritic cells, thus culminating in a reduction in inflammation (Kim, 2005). In addition, curcumin administration to dendritic cells reduced the expression of pro-inflammatory cytokines (IL-1, IL-6, TNF-α), and notably that of IL-12, thus reducing the ability of these dendritic cells to promote Th1-type responses (Kim, 2005). Furthermore, curcumin can diminish the expression of intercellular adhesion molecule-1 (ICAM-1) and CD11c in dendritic cells, being these molecules associated with both cellular adhesion and T-cell activation (Abdollahi, 2018), most probably through an AP-1-dependent pathway. However, the most crucial effect of curcumin on dendritic cells is probably to impede their maturation by suppressing indoleamine 2,3-dioxygenase (IDO), thereby exerting an anti-inflammatory effect akin to that of corticosteroids (Jung, 2010; Kim, 2005).

Additionally, curcumin has been also shown to obstruct T cell-mediated immune functions, which play a pivotal role in chronic intestinal inflammatory conditions (Gao, 2004). Among these, curcumin has the capacity to reduce the proliferative response of lymphocytes, which has been associated with its antioxidant properties, thus decreasing ROS-related proliferation, as well as with the inhibition of ribonucleotide reductase and DNA polymerase activation, which are crucial in the T cell cycle (Gao, 2004). Furthermore, curcumin has been shown to suppress NF-κB-induced T cell-initiated cytokine production (Gao, 2004), encompassing Th1-type cytokines, IL-2, and IFN-γ, thus suppressing the proliferation of lymphocytes (Gao, 2004; Jagetia & Aggarwal, 2007). In CD, Th1 cells are the predominant type and are thought to fuel much of the inflammation related to the adaptive immune system (Parronchi et al., 1997). Curcumin can obstruct the production of the Th1 subset by dampening macrophage production of IL-12, while also promoting the proliferation of the Th2 subclass (Kang, 1999; Zhang et al., 2006), characterized by a more anti-inflammatory cytokine profile. For instance, in a rodent model of TNBS-induced colitis, curcumin augmented Th2 synthesis and suppressed Th1 proliferation, resulting in a less inflammatory T helper profile (Zhang et al., 2006).

As commented above, curcumin has been demonstrated to inhibit TLR4 activation (Lubbad et al., 2009), which has been associated with the ability of curcumin to boost the production of TGF-β1, particularly in regions of active inflammation where a disrupted intestinal barrier can occur. In rat macrophages, curcumin treatment diminishes the ability of cells to generate ROS and secrete lysosomal breakdown enzymes (Joe & Lokesh, 2000), potentially leading to a decline in mucosal inflammation. Curcumin can also hinder NF-κB-induced macrophage and monocyte production of IL-12, IFN-γ, iNOS, MIP-2, IL-1β, IL-8, MCP-1, MIP-1α, and TNF-α (Billerey-Larmonier, 2008; Chan, 1998; Gao, 2004; Kang, 1999). Inhibition of IL-12 is particularly crucial in the context of adaptive immune cell differentiation and the further advancement of intestinal inflammation. Moreover, several studies have indicated that treatment with curcumin enhances the phagocytic activity of macrophages (Antony et al., 1999; Bai, 2016; Li & Liu, 2005; Mimche et al., 2012).

Intestinal inflammatory diseases are characterized by the recruitment and activation of neutrophils at the site(s) of injury, an early step providing a major source of ROS (Martin, 2004) for further mucosal and epithelial degradation. Curcumin is acknowledged to impede neutrophil recruitment (Martin, 2004), primarily achieved through the downregulation of NF-κB- and PI3K-Akt-induced chemotaxis (Larmonier et al., 2011; Larmonier, 2008), as well as a reduction in superoxide release (Billerey-Larmonier, 2008). Additionally, curcumin restrains neutrophils from aggregating, degranulating, and generating superoxide radicals (Srivastava, 1989). In both B cells and natural killer cells, curcumin has been demonstrated to bolster activity (Churchill, 2000; Okazaki et al., 2010; Yadav et al., 2005) or suppress activation (Decote-Ricardo, 2009; Gao, 2004) contingent upon the dosage and context (Jagetia & Aggarwal, 2007). Both B cells and natural killer cells have been identified as potential general sources of inflammation in the intestine, particularly in the context of microbial infection(Harrington et al., 2007).

Apart from its effects on specific immune cells, curcumin also influences the broad production of cytokines throughout the entire intestinal immune system. It reduces the levels of pro-inflammatory cytokines such as TNF-α, IFN-γ, IL-1, IL-2, IL-6, and IL-8, while simultaneously increasing the levels of anti-inflammatory cytokines like IL-10 and TGF-β (Zhang, 2016). For instance, in various models including rat methotrexate-colitis and LPS-treated IEC-6 cells, curcumin has been shown to decrease TNF-α and IL-1β levels while enhancing IL-10 levels (Kim, 2005). Additionally, in vitro studies using HT29 IECs demonstrated that curcumin suppressed TNF-α- and IL-1β-induced IL-8 release (Moon, 2006). Moreover, curcumin targets COX-2, an inflammatory enzyme implicated in prostaglandin synthesis and upregulated in intestinal inflammatory diseases. Through inhibition of signaling pathways like NF-κB and AP-1, curcumin effectively reduces COX-2 expression. Clinical trials have demonstrated the efficacy of curcumin in treating conditions like UC, likely due to its potent anti-inflammatory properties, although further research is needed to fully elucidate its mechanisms of action. For example, in trials involving quiescent UC patients, 2 g of curcumin effectively maintained remission, while in patients with mild-to-moderate disease, a combination of 3 g of curcumin with the anti-inflammatory drug mesalamine induced remission in over half of the participants (Lang, 2015).

Curcumin is characterized by its well-known antioxidant properties, being able to ameliorate the situation of oxidative stress that has been considered a pivotal player in the onset and progression of intestinal inflammatory diseases. Curcumin is able to effectively scavenge oxygen- and nitrogen-derived reactive species. Among the latter, it has been also reported the inhibitory effect exerted by this compound on NO production, which is associated to the modulation in the expression of iNOS and the protection of the intestinal tissue from oxidative damage (Venkataranganna et al., 2007). Additionally, curcumin interferes with the inflammatory cascade by reducing myeloperoxidase (MPO) activity, thus mitigating neutrophil-mediated tissue injury, and inhibiting matrix metalloproteinases (MMPs) responsible for extracellular matrix degradation (Lubbad et al., 2009). Moreover, curcumin orchestrates a defensive mechanism by enhancing the activity of phase II enzymes and bolstering the expression of antioxidant proteins such as nuclear factor (erythroid-derived 2)-related factor (Nrf2) and heme oxygenase-1 (HO-1) (McNally et al., 2007).

This multifaceted profile of curcumin, which includes its ability to counteract the oxidative-induced damage, preservation of the intestinal integrity, and inhibition of the associated inflammatory process, makes it a real promising therapeutic agent in the management of intestinal inflammatory disorders.

Quercetin

Oxidative stress is recognized as a factor that exacerbates symptoms of IBD and contributes to the development of IBD. The published reports have revealed that quercetin can significantly decrease MPO levels and inhibit ROS overproduction in colitis models, thereby blocking the following tissue damage (Dong, 2020; Hong & Piao, 2018). In normal physiological circumstances, a small amount NO provides protection against inflammatory injury. On the contrary, in the inflammatory state, the overexpression of iNOS results in excessive NO production, causing an overproduction of peroxynitrites and followed by intestinal destruction under the action of superoxide anions (Veljaca et al., 1995). Recent studies have shown that quercetin can curb the expression of iNOS, leading to a noticeable decrease in NO production (Khater, 2022).

As commented before, the loss of the epithelial barrier is a common event in the development of IBD. Thus preserving the integrity of the intestinal epithelial barrier is a critical step in preventing the progression of IBD. Studies have shown that quercetin may have a beneficial effect on the epithelial barrier (Suzuki & Hara, 2011). Oral administration of quercetin in mice colitis model induced by dextran sulfate sodium (DSS) have restored the loss of epithelial integrity by the induction of TJ proteins (Riemschneider, 2021).

The imbalance of gut microbiota, marked by a decline in microbial diversity and decreased resilience, plays a crucial role in the onset of IBD disrupting the host homeostasis. Recent research indicates that dietary quercetin holds promise in shielding against IBD by ameliorating intestinal dysbiosis. For instance, Lin et al. administered dietary quercetin to mice infected with *Citrobacter rodentium*, observing a notable reduction in colitis-induced inflammation. Moreover, quercetin was observed to bolster populations of beneficial bacteria such as *Bacteroidota*, *Bifidobacterium*, *Lactobacillus*, and *Clostridia*, while reducing those of harmful species like *Fusobacterium* and *Enterococcus*. Hence, the mitigation of colitis in mice is likely attributable to quercetin's ability to modulate gut microbiota composition (Lin et al., 2019). Additionally, Zhu et al. noted a decline in the population of *Bacillota* (previously named *Firmicutes*) and an increase in *Proteobacteriodota* in mice with DSS-induced colitis. Quercetin was found to reverse these alterations, thereby influencing the outcomes of microbiota-related disorders (Hong & Piao, 2018).

IBD is closely linked with an aberrant response of the intestinal immune system. In this scenario, quercetin has been demonstrated to reduce the infiltration of macrophages, neutrophils, and Th17 cells, while simultaneously enhancing the proportions of Treg cells in experimental colitis (Riemschneider, 2021). Additionally, in experimental models of colitis quercetin, supplementations have demonstrated higher ability to inhibit the production of proinflammatory markers, including IL-17,

TNF-α, and IL-6, while concurrently enhancing the release of anti-inflammatory cytokines such as IL-10 (Lin et al., 2019). Specifically, quercetin suppresses TNF-α, IL-1β and IL-6 production by modulating the miR-369−3p/C/EBP-β axis in dendritic cells and interfering with the MAPK and NF-κB signaling pathways in macrophages, thus showcasing its anti-inflammatory properties (Cho et al., 2003; Galleggiante, 2019). This consistent effect has been also evident in human IECs, where the inhibition of C/EBP-β signaling by quercetin elucidates its protective role against IL-6 production induced by the heat shock response in the intestinal mucosa (Hungness, 2002).

Resveratrol

Resveratrol (3,4′,5-trihydroxystilbene) is a naturally occurring stilbene found in a variety of plant species (e.g., grapes, groundnuts, and several species of berries, such as blueberries, bilberries, and cranberries). Oral administration of resveratrol, or dietary supplementation, has been shown to prevent and ameliorate intestinal inflammation as well as to mitigate the severity of colitis. Although, human clinical trials of the potential therapeutic, or, better, ameliorating effects of resveratrol in IBD are few, a randomized, double-blind, and placebo-controlled pilot study has confirmed that resveratrol capsules treatment increased anti-oxidative capacity, decreased serum malondialdehyde (MDA) level and disease activity, and increased quality of life in patients with UC. Administration of resveratrol, primarily through oral gavage or dietary supplementation, has been shown to mitigate the severity of colitis, as evidenced by reductions in various macroscopic and microscopic inflammatory markers and colon damage in animal models of experimental IBD (Alrafas, 2020). Moreover, a randomized, double-blind, placebo-controlled clinical trial evaluating the effects of resveratrol supplementation on inflammatory biomarkers and the quality of life of patients with UC have demonstrated that a dose of 0.5 g/day of resveratrol for 6 weeks diminish the plasma levels of high-sensitivity C-reactive protein, TNF-α and NF-κB p65 levels and significantly improve the clinical colitis activity index score compared with the placebo group (Samsami-Kor et al., 2015). The antiinflammatory effect of the resveratrol in UC patients was also demonstrated reducing TNF-α and IL-1β and increasing IL-11 production (Sabzevary-Ghahfarokhi, 2020). Additionally, the improvement of the oxidative stress status of patients with mild to moderate UC treated with resveratrol markedly improved as evidenced by increases of superoxide dismutase and total antioxidant activity and reductions in antioxidant MDA levels (Nunes, 2018; Samsami-Kor et al., 2015). Collectively, these studies indicate that resveratrol has the potential to enhance the quality of life for individuals with IBD by reducing inflammation and oxidative stress. The mechanisms involved in this beneficial clinical effect have been extensively studied in vitro and in vivo.

Several groups have previously characterized the anti-inflammatory effects of resveratrol in intestinal cells. Thus one study has revealed the protective effects of moderate to high concentrations of resveratrol (10–50 μM) in Caco-2 cells Caco-2 cells exposed to bacterial-derived LPS. The results demonstrated that resveratrol-treated cells exhibited reduced levels of PGE2, accompanied by a decrease in COX-2 expression. Moreover, resveratrol inhibited NF-κB activation by attenuating the degradation rate of IκB, a natural inhibitor of NF-κB (Cianciulli, 2012). Similarly, in Caco-2 cells and SW480 human colon adenocarcinoma cells treated with lipopolysaccharide, high concentrations of resveratrol (40 μM) suppressed inflammatory responses by reducing TLR 4 and inducible NO synthase expression, as well as inhibiting IκB-α degradation (Panaro, 2012). Conversely, in another study utilizing Caco-2 cells, treatment with resveratrol (50 μM) did not inhibit NF-κB activation; instead, it exacerbated inflammatory responses (Romier, 2008). It is well described that mitochondrial dysfunction plays a crucial role in the pathogenesis of IBD, as oxidative stress and compromised ATP production are significant contributors to the onset and advancement of the condition. applied at extremely high concentrations (440 μM), protected Caco-2 cells against indomethacin-induced mitochondrial dysfunction (Carrasco-Pozo, 2012a, 2012b). It has been postulated that resveratrol's protective mechanism against mitochondrial dysfunction involves binding to the ubiquinone site of complex I, akin to rotenone (a complex I inhibitor), thereby impeding the interaction between indomethacin and rotenone (Carrasco-Pozo, 2012a). Consistently, resveratrol has been demonstrated to inhibit the mobilization of Ca^{2+}, the release of caspase-3, -9, and cytochrome c from the mitochondria, and the initiation of apoptosis in response to cytotoxic concentrations of indomethacin (Carrasco-Pozo, 2012b). Therefore resveratrol appears to potently inhibit mitochondrial dysfunction (Ungvari, 2011) and prevent the onset of programmed cell death. In conclusion, in vitro studies show that resveratrol, at moderate to high concentrations, can modulate inflammatory responses within intestinal cells by downregulating NF-κB activation and preventing mitochondrial dysfunction. Additionally, in vivo studies using models for inflammatory intestinal chronic diseases have contributed to understanding of the effects of resveratrol in IBD. In a rat model of chronic inflammation induced by TNBS, Martín et al. demonstrated that resveratrol mitigates intestinal mucosal damage by diminishing the recruitment of neutrophils and the secretion of TNF-α (Martin, 2006). Moreover, resveratrol normalized the production of PGE2 and PGD2, reduced COX-2 expression, and induced apoptosis in colon mucosa during the early stages of the disease (Martin, 2006). Previously, these researchers also revealed the effectiveness of resveratrol in ameliorating early colon inflammation induced by TNBS in rats (Martin, 2004). Similarly, other studies have shown that resveratrol hindered the activity of intercellular adhesion molecule-1 and vascular cell adhesion protein-1, thereby impeding neutrophil infiltration through attenuating the interaction between leukocytes and endothelial cells in TNBS rat model (Abdallah & Ismael, 2011). Additionally, it has been demonstrated that pre-treatment with intraperitoneal resveratrol for 5 days prior to the induction of UC by TNBS significantly reduced the microscopy injury score and MDA levels, while boosting glutathione peroxidase activity (Yildiz, 2015). It has been also reported in DSS-induced colitis rat model resveratrol exerts anti-inflammatory effects in vivo at a feasible dietary dosage (Larrosa, 2009). In fact, the analysis via microarray revealed that resveratrol treatment significantly regulated 6% of the total genes examined in distal colon mucosa, encompassing those implicated in IL-6 signaling, apoptosis, mitochondrial fatty acid oxidation, and Wnt signaling. These findings underscore resveratrol's significance as a multi-target anti-inflammatory agent, implying that resveratrol is not solely a promising anti-inflammatory compound at pharmacological doses but also at doses attainable through the normal consumption of resveratrol-rich foods.

Several researchers groups have dilapidated the resveratrol's antioxidative effects in vivo, including a study employing a DSS-induced mice model. Here, researchers observed that resveratrol alleviated colitis, diminished mucosal inflammation, and heightened the activity of superoxide dismutase and glutathione peroxidase, indicating resveratrol's ability to modulate antioxidant enzyme expression. In these mice, myeloperoxidase activity and production of IL-8, TNF-α, and interferon-γ were significantly reduced in a dose-dependent manner following resveratrol administration (Yao, 2010). Furthermore, resveratrol treatment ameliorated clinical symptoms associated with DSS-induced chronic inflammation, such as weight loss, diarrhea, and rectal bleeding. Moreover, the mortality rate markedly decreased in mice subjected to both DSS and resveratrol compared to those exposed solely to DSS (Sanchez-Fidalgo, 2010). Similarly, oral resveratrol administration prevented the onset of DSS-induced mouse colitis by suppressing the phosphorylation of extracellular signal-regulated kinase and STAT3, pivotal mediators of NF-κB signaling and inflammation (Youn, 2009). Moreover, it has been confirmed that the administration of moderate to high doses of resveratrol (50 and 100 mg/kg) regulates the Th17 and regulatory T (Treg) cell imbalance by reducing Th17 cell numbers and upregulating Treg cell numbers in a DSS mice model (Yao et al., 2015). Interestingly, it has been uncovered sex-based disparities in the biological effects of resveratrol in mice treated with DSS. In DSS-exposed mice receiving resveratrol, females exhibited lower colon myeloperoxidase activity compared to males, likely linked to sex hormone modulation of neutrophil activity (Wagnerova, 2017).

Elevated levels of TNF-α-converting enzyme (TACE) have been observed in the colon tissue of IBD patients. TACE belongs to a large group of type I integral membrane proteins, known as ADAMs (a disintegrin and metalloproteinase), which are involved in a variety of cellular processes, including ectodomain shedding of membrane-bound proteins. Due to the firm evidence for TACE as the major TNF-α convertase, this enzyme has attracted considerable interest as a specific therapeutic target in diseases known to benefit from anti-TNF-α treatment including CD and UC((Baert, 1999; D'Haens, 1999; Evans et al., 1997; Føgh, 1999; Moss, 2001; Rutgeerts, 1999). Correspondingly, many evidence has suggested that resveratrol can mitigate intestinal inflammation by inhibiting TACE in C57BL/6 mice with DSS-induced colitis (Sharma, 2014).

The IL-10−/− mouse model, lacking a functional version of the anti-inflammatory cytokine IL-10 (Kuhn, 1993), serves as a widely utilized IBD model. In this sense, it has been shown that oral resveratrol administration ameliorated chronic colitis in IL-10−/− mice by inducing myeloid-derived suppressor cells (MDSC) and reducing mucosal and systemic inflammatory cytokine responses (Singh, 2012). MDSCs possess robust immunosuppressive properties (Haile, 2008), and their modulation holds therapeutic potential for inflammatory conditions like IBD.

In summary, a number of in vitro and animal studies have shown that resveratrol may reduce the severity of intestinal inflammation in IBD through the inhibition of NF-κB activation, modulation of the immune system, the reduction of the COX-2 expression, PGE_2 levels, and proinflammatory cytokines as well as the attenuation of reactive species production.

Catechins

Catechins are the most active constituent polyphenols in the green tea (*Camelia sinensis*) (Biasi et al., 2011). Catechins mainly include catechin, epicatechin (EC), epicatechin gallate (ECG), epigallocatechin (EGC) and its stereoisomer gallocatechin (GC), EGCG and its stereoisomer gallocatechin gallate (GCG). EGCG, which accounts for almost 50% of the total catechins content of green tea extracts has the strongest chemopreventive potential containing anti-inflammatory, anti-mutagenic, and anti-carcinogenic effects among the green tea catechins which contain EC, EGC, and EGCG (Biasi et al., 2011; Lee & Lee, 2006; Pan et al., 2011). In recent years, catechins have garnered considerable attention due to their potential involvement in various physiological functions, including anti-hypertensive, antibacterial, anti-inflammatory, and antioxidative properties as well as their role in protecting against atherosclerosis (da Silva et al., 2010; Fathima & Rao, 2016; Fraga, 2010; Melgarejo, 2010; Sánchez-Fidalgo, 2013; Scalia et al., 2013). A double-blinded randomized clinical trial has revealed that patients with mild to moderate UC treated for 56 days with polyphenon E (a standardized green tea preparation consisting of 65% EGCG and 22% other catechins), 400 mg or 800 mg of total EGCG daily, administered in split doses, induce the improvement in response and in the remission rate in mild to moderate, but refractory, UC patients. Thus administration of EGCG resulted in a therapeutic benefit for patients who were refractory to 5-aminosalicylic and/or azathioprine (Dryden et al., 2013). Only one IBD human clinical trial with catechins has been published; however, many preclinical studies have revealed potential mechanisms for the beneficial use of catechins in IBD. Remarkably, it has not yet been entirely made clear that the mechanisms by which catechins benefit in IBD, however, there is abundant evidence showing that catechins may work through a combination of oxidation inhibition, alteration of cellular signaling, and regulation of intestinal microbiota (Chiou, 2012; Liu et al., 2014; Sanchez-Fidalgo, 2013). Thus the impact of the catechins on oxidative stress has been widely reported. The dysregulation between oxidative reactions and antioxidative defenses is a crucial factor in the development of IBD (Alzoghaibi, 2013; Biasi et al., 2011; Vasconcelos, 2012). IBD is recognized as one of the primary conditions characterized by an overload of oxyradicals, contributing to a phenotype prone to cancer (Yum, 2013). It has been documented that an excessive production of ROS and radical nitrogen metabolites (RNS) in the gut of IBD patients (Kruidenier et al., 2003). In this sense, it has been demonstrated that treating lymphocytes from IBD patients with catechins in vitro provide protection against oxidative stress induced by 2-amino-3-methylimidazo[4,5-f]-quinolone (Najafzadeh et al., 2009). Renato et al. have also revealed the dose-dependent antioxidant capacity of (+)-catechin within the range of 0−100 μM (Pereira, 2013). Studies on the antioxidant effects of catechins have shown that EGCG, ECG, and peracetylated (−)-epigallocatechin-3-gallate (AcEGCG) can significantly reduce ROS levels induced by LPS in RAW264.7 cells. Additionally, it has been suggested that high doses of EGCG can induce ROS production and act as a prooxidant, leading to the activation of the transcription factor nuclear factor (erythroid-derived 2)-like 2 (Nrf2) pathway (Chiou, 2016). Catechins exhibit both "direct antioxidant effects" and "indirect antioxidant effects" as antioxidative agents. Catechins can scavenge free radicals and chelate redox-active metals, acting as direct antioxidants. Additionally, as indirect antioxidants, catechins regulate protein synthesis activities and signaling pathways, including the modulation of prooxidant enzyme properties. These effects are concentration-dependent, with high doses demonstrating more robust antioxidant capacity in the digestive tract (Dryden et al., 2006; Fraga & Oteiza, 2011; Fraga, 2010). Catechins have demonstrated direct antioxidant effects through scavenging free radicals and sequestering redox-active metals in the digestive tract. Despite their various diastereoisomers such as EC, ECG, EGC, and EGCG, catechins share a similar chemical structure that stabilizes free radicals through phenolic groups (Fraga, 2010; Silva et al., 2010). A study evaluating the effect of (+)-catechin on LPS-induced murine peritoneal macrophages indicated that (+)-catechin significantly inhibited intracellular ROS and NO production, demonstrating antioxidant ability and confirming its efficacy against ROS and RNS activity in RAW264.7 macrophages or in vitro (Sanchez-Fidalgo, 2013). As scavengers of free radicals, catechins possess the ability to interact with free radicals, thereby interrupting their chain reactions, such as the common oxidative process of lipid oxidation. There is no convincing experimental evidence to compare the reducibility between catechins, which exhibit extensive antioxidant activity in vitro, and GSH, an endogenous antioxidant system present in both reduced (GSH) and oxidized (GSSG) forms. However, when pretreated with Aβ, the GSH content decreases, but upon addition of EGCG, the GSH content is restored in BV2 cells (Kim, 2009), indicating that EGCG's reduction potential is greater than that of GSH. Nonetheless, in general, catechins increase GSH activity by forming complexes with greater reduction potential in the B ring. However, further precise and specific experiments, such as employing chromogenic compounds to assess the antioxidant's ability to trap free radicals, are needed to compare the reducibility between catechins and GSH (Bao, 2013). Moreover, catechins at low concentrations possess an indirect antioxidative capacity by hindering the modulation of enzymes involved in oxidant generation, such as NADPH-oxidase (NOX), and altering oxidant production by preventing the binding of ligands to receptors, such as TNF-α to TNF-αR, in intestinal cells (Fraga & Oteiza, 2011). It has been suggested that EGCG could suppress NOX and the subsequent increase in ROS (Biasi et al., 2011). It has been suggested that EC and its metabolites might inhibit NOX, leading to a reduction in superoxide production. One of EC's metabolites, O-metabolite, shares a similar structure with apocynin, a classical NOX inhibitor (Fraga & Oteiza, 2011; Fraga, 2010). EC could inhibit NOX through the following mechanisms: (i) direct binding of EC with NOX; (ii) alteration of calcium influx, thereby inhibiting NOX activation; and (iii) regulation of upstream NOX signaling, such as blocking ligand-receptor interactions that promote NOX activity (Fraga & Oteiza, 2011).

GSH produced by mucosal cells acts as a crucial antioxidant defending cellular integrity and the excessive generation of reactive oxygen- nitrogen species (RONS) can deplete intracellular GSH levels. This depletion may trigger the activation of Nrf2 signaling, stimulating the upregulation of HO-1 and GSH to maintain cellular homeostasis. However, when RONS-induced damage surpasses the protective capacity of antioxidants, it can lead to IBD onset. In this sense, it has been demonstrated through flow cytometry that catechin analogs (ECG, EGCG, and AcEGCG) could elevate GSH levels in LPS-induced RAW264.7 cells, albeit not in a dose-dependent manner. Interestingly, lower doses of catechins exhibited more potent effects compared to higher doses, possibly by inducing the expression of Nrf2-related GSH-related enzymes, such as glutathione peroxidase (GPO), thereby regulating GSH expression. Additionally, EC administration at 10 mg/kg induced higher GSH levels in both acute and chronic colitis rat models, suggesting an increase in GSH expression and/or a decrease in depletion induced by EC (Chiou, 2016; Pereira, 2013). Conversely, in response to mild oxidative stress, defense mechanisms are activated, leading to increased production of enzymatic antioxidants and other low molecular weight antioxidants. Prolonged exposure to oxidative stress from ROS overproduction can overwhelm these defense mechanisms, resulting in insufficient production of antioxidant enzymes such as catalase, GPO, and SOD. It has been observed that ROS can inactivate several antioxidant enzymes, potentially contributing to the development of IBD. Catechins have been demonstrated to enhance the activity of various protective enzymes, conferring a protective effect. In a murine model of colitis induced by oral administration of DSS, the combination of EGCG and piperine exhibited positive antioxidative effects on GPO and SOD. Additionally, EGCG and EC increased catalase activity, potentially through the reduction of ROS, MDA, and protein carbonyls. However, the precise mechanisms underlying the effects of EGCG on antioxidant enzymes remain unclear.

The key players in IBD encompass neutrophils, lymphocytes, monocytes, and macrophages. Their migration and activation within inflamed tissue rely on cytokines, chemokines, and adhesion molecules. These extracellular signaling molecules facilitate various cellular processes such as proliferation, differentiation, gene expression, migration, immunity, and inflammation. IBD's inflammatory cascade involves substantial infiltration of T cells, polymorpho-, and mononuclear phagocytic leukocytes, intensifying local inflammation and oxidative stress. This process exacerbates mucosal damage by producing copious pro-inflammatory cytokines, chemokines, and reactive oxygen intermediates, further enhancing cell infiltration and perpetuating inflammation. EGCG exhibits an inhibitory effect on endothelial VACM-1, reducing monocyte adhesion to endothelial cells and their migration into inflamed tissue. Furthermore, EGCG treatment decreases the gene and protein expression of monocyte chemotactic protein 1 (MCP-1) and its receptor (CCR2), inhibiting excessive monocyte migration. Neutrophil infiltration levels can be assessed by myeloperoxidase (MPO) activity, which is abundantly present in neutrophil granulocytes. EGCG has been reported to reduce MPO activity, indicative of decreased neutrophil infiltration. In a study evaluating EGCG's effects on DSS-induced murine colitis, MPO levels were reduced in colon tissue with piperine plus EGCG treatment (Abboud et al., 2008). Similarly, EC did not decrease neutrophil infiltration in TNBS-induced colitis rat models via MPO modulation, potentially attributed to sustained high MPO concentrations(Vasconcelos, 2012). Another study has demonstrated EGCG's significant reduction of MPO and histamine levels in the distal gut mucosa of TNBS-induced rat colitis. EGCG's ability to suppress macrophage migration leads to inhibition of neutrophil migration, thereby attenuating neutrophil infiltration owing to histamine's chemoattractant properties (Mochizuki & Hasegawa, 2010). It is important that EGCG can suppress the macrophage migration, causing the inhibition of neutrophil migration. The experiment exhibited the increasing mast cells and the high concentration of histamine, both of which could be found in the mucosa of IBD, and were also ameliorated by EGCG in a dose-dependent manner. Thus the neutrophil infiltration could be inhibited by EGCG, owing to histamine being a kind of chemoattractant of neutrophils (Melgarejo, 2010; Mochizuki & Hasegawa, 2010).

Lastly, TJs, a dynamic and versatile complex that creates a seal between neighboring epithelial cells, rely heavily on the zonula occludens-(ZO-) 1 protein to function effectively, facilitating connections between occludin and claudin proteins and other cellular components. This intricate network plays a crucial role in maintaining the integrity of the intestinal barrier (Contreras, 2015; Kaulmann & Bohn, 2016; Steed et al., 2010). Alterations in TJ regulators and components, and subsequent changes in barrier permeability, can exacerbate tissue damage in IBD (Salim & Söderholm, 2011). Multiple cellular studies indicate that catechins may offer therapeutic benefits in IBD by enhancing the stability of TJs in epithelial cells (Carrasco-Pozo et al., 2013; Carrasco-Pozo, 2012b; Contreras, 2015; Kaulmann & Bohn, 2016). Thus it has been demonstrated that EC could prevent the exacerbation of IBD by inhibiting TNF-α-induced permeabilization of Caco-2 cell monolayers. This effect was achieved by preserving the levels and proper distribution of ZO-1 protein (Contreras, 2015). Furthermore, Carrasco-Pozo et al. established that various polyphenols, including EGCG, exerted similar protective effects by thwarting the redistribution of ZO-1 protein and the downregulation of ZO-1 and occludin expression induced by indomethacin in a Caco-2-based model (Carrasco-Pozo et al., 2013; Kaulmann & Bohn, 2016).

In summary, catechins show potential for ameliorating IBD and associated conditions due to their known anti-inflammatory, antioxidative, and antibacterial properties. As antioxidants, catechins mitigate cellular and tissue damage caused by oxidative imbalance by enhancing antioxidant enzymes and exerting direct or indirect antioxidative effects. Catechins inhibit immune cell infiltration and proliferation, regulate inflammation and oxidative reactions by interacting with various pathways like NF-κB, MAPKs and Nrf2, leading to reduced production of cytokines, chemokines, and inflammatory enzymes. Catechins may also prevent gastrointestinal lesions from by regulating cell gap junctions and improving epithelial

TJ expression, thus benefiting IBD-related inflammation. Moreover, they might stabilize intestinal flora, aiding IBD recovery. However, limited studies in animal models or IBD patients exist to validate these biological activities. Further research is necessary to elucidate precise mechanisms, potentially at the molecular level.

Conclusion

Research into the intestinal anti-inflammatory properties of polyphenols indicates their potential effectiveness as treatments for IBD. Exploring alternative therapeutic approaches with diverse targets offers hope for improving disease management and enhancing patients' well-being. However, conducting thorough clinical studies is imperative to establish the safety and efficacy profiles of polyphenols in addressing these intestinal inflammations, thereby cementing their status as dependable treatment options. Numerous studies investigating the intestinal anti-inflammatory effects of polyphenols offer compelling evidence of their promise as treatments for IBD in humans. Consistent findings from experimental models highlight their efficacy, aligning with the mechanisms of action of current pharmaceuticals used in human therapy. Moreover, their long-standing consumption in edible plants underscores their safety. Nonetheless, conducting clinical trials is essential to conclusively ascertain their role in managing these inflammatory conditions. While human studies primarily focus on interventions rich in polyphenol-containing plants, preliminary results indicate favorable outcomes in inducing clinical remission and response among IBD patients. Nonetheless, additional clinical investigation is warranted.

References

Abboud., et al. (2008). Therapeutic effect of epigallocatechin-3-gallate in a mouse model of colitis. *European Journal of Pharmacology*, *579*(1-3), 411–417.

Abdallah, D. M., & Ismael, N. R. (2011). Resveratrol abrogates adhesion molecules and protects against TNBS-induced ulcerative colitis in rats. *Canadian Journal of Physiology and Pharmacology*, *89*(11), 811–818. Available from https://doi.org/10.1139/Y11-080.

Abdollahi, E., et al. (2018). Therapeutic effects of curcumin in inflammatory and immune-mediated diseases: A nature-made jack-of-all-trades? *Journal of Cellular Physiology*, *233*(2), 830–848.

Alrafas, H. R., et al. (2020). Resveratrol downregulates miR-31 to promote T regulatory cells during prevention of TNBS-induced colitis. *Molecular Nutrition & Food Research, 64*(1), e1900633.

Alzoghaibi, M. A. (2013). Concepts of oxidative stress and antioxidant defense in Crohn's disease. *World Journal of Gastroenterology, 19*(39), 6540–6547. Available from https://doi.org/10.3748/wjg.v19.i39.6540, http://www.wjgnet.com/1007-9327/pdf/v19/i39/6540.pdf.

Antony, S., Kuttan, R., & Kuttan, G. (1999). Immunomodulatory activity of Curcumin. *Immunological Investigations, 28*(5-6), 291–303. Available from https://doi.org/10.3109/08820139909062263.

Atreya, I., Atreya, R., & Neurath, M. F. (2008). NF-kappaB in inflammatory bowel disease. *Journal of Internal Medicine, 263*(6), 591–596.

Baert, F. J., et al. (1999). Tumor necrosis factor alpha antibody (infliximab) therapy profoundly down-regulates the inflammation in Crohn's ileocolitis. *Gastroenterology, 116*(1), 22–28.

Bai, X., et al. (2016). Curcumin enhances human macrophage control of Mycobacterium tuberculosis infection. *Respirology, 21*(5), 951–957.

Baliga, M. S., et al. (2012). Curcumin, an active component of turmeric in the prevention and treatment of ulcerative colitis: Preclinical and clinical observations. *Food Funct, 3* (11), 1109–1117.

Bao, G. H., et al. (2013). EGCG inhibit chemical reactivity of iron through forming an Ngal-EGCG-iron complex. *Biometals: An International Journal on the Role of Metal Ions in Biology, Biochemistry, and Medicine, 26*(6), 1041–1050.

Baumgart, D. C., & Sandborn, W. J. (2007). Inflammatory bowel disease: Clinical aspects and established and evolving therapies. *Lancet, 369*(9573), 1641–1657.

Biasi, F., Astegiano, M., Maina, M., Leonarduzzi, G., & Poli, G. (2011). Polyphenol supplementation as a complementary medicinal approach to treating inflammatory bowel disease. *Current Medicinal Chemistry, 18*(31), 4851–4865. Available from https://doi.org/10.2174/092986711797535263.

Billerey-Larmonier, C., et al. (2008). Protective effects of dietary curcumin in mouse model of chemically induced colitis are strain dependent. *Inflammatory Bowel Diseases, 14*(6), 780–793.

Boone, D. L., & Ma, A. (2003). Connecting the dots from Toll-like receptors to innate immune cells and inflammatory bowel disease. *The Journal of Clinical Investigation, 111*(9), 1284–1286.

Boozari, M., Butler, A. E., & Sahebkar, A. (2019). Impact of curcumin on toll-like receptors. *Journal of Cellular Physiology, 234*(8), 12471–12482.

Burge, K., et al. (2019). Curcumin and intestinal inflammatory diseases: molecular mechanisms of protection. *International Journal of Molecular Science, 20*(8), 1912. Available from https://doi.org/10.3390/ijms20081912.

Camacho-Barquero, L., Villegas, I., Sánchez-Calvo, J. M., Talero, E., Sánchez-Fidalgo, S., Motilva, V., & Lastra, C. A. de la (2007). Curcumin, a Curcuma longa constituent, acts on MAPK p38 pathway modulating COX-2 and iNOS expression in chronic experimental colitis. *International Immunopharmacology, 7*(3), 333–342. Available from https://doi.org/10.1016/j.intimp.2006.11.006.

Carrasco-Pozo, C., et al. (2012a). Differential protective effects of quercetin, resveratrol, rutin and epigallocatechin gallate against mitochondrial dysfunction induced by indomethacin in Caco-2 cells. *Chemico-Biological Interactions, 195*(3), 199–205.

Carrasco-Pozo, C., et al. (2012b). Stimulation of cytosolic and mitochondrial calcium mobilization by indomethacin in Caco-2 cells: modulation by the polyphenols quercetin, resveratrol and rutin. *Biochimica et Biophysica Acta, 1820*(12), 2052–2061.

Carrasco-Pozo, C., Morales, P., & Gotteland, M. (2013). Polyphenols protect the epithelial barrier function of Caco-2 cells exposed to indomethacin through the modulation of occludin and zonula occludens-1 expression. *Journal of Agricultural and Food Chemistry*, *61*(22), 5291–5297.

Chan, M. M., et al. (1998). In vivo inhibition of nitric oxide synthase gene expression by curcumin, a cancer preventive natural product with anti-inflammatory properties. *Biochemical Pharmacology*, *55*(12), 1955–1962.

Chiou, Y. S., et al. (2012). Peracetylated (-)-epigallocatechin-3-gallate (AcEGCG) potently suppresses dextran sulfate sodium-induced colitis and colon tumorigenesis in mice. *Journal of Agricultural and Food Chemistry*, *60*(13), 3441–3451.

Chiou, Y. S., et al. (2016). Directly interact with Keap1 and LPS is involved in the anti-inflammatory mechanisms of (-)-epicatechin-3-gallate in LPS-induced macrophages and endotoxemia. *Free Radical Biology & Medicine*, *94*, 1–16.

Cho, S. Y., Park, S. J., Kwon, M. J., Jeong, T. S., Bok, S. H., Choi, W. Y., Jeong, W. I., Ryu, S. Y., Do, S. H., Lee, C. S., Song, J. C., & Jeong, K. S. (2003). Quercetin suppresses proinflammatory cytokines production through MAP kinases and NF-κB pathway in lipopolysaccharide-stimulated macrophage. *Molecular and Cellular Biochemistry*, *243*(1-2), 153–160. Available from https://doi.org/10.1023/A:1021624520740.

Churchill, M., et al. (2000). Inhibition of intestinal tumors by curcumin is associated with changes in the intestinal immune cell profile. *The Journal of Surgical Research*, *89*(2), 169–175.

Cianciulli, A., et al. (2012). Modulation of NF-kappaB activation by resveratrol in LPS treated human intestinal cells results in downregulation of PGE2 production and COX-2 expression. *Toxicology in Vitro: An International Journal Published in Association with BIBRA*, *26*(7), 1122–1128.

Cong, Y., et al. (2009). Curcumin induces the tolerogenic dendritic cell that promotes differentiation of intestine-protective regulatory T cells. *European Journal of Immunology*, *39*(11), 3134–3146.

Contreras, T. C., et al. (2015). -)-Epicatechin in the prevention of tumor necrosis alpha-induced loss of Caco-2 cell barrier integrity. *Archives of Biochemistry and Biophysics*, *573*, 84–91.

D'Haens, G., et al. (1999). Endoscopic and histological healing with infliximab anti-tumor necrosis factor antibodies in Crohn's disease: A European multicenter trial. *Gastroenterology*, *116*(5), 1029–1034.

Decote-Ricardo, D., et al. (2009). Modulation of in vitro murine B-lymphocyte response by curcumin. *Phytomedicine: International Journal of Phytotherapy and Phytopharmacology*, *16*(10), 982–988.

Dong, Y., et al. (2020). Quercetin alleviates intestinal oxidative damage induced by H(2) O(2) via modulation of GSH: In vitro screening and in vivo evaluation in a colitis model of mice. *ACS Omega*, *5*(14), 8334–8346.

Dryden, G. W., Lam, A., Beatty, K., Qazzaz, H. H., & McClain, C. J. (2013). A pilot study to evaluate the safety and efficacy of an oral dose of (-)-Epigallocatechin-3-gallate-rich polyphenon E in patients with mild to moderate ulcerative colitis. *Inflammatory Bowel Diseases*, *19*(9), 1904–1912. Available from https://doi.org/10.1097/MIB.0b013e31828f5198.

Dryden, G. W., Song, M., & McClain, C. (2006). Polyphenols and gastrointestinal diseases. *Current Opinion in Gastroenterology*, *22*(2), 165–170.

Elhag, D. A., et al. (2022). Inflammatory bowel disease treatments and predictive biomarkers of therapeutic response. *International Journal of Molecular Science*, *23*(13), 6966.

Evans, R. C., Clarke, L., Heath, P., Stephens, S., Morris, A. I., & Rhodes, J. M. (1997). Treatment of ulcerative colitis with an engineered human anti-TNFα antibody

CDP571. *Alimentary Pharmacology & Therapeutics, 11*(6), 1031–1035. Available from https://doi.org/10.1046/j.1365-2036.1997.00251.x.

Fathima, A., & Rao, J. R. (2016). Selective toxicity of Catechin-a natural flavonoid towards bacteria. *Applied Microbiology and Biotechnology, 100*(14), 6395–6402.

Fraga, C. G., et al. (2010). Basic biochemical mechanisms behind the health benefits of polyphenols. *Molecular Aspects of Medicine, 31*(6), 435–445.

Fraga, C. G., & Oteiza, P. I. (2011). Dietary flavonoids: Role of (-)-epicatechin and related procyanidins in cell signaling. *Free Radical Biology & Medicine, 51*(4), 813–823.

Føgh, P., et al. (1999). TNF-alpha-converting enzyme activity in colonic biopsy specimens from patients with inflammatory bowel disease revealed by mRNA and in vitro assay. *Annals of the New York Academy of Sciences, 878*, 692–695.

Galleggiante, V., et al. (2019). Quercetin-induced miR-369-3p suppresses chronic inflammatory response targeting C/EBP-β. *Molecular Nutrition & Food Research, 63*(19), e1801390.

Gao, X., et al. (2004). Immunomodulatory activity of curcumin: suppression of lymphocyte proliferation, development of cell-mediated cytotoxicity, and cytokine production in vitro. *Biochemical Pharmacology, 68*(1), 51–61.

Gradisar, H., et al. (2007). MD-2 as the target of curcumin in the inhibition of response to LPS. *Journal of Leukocyte Biology, 82*(4), 968–974.

Gupta, S. C., Patchva, S., & Aggarwal, B. B. (2013). Therapeutic roles of curcumin: lessons learned from clinical trials. *The AAPS Journal, 15*(1), 195–218.

Haile, L. A., et al. (2008). Myeloid-derived suppressor cells in inflammatory bowel disease: a new immunoregulatory pathway. *Gastroenterology, 135*(3), 871–881, 881.e1-5.

Hanai, H., et al. (2006). Curcumin maintenance therapy for ulcerative colitis: Randomized, multicenter, double-blind, placebo-controlled trial. *Clinical Gastroenterology and Hepatology: the Official Clinical Practice Journal of the American Gastroenterological Association, 4*(12), 1502–1506.

Harrington, L., Srikanth, C. V., Antony, R., Shi, H. N., & Cherayil, B. J. (2007). A role for natural killer cells in intestinal inflammation caused by infection with Salmonella enterica serovar Typhimurium. *FEMS Immunology & Medical Microbiology, 51*(2), 372–380. Available from https://doi.org/10.1111/j.1574-695x.2007.00313.x.

Holt, P. R., Katz, S., & Kirshoff, R. (2005). Curcumin therapy in inflammatory bowel disease: A pilot study. *Digestive Diseases and Sciences, 50*(11), 2191–2193.

Hong, Z., & Piao, M. (2018). Effect of quercetin monoglycosides on oxidative stress and gut microbiota diversity in mice with dextran sodium sulphate-induced colitis. *Biomed Research International, 2018*, 8343052.

Hungness, E. S., et al. (2002). Proteasome inhibitors activate the transcription factors C/EBP-beta and delta in human intestinal epithelial cells. *Biochemical and Biophysical Research Communications, 290*(1), 469–474.

Hwang, S. W., et al. (2018). Intestinal alkaline phosphatase ameliorates experimental colitis via toll-like receptor 4-dependent pathway. *European Journal of Pharmacology, 820*, 156–166.

Jagetia, G. C., & Aggarwal, B. B. (2007). "Spicing up" of the immune system by curcumin. *Journal of Clinical Immunology, 27*(1), 19–35.

Jian, Y. T., Mai, G. F., Wang, J. D., Zhang, Y. L., Luo, R. C., & Fang, Y. X. (2005). Preventive and therapeutic effects of NF-kappaB inhibitor curcumin in rats colitis induced by trinitrobenzene sulfonic acid. *World Journal of Gastroenterology, 11*(12), 1747–1752. Available from https://doi.org/10.3748/wjg.v11.i12.1747, https://www.wjgnet.com/1007-9327.

Joe, B., & Lokesh, B. R. (2000). Dietary n-3 fatty acids, curcumin and capsaicin lower the release of lysosomal enzymes and eicosanoids in rat peritoneal macrophages. *Molecular and Cellular Biochemistry*, *203*(1-2), 153–161. Available from https://doi.org/10.1023/a:1007005605869, http://www.kluweronline.com/issn/0300-8177/.

Juillerat, P., et al. (2022). Positioning biologics in the treatment of IBD: A practical guide - Which mechanism of action for whom? *Current Research in Pharmacology and Drug Discovery*, *3*, 100104.

Jung, I. D., et al. (2010). COX-2 and PGE2 signaling is essential for the regulation of IDO expression by curcumin in murine bone marrow-derived dendritic cells. *International Immunopharmacology*, *10*(7), 760–768.

Kang, B. Y., et al. (1999). Inhibition of interleukin-12 production in lipopolysaccharide-activated macrophages by curcumin. *European Journal of Pharmacology*, *384*(2-3), 191–195.

Kaulmann, A., & Bohn, T. (2016). Bioactivity of polyphenols: Preventive and adjuvant strategies toward reducing inflammatory bowel diseases-promises, perspectives, and pitfalls. *Oxidative Medicine and Cellular Longevity*, *2016*, 9346470.

Khater, S. I., et al. (2022). Therapeutic potential of quercetin loaded nanoparticles: Novel insights in alleviating colitis in an experimental DSS induced colitis model. *Biomedicines*, *10*(7), 1654.

Kim, C. Y., et al. (2009). Neuroprotective effect of epigallocatechin-3-gallate against beta-amyloid-induced oxidative and nitrosative cell death via augmentation of antioxidant defense capacity. *Archives of Pharmacal Research*, *32*(6), 869–881.

Kim, G. Y., et al. (2005). Curcumin inhibits immunostimulatory function of dendritic cells: MAPKs and translocation of NF-kappa B as potential targets. *Journal of Immunology*, *174*(12), 8116–8124.

Kruidenier, L., Kuiper, I., Lamers, C. B. H. W., & Verspaget, H. W. (2003). Intestinal oxidative damage in inflammatory bowel disease: Semi-quantification, localization, and association with mucosal antioxidants. *The Journal of Pathology*, *201*(1), 28–36. Available from https://doi.org/10.1002/path.1409.

Kuhn, R., et al. (1993). Interleukin-10-deficient mice develop chronic enterocolitis. *Cell*, *75*(2), 263–274.

Lang, A., et al. (2015). Curcumin in combination with mesalamine induces remission in patients with mild-to-moderate ulcerative colitis in a randomized controlled trial. *Clinical Gastroenterology and Hepatology: The Official Clinical Practice Journal of the American Gastroenterological Association*, *13*(8), 1444–1449, e1.

Larmonier, C. B., et al. (2008). Limited effects of dietary curcumin on Th-1 driven colitis in IL-10 deficient mice suggest an IL-10-dependent mechanism of protection. *American Journal of Physiology. Gastrointestinal and Liver Physiology*, *295*(5), G1079–G1091.

Larmonier, C. B., Midura-Kiela, M. T., Ramalingam, R., Laubitz, D., Janikashvili, N., Larmonier, N., Ghishan, F. K., & Kiela, P. R. (2011). Modulation of neutrophil motility by curcumin: Implications for inflammatory bowel disease. *Inflammatory Bowel Diseases*, *17*(2), 503–515. Available from https://doi.org/10.1002/ibd.21391.

Larrosa, M., et al. (2009). Effect of a low dose of dietary resveratrol on colon microbiota, inflammation and tissue damage in a DSS-induced colitis rat model. *Journal of Agricultural and Food Chemistry*, *57*(6), 2211–2220.

Lee, K. W., & Lee, H. J. (2006). The roles of polyphenols in cancer chemoprevention. *Biofactors*, *26*(2), 105–121.

Li, X., & Liu, X. (2005). Effect of curcumin on immune function of mice. *Journal of Huazhong University of Science and Technology. Medical Sciences = Hua Zhong ke ji da xue xue bao. Yi xue Ying De wen ban = Huazhong Keji Daxue Xuebao. Yixue Yingdewen ban*, *25*(2), 137–140.

Lin, R., Piao, M., & Song, Y. (2019). Dietary quercetin increases colonic microbial diversity and attenuates colitis severity in citrobacter rodentium-infected mice. *Frontier Microbiology*, *10*, 1092.

Lin, S. C., & Cheifetz, A. S. (2018). The use of complementary and alternative medicine in patients with inflammatory bowel disease. *Gastroenterology and Hepatology*, *14*(7), 415–425. Available from http://www.gastroenterologyandhepatology.net/files/2018/07/gh0718LinCheifetz-1.pdf.

Liu, F., et al. (2021). Polyphenols intervention is an effective strategy to ameliorate inflammatory bowel disease: a systematic review and meta-analysis. *International Journal of Food Sciences and Nutrition*, *72*(1), 14–25.

Liu, S. H., Lu, T. H., Su, C. C., Lay, I. S., Lin, H. Y., Fang, K. M., Ho, T. J., Chen, K. L., Su, Y. C., Chiang, W. C., & Chen, Y. W. (2014). Lotus leaf (nelumbo nucifera) and its active constituents prevent inflammatory responses in macrophages via JNK/NF-κB signaling pathway. *American Journal of Chinese Medicine*, *42*(4), 869–889. Available from https://doi.org/10.1142/S0192415X14500554, http://www.worldscinet.com/ajcm/ajcm.shtml.

Lubbad, A., Oriowo, M. A., & Khan, I. (2009). Curcumin attenuates inflammation through inhibition of TLR-4 receptor in experimental colitis. *Molecular and Cellular Biochemistry*, *322*(1-2), 127–135. Available from https://doi.org/10.1007/s11010-008-9949-4.

Martin, A. R., et al. (2004). Resveratrol, a polyphenol found in grapes, suppresses oxidative damage and stimulates apoptosis during early colonic inflammation in rats. *Biochemical Pharmacology*, *67*(7), 1399–1410.

Martin, A. R., et al. (2006). The effects of resveratrol, a phytoalexin derived from red wines, on chronic inflammation induced in an experimentally induced colitis model. *British Journal of Pharmacology*, *147*(8), 873–885.

McNally, S. J., Harrison, E. M., Ross, J. A., Garden, O. J., & Wigmore, S. J. (2007). Curcumin induces heme oxygenase 1 through generation of reactive oxygen species, p38 activation and phosphatase inhibition. *International Journal of Molecular Medicine*, *19*(1), 165–172. Available from https://doi.org/10.3892/ijmm.19.1.165, http://www.spandidos-publications.com/serveFile/ijmm_19_1_165_PDF.pdf?type = article&article_id = ijmm_19_1_165&item = PDF.

Melgarejo, E., et al. (2010). Targeting of histamine producing cells by EGCG: a green dart against inflammation? *Journal of Physiology and Biochemistry*, *66*(3), 265–270.

Mimche, P. N., Thompson, E., Taramelli, D., & Vivas, L. (2012). Curcumin enhances non-opsonic phagocytosis of Plasmodium falciparum through up-regulation of CD36 surface expression on monocytes/macrophages. *Journal of Antimicrobial Chemotherapy*, *67*(8), 1895–1904. Available from https://doi.org/10.1093/jac/dks132.

Mochizuki, M., & Hasegawa, N. (2010). (-)-Epigallocatechin-3-gallate reduces experimental colon injury in rats by regulating macrophage and mast cell. *Phytotherapy Research: PTR*, *24*(Suppl 1), S120–S122.

Moon, D. O., et al. (2006). Curcumin decreases binding of Shiga-like toxin-1B on human intestinal epithelial cell line HT29 stimulated with TNF-alpha and IL-1beta: suppression of p38, JNK and NF-kappaB p65 as potential targets. *Biological & Pharmaceutical Bulletin*, *29*(7), 1470–1475.

Moss, M. L., et al. (2001). TACE and other ADAM proteases as targets for drug discovery. *Drug Discovery Today*, *6*(8), 417–426.

Murano, M., Maemura, K., Hirata, I., Toshina, K., Nishikawa, T., Hamamoto, N., Sasaki, S., Saitoh, O., & Katsu, K. (2000). Therapeutic effect of intracolonically administered nuclear factor κ B (p65) antisense oligonucleotide on mouse dextran sulphate sodium (DSS)-induced colitis. *Clinical and Experimental Immunology*, *120*(1), 51–58. Available from https://doi.org/10.1046/j.1365-2249.2000.01183.x.

Najafzadeh, M., Reynolds, P. D., Baumgartner, A., & Anderson, D. (2009). Flavonoids inhibit the genotoxicity of hydrogen peroxide (H2O2) and of the food mutagen 2-amino-3-methylimadazo[4,5-f]-quinoline (IQ) in lymphocytes from patients with inflammatory bowel disease (IBD). *Mutagenesis*, *24*(5), 405–411. Available from https://doi.org/10.1093/mutage/gep016.

Nunes, S., et al. (2018). Resveratrol and inflammatory bowel disease: the evidence so far. *Nutrition Research Reviews*, *31*(1), 85–97.

Okazaki, Y., Han, Y., Kayahara, M., Watanabe, T., Arishige, H., & Kato, N. (2010). Consumption of curcumin elevates fecal immunoglobulin a, an index of intestinal immune function, in rats fed a high-fat diet. *Journal of Nutritional Science and Vitaminology*, *56*(1), 68–71. Available from https://doi.org/10.3177/jnsv.56.68.

Pan, M. H., Chiou, Y. S., Wang, Y. J., Ho, C. T., & Lin, J. K. (2011). Multistage carcinogenesis process as molecular targets in cancer chemoprevention by epicatechin-3-gallate. *Food and Function*, *2*(2), 101–110. Available from https://doi.org/10.1039/c0fo00174k.

Pan, M. H., Lin-Shiau, S. Y., & Lin, J. K. (2000). Comparative studies on the suppression of nitric oxide synthase by curcumin and its hydrogenated metabolites through down-regulation of IkappaB kinase and NFkappaB activation in macrophages. *Biochemical Pharmacology*, *60*(11), 1665–1676.

Panaro, M. A., et al. (2012). Anti-inflammatory effects of resveratrol occur via inhibition of lipopolysaccharide-induced NF-kappaB activation in Caco-2 and SW480 human colon cancer cells. *The British Journal of Nutrition*, *108*(9), 1623–1632.

Parronchi, P., Romagnani, P., Annunziato, F., Sampognaro, S., Becchio, A., Giannarini, L., Maggi, E., Pupilli, C., Tonelli, F., & Romagnani, S. (1997). Type 1 T-helper cell predominance and lnterleukin-12 expression in the gut of patients with Crohn's disease. *American Journal of Pathology*, *150*(3), 823–832.

Pereira, R. B., et al. (2013). Glutathione and the antioxidant potential of binary mixtures with flavonoids: Synergisms and antagonisms. *Molecules (Basel, Switzerland)*, *18*(8), 8858–8872.

Piovani, D., et al. (2020). Inflammatory bowel disease: estimates from the global burden of disease 2017 study. *Alimentary Pharmacology & Therapeutics*, *51*(2), 261–270.

Quezada, S. M., McLean, L. P., & Cross, R. K. (2018). Adverse events in IBD therapy: the 2018 update. *Expert Review of Gastroenterology & Hepatology*, *12*(12), 1183–1191.

Ramos, G. P., & Papadakis, K. A. (2019). Mechanisms of disease: Inflammatory bowel diseases. *Mayo Clinic Proceedings. Mayo Clinic*, *94*(1), 155–165.

Riemschneider, S., et al. (2021). Indol-3-carbinol and quercetin ameliorate chronic DSS-induced colitis in C57BL/6 mice by AhR-mediated anti-inflammatory mechanisms. *International Journal of Environmental Research and Public Health*, *18*(5), 2262.

Del Rio, D., Rodriguez-Mateos, A., Spencer, J. P. E., Tognolini, M., Borges, G., & Crozier, A. (2013). Dietary (poly)phenolics in human health: structures, bioavailability, and evidence of protective effects against chronic diseases. *Antioxidants & Redox Signaling*, *18*(14), 1818–1892. Available from https://doi.org/10.1089/ars.2012.4581.

Romier, B., et al. (2008). Modulation of signalling nuclear factor-kappaB activation pathway by polyphenols in human intestinal Caco-2 cells. *The British Journal of Nutrition*, *100*(3), 542–551.

Rutgeerts, P., et al. (1999). Efficacy and safety of retreatment with anti-tumor necrosis factor antibody (infliximab) to maintain remission in Crohn's disease. *Gastroenterology*, *117*(4), 761–769.

Sabzevary-Ghahfarokhi, M., et al. (2020). The protective effects of resveratrol on ulcerative colitis via changing the profile of Nrf2 and IL-1beta protein. *Molecular Biology Reports*, *47*(9), 6941–6947.

Salim, S. Y., & Söderholm, J. D. (2011). Importance of disrupted intestinal barrier in inflammatory bowel diseases. *Inflammatory Bowel Diseases*, *17*(1), 362–381.

Samsami-Kor, M., Daryani, N. E., Asl, P. R., & Hekmatdoost, A. (2015). Anti-inflammatory effects of resveratrol in patients with ulcerative colitis: A randomized, double-blind, placebo-controlled pilot study. *Archives of Medical Research*, *46*(4), 280–285. Available from https://doi.org/10.1016/j.arcmed.2015.05.005.

Sanchez-Fidalgo, S., et al. (2010). Dietary supplementation of resveratrol attenuates chronic colonic inflammation in mice. *European Journal of Pharmacology*, *633*(1-3), 78–84.

Sanchez-Fidalgo, S., et al. (2013). Abarema cochliacarpos reduces LPS-induced inflammatory response in murine peritoneal macrophages regulating ROS-MAPK signal pathway. *Journal of Ethnopharmacology*, *149*(1), 140–147.

Scalia, S., Marchetti, N., & Bianchi, A. (2013). Comparative evaluation of different co-antioxidants on the photochemical- and functional-stability of epigallocatechin-3-gallate in topical creams exposed to simulated sunlight. *Molecules*, *18*(1), 574–587.

Schreiber, S., Nikolaus, S., & Hampe, J. (1998). Activation of nuclear factor kappa B in inflammatory bowel disease. *Gut*, *42*(4), 477–484. Available from https://doi.org/10.1136/gut.42.4.477.

Sharma, M., et al. (2014). Involvement of TACE in colon inflammation: A novel mechanism of regulation via SIRT-1 activation. *Cytokine*, *66*(1), 30–39.

Silva, M. S. da, Almeida, A. C. A. de, Faria, F. M. de, Luiz-Ferreira, A., Silva, M. A. da, Vilegas, W., Pellizzon, C. H., & Brito, A. R. M. S. (2010). Abarema cochliacarpos: Gastroprotective and ulcer-healing activities. *Journal of Ethnopharmacology*, *132*(1), 134–142. Available from https://doi.org/10.1016/j.jep.2010.08.001.

Singh, U. P., et al. (2012). Role of resveratrol-induced CD11b(+) Gr-1(+) myeloid derived suppressor cells (MDSCs) in the reduction of CXCR3(+) T cells and amelioration of chronic colitis in IL-10(-/-) mice. *Brain, Behavior, and Immunity*, *26*(1), 72–82.

Srivastava, R. (1989). Inhibition of neutrophil response by curcumin. *Agents and Actions*, *28*(3-4), 298–303. Available from https://doi.org/10.1007/bf01967418.

Steed, E., Balda, M. S., & Matter, K. (2010). Dynamics and functions of tight junctions. *Trends in Cell Biology*, *20*(3), 142–149.

Suzuki, T., & Hara, H. (2011). Role of flavonoids in intestinal tight junction regulation. *The Journal of Nutritional Biochemistry*, *22*(5), 401–408.

Sánchez-Fidalgo, S., et al. (2013). Abarema cochliacarpos reduces LPS-induced inflammatory response in murine peritoneal macrophages regulating ROS-MAPK signal pathway. *Journal of Ethnopharmacology*, *149*(1), 140–147.

Ungvari, Z., et al. (2011). Mitochondrial protection by resveratrol. *Exercise and Sport Sciences Reviews*, *39*(3), 128–132.

Vasconcelos, P. C., et al. (2012). Epicatechin used in the treatment of intestinal inflammatory disease: an analysis by experimental models. *Evidence-Based Complementary and Alternative Medicine: eCAM*, *2012*, 508902.

Veljaca, M., Lesch, C. A., Pllana, R., Sanchez, B., Chan, K., & Guglietta, A. (1995). BPC-15 reduces trinitrobenzene sulfonic acid-induced colonic damage in rats. *Journal of Pharmacology and Experimental Therapeutics*, *272*(1), 417–422.

Venkataranganna, M. V., Rafiq, M., Gopumadhavan, S., Peer, G., Babu, U. V., & Mitra, S. K. (2007). NCB-02 (standardized Curcumin preparation) protects dinitrocholorobenzene-induced colitis through down-regulation of NFκ-B and iNOS. *World Journal of Gastroenterology*, *13*(7), 1103–1107. Available from https://doi.org/10.3748/wjg.v13.i7.1103, https://www.wjgnet.com/1007-9327.

Vucelic, B. (2009). Inflammatory bowel diseases: Controversies in the use of diagnostic procedures. *Digestive Diseases*, *27*(3), 269–277.

Waetzig, G. H., et al. (2002). p38 mitogen-activated protein kinase is activated and linked to TNF-alpha signaling in inflammatory bowel disease. *Journal of Immunology*, *168*(10), 5342–5351.

Wagnerova, A., et al. (2017). Sex differences in the effect of resveratrol on DSS-induced colitis in mice. *Gastroenterology Research and Practice*, *2017*, 8051870.

Wang, J., et al. (2024). Global trends and hotspots of ulcerative colitis based on bibliometric and visual analysis from 1993 to 2022. *Medicine*, *103*(5), e37095.

Wang, J., Ghosh, S. S., & Ghosh, S. (2017). Curcumin improves intestinal barrier function: Modulation of intracellular signaling, and organization of tight junctions. *American Journal of Physiology. Cell Physiology*, *312*(4), C438–C445.

Yadav, V. S., Mishra, K. P., Singh, D. P., Mehrotra, S., & Singh, V. K. (2005). Immunomodulatory effects of curcumin. *Immunopharmacology and Immunotoxicology*, *27*(3), 485–497. Available from https://doi.org/10.1080/08923970500242244.

Yao, J., et al. (2010). Anti-oxidant effects of resveratrol on mice with DSS-induced ulcerative colitis. *Archives of Medical Research*, *41*(4), 288–294.

Yao, J., Wei, C., Wang, J. Y., Zhang, R., Li, Y. X., & Wang, L. S. (2015). Effect of resveratrol on Treg/Th17 signaling and ulcerative colitis treatment in mice. *World Journal of Gastroenterology*, *21*(21), 6572–6581. Available from https://doi.org/10.3748/wjg.v21.i21.6572, http://www.wjgnet.com/1007-9327/pdf/v21/i21/6572.pdf.

Yildiz, G., et al. (2015). Resveratrol pretreatment Ameliorates TNBS colitis in rats. *Recent Patents on Endocrine, Metabolic & Immune Drug*, *9*(2), 134–140.

Youn, J., et al. (2009). Resveratrol and piceatannol inhibit iNOS expression and NF-kappaB activation in dextran sulfate sodium-induced mouse colitis. *Nutrition and Cancer*, *61*(6), 847–854.

Yum, H. W., et al. (2013). Oligonol inhibits dextran sulfate sodium-induced colitis and colonic adenoma formation in mice. *Antioxidants & Redox Signaling*, *19*(2), 102–114.

Zhang, M., Deng, C.-S., Zheng, J.-J., & Xia, J. (2006). Curcumin regulated shift from Th1 to Th2 in trinitrobenzene sulphonic acid-induced chronic colitis. *Acta Pharmacologica Sinica*, *27*(8), 1071–1077. Available from https://doi.org/10.1111/j.1745-7254.2006. 00322.x.

Zhang, X., et al. (2016). Protective effect of curcumin on TNBS-induced intestinal inflammation is mediated through the JAK/STAT pathway. *BMC Complementary and Alternative Medicine*, *16*(1), 299.

Zhang, Y. Z., & Li, Y. Y. (2014). Inflammatory bowel disease: Pathogenesis. *World Journal of Gastroenterology*, *20*(1), 91–99. Available from https://doi.org/10.3748/wjg.v20.i1.91, http://www.wjgnet.com/1007-9327/pdf/v20/i1/91.pdf.

CHAPTER FOUR

Urinary-related disorders: role of polyphenols

María Teresa Melgarejo Segura[1,*], Yaiza María Yáñez Castillo[1,*] and Francisco Javier Leyva-Jimenez[2,3]

[1]Department of Urology, San Cecilio University Hospital, Granada, Spain
[2]Regional Institute for Applied Scientific Research (IRICA), University of Castilla-La Mancha, Ciudad Real, Spain
[3]Area of Food Science and Technology, Faculty of Chemical Sciences and Technologies, University of Castilla-La Mancha, Ciudad Real, Spain

Introduction

Urology is the medical specialty responsible for diagnosing and treating urinary-related disorders. These disorders can range from oncological diseases, such as prostate, bladder, and renal cancer, to benign diseases, such as urolithiasis, benign prostatic hyperplasia (BPH), or urinary tract infections (UTIs) (¿Qué es la Urología?, 2024).

Urological diseases have a high prevalence. For example, prostate cancer (PCa) is the most common non-cutaneous tumor in men in the world, with an incidence of 1,600,000 new cases and 366,000 deaths annually (Wang, Zhao, et al., 2018). On the other hand, UTIs are the most common bacterial infections in the world, affecting 150 million people. The social costs of UTIs, taking into account healthcare expenses and work absenteeism, are approximately $3.5 billion per year only in the United States (Flores-Mireles et al., 2015).

Various factors are involved in the development of urological diseases, such as old age in some urological cancers (Daniyal et al., 2014). Life habits can also influence their development, with smoking as one of the main risk factors for bladder and renal cancer (Capitanio & Montorsi, 2016), and excessive consumption of salt and fat increasing the probability of suffering from urinary stones and low water intake in UTIs (Morales-Martínez et al., 2021). Diseases such as high blood pressure may be related to a higher incidence of renal cancer (Capitanio & Montorsi, 2016).

[*]Both authors contributed equally to the study.

Bioactive Polyphenols for Health and Pathology Treatment.
DOI: https://doi.org/10.1016/B978-0-443-13821-8.00004-4

Due to the extensive complexity of these urinary disorders, the available treatments can be diverse, involve high healthcare costs, and are not free of side effects.

On the one hand, the treatments available for BPH are associated with dizziness and sexual dysfunction (Sausville & Naslund, 2010). On the other hand, medical treatments for different types of urinary stones can produce metabolic alterations and gastrointestinal dysfunction (Martínez et al., 2021). Finally, the use of antibiotics in UTIs can be associated with bacterial resistance, representing an important healthcare problem (Maisto et al., 2023).

For all these reasons, in recent years, research has intensified on bioactive compounds from natural sources that serve as an alternative for the prevention and/or treatment of different urological diseases and that, in addition, reduce the side effects of traditional pharmacological treatment, improving thereby the prognosis of the disease. In this scenario, phenolic compounds are gaining the attention of the scientific community as an interesting option in the treatment of urinary disorders.

Phenolic compounds are secondary metabolites of plants used for protective or reproductive functions, among others. The main characteristic of these compounds is that they all have one or more aromatic rings with one or more hydroxyl substituents, giving rise to more than 10,000 different structures. One of the most relevant properties of these compounds lies in their antioxidant and anti-inflammatory capacity (Leyva-Jiménez et al., 2019). In the case of the urinary system, physiological imbalances in oxidative stress (OS) and the maintenance of a proinflammatory state can lead to the development or aggravation of urinary disorders, worsening the symptoms and prognosis of the disease (Minciullo et al., 2015).

The objective of this chapter is to summarize the most recent information about the use of phenolic compounds in the treatment and prevention of different urinary disorders and highlight their effectiveness in different urological diseases. These compounds represent an interesting complementary therapeutic alternative to commonly used treatments.

Benign prostatic hyperplasia

BPH is a benign enlargement of the transitional zone of the prostate gland caused by a proliferative increase in epithelial cells, smooth muscle, and stroma. These changes cause compression of the urethra and are responsible for the majority of lower urinary tract symptoms (LUTS) in men, which include storage symptoms (nocturia, frequency, and urgency) and voiding symptoms (weak urinary flow, incomplete bladder emptying, choppy flow, terminal drip). Although it is not a malignant pathology, LUTS derived from BPH negatively affect patients' quality of life (QoL). The prevalence increases with age, with symptoms starting around 40 years of age. In men aged 50–60 years, it reaches a prevalence of 40%–50%, rising to 80% in men over 80 years old (EAU Guidelines on the Management of Non-neurogenic Male LUTS, 2023).

BPH has a complex and multifactorial etiology, influenced by various factors, such as aging, hormonal modifications, and other elements, such as obesity and type II diabetes mellitus (Chen et al., 2015; Minciullo et al., 2015). During the aging process, a physiological imbalance occurs in the relationship between estrogens and androgens (testosterone), which stimulates cell proliferation. Besides, chronic inflammation of the prostate has also been identified as another factor involved in the etiology of BPH (Di Silverio et al., 2003; Stewart & Lephart, 2023).

Currently, there are two main drugs used for the treatment of BPH: (1) α1-adrenergic receptor blockers (such as doxazosin, terazosin, and tamsulosin) that relieve LUTS by relaxing the smooth muscles of the prostate and bladder neck, and (2) 5α-reductase enzyme inhibitors, which reduce dihydrotestosterone (DHT) production and inhibit the growth of BPH. Despite having proven effectiveness, these medications are not free of side effects. α1-adrenergic blockers can cause dizziness, disorientation, drowsiness, nasal congestion, and disturbances in ejaculation, in addition to causing confusion and depression. Furthermore, 5α-reductase inhibitors can lead to problems such as erectile dysfunction, ejaculation disorders, breast enlargement in men (gynecomastia), depression, anxiety, and an increased risk of developing PCa (Thompson et al., 2003).

Another alternative is surgery; however, the high costs and associated risks exclude it as a routine treatment alternative (EAU Guidelines on the Management of Non-neurogenic Male LUTS, 2023). All these factors have driven the search for alternative methods to control this disease, with phenolic compounds receiving special attention.

Epigallocatechin-3-gallate

Both in vitro and in vivo studies demonstrate the effects of epigallocatechin-3-gallate (EGCG) from green tea on BPH cells.

In in vitro studies, it was observed that EGCG reduced cell viability and migration by inhibiting the phosphorylation of focal adhesion kinase (FAK), reduced the levels of cell division cycle 42 (Cdc42), and the protein levels of the homolog of Ras (Rho) (Tepedelen et al., 2017). These anti-cell growth effects were also confirmed in in vivo studies in rats, where EGCG improved BPH and testosterone-induced fibrosis. The underlying mechanisms identified include the decrease in prostate levels of cyclooxygenase (COX-2), interleukins (IL-1B, IL-6), and tumor necrosis factor (TNF-α), in addition to the inhibition of overexpression of androgen receptors (ARs), estrogen receptor α (ER-α), hypoxia-inducible factor 1 alpha subunit (HIF-1α), transforming growth factor (TGF-β1, TGF-βRI) and p-SMAD3 (an intracellular signaling protein), as well as increased expression of ER-β (Zhou et al., 2018). It has also been proven that EGCG has protective effects against testosterone-induced BPH in rats with metabolic syndrome. A decrease in insulin-like growth factor (IGF-1 and IGF-2) and other inflammatory markers such as IL-Iβ, IL-I6, and TNF-α were observed. Furthermore, in the prostate tissue of these rats, an increase in the expression of IGF-binding protein 3 (IGFBP-3) and the peroxisome proliferator-activated receptor (PPAR) was observed (Jinglou Chen & Song, 2016), both with negative effects on the development of BPH (Refaie et al., 2018; Sreenivasulu et al., 2018).

Although there are few in vivo clinical trials in humans, it has been shown that administering a mixture of green and black tea extracts containing EGCG to patients with LUTS achieves clinical improvement and QoL after 6 weeks of treatment (Katz et al., 2014). Despite these promising results have been obtained, clinical trials with a larger population, as well as isolated EGCG, are required to determine the safety and clinical efficacy of EGCG in the treatment of BPH.

Resveratrol

Resveratrol (RSV) is a stilbene found in the skin of grapes, peanuts, and various woody plants (Vestergaard & Ingmer, 2019). The ability of RSV to induce apoptosis of human BPH cells (BPH-1) and inhibit cell proliferation in a concentration-dependent manner has been demonstrated through in vitro studies. This function is achieved by arresting cells in the S phase, increasing the phosphorylation of p38 MAPK, and repressing FOXO3a, producing the accumulation of reactive oxygen species (ROS) leading to cell apoptosis (Li et al., 2019). Its proapoptotic effects have also been studied *in vivo*, observing that the mechanisms underlying its antiproliferative capacity are found in the increase in the proapoptotic protein Bax and the decrease in Bcl-2 and Bcl-XL. It also has anti-inflammatory effects by reducing the expression of COX-2 and Inos (Chung et al., 2015). In other in vivo studies, its effect on obesity-induced BPH is demonstrated by improving the production of ROS and increasing the expression of phosphorus-AKT induced by insulin (Calmasini et al., 2018).

In a placebo-controlled clinical trial by Kjaer and colleagues, the effect of administering 1 gram of RSV for 4 months in middle-aged men with metabolic syndrome was investigated. It was observed that RSV significantly reduced serum levels of the androgen precursors androstenedione, DHEA, and DHEAS. However, no changes were found in prostate size or circulating levels of PSA, testosterone, free testosterone, and DHT. Therefore it was concluded that there is insufficient support for using RSV in treating BPH (Kjaer et al., 2015).

Isoflavones

Isoflavones are flavonoid polyphenols with structural similarity to human estrogens. In vitro, it has been shown that Equol, derived from the intestinal metabolism of isoflavones such as soy, has different effects on the prostate: (1) it binds to 5α-DHT, inhibiting the aromatase enzyme responsible for the biosynthesis of estrogens and cell proliferation; (2) binds to the estrogen-related receptor gamma (ERR-γ), slowing cell proliferation; and finally, (3) binds to the estrogen receptor beta (ER-β) acting as a selective modulator of estrogen receptors (Lephart, 2014).

In vivo studies on rats fed with red clover isoflavones showed an increase in the expression of the estrogen receptor-β in the prostate epithelium, in addition to inhibiting cell proliferation and the adhesion protein E-cadherin, decreasing the beta growth factor (TGF-β) (Slater et al., 2002). An inhibition of prostate proliferation and an increase in the expression of nitric oxide (NO) and NO synthase were observed when the rats were feeding with soy isoflavone (Yang et al., 2009). The administration of Seoritae, a black soybean with a high content of isoflavones and anthocyanins, caused a decrease in prostate weight, OS, apoptosis, and 5-α reductase activity (Jang et al., 2014). Finally, consumption of a diet rich in isoflavones for 5 weeks also caused a decrease in prostate weight in rats but without changes in 5-α-reductase activity or testosterone levels (Weber et al., 2001).

In human studies, the daily consumption of 40 mg of soy isoflavones was evaluated for 12 months in 176 patients with BPH and LUTS. They found that, compared to placebo, those who received isoflavones obtained better results in the score of the IPSS questionnaire (*International Prostate Symptoms Score*, a standardized questionnaire used to objectify LUTS) and flowmetry results (Wong et al., 2012). Likewise, the administration of the combination of daidzein with isolase and zinc in patients with LUTS due to BPH for 6 months caused an improvement in the symptoms and QoL of the patients. In another pilot clinical trial of 18 middle-aged men with moderate or severe LUTS secondary to BPH, treatment for 4 weeks with Equol improved IPSS and QoL values. Furthermore, a decrease in 5α-DHT was detected in 21% of men with severe symptoms (Lephart, 2013).

Urinary tract infections

UTIs constitute a significant health problem due to their high prevalence and are the most frequent infections in our environment (Flores-Mireles et al., 2015). They are more common in women due to the shorter urethra length. It is estimated that approximately 40%–50% of adult women will experience a UTI at some point in their lives, and around 25% will experience recurrences (Scholes et al., 2000). From a clinical point of view, recurrent UTIs constitute a serious public health problem and are diagnosed when they occur more than 2 times in 6 months (European Association of Urology, 2024). The main microorganisms that cause UTIs are Gram-negative and Gram-positive, especially *Escherichia coli*, *Pseudomonas aeruginosa*, *Proteus mirabilis*, and *Enterococcus faecalis* (Ortega-Lozano et al., 2023). In total, 85% of cystitis registered worldwide are caused by *E. coli* (González de Llano et al., 2020). UTIs associated with urinary catheters constitute a frequent problem of nosocomial infections. *E. coli, P. aeruginosa*, and *Proteus spp*., which have the capacity to form biofilm on catheters, mainly produce them. The biofilm consists of a community of the same or different groups of bacteria that bind to each other and a substrate (bladder mucosa, urinary catheters) irreversibly. This assembly adheres to a hydrated exopolymeric matrix (MEP) (Jagannathan & Viswanathan, 2018). Controlling this process is important to reduce the incidence of this type of infection since the biofilm results in a chronic persistent infection that is difficult to eradicate with simple antimicrobial agents because the MEP prevents the passage of antibiotics (Ortega-Lozano et al., 2023).

The primary treatment for these pathologies has been antibiotics. However, the emergence of antibiotic resistance (*Pseudomonas spp.* resistant to carbapenems and fluoroquinolones, *Enterococcus spp.* resistant to vancomycin, Enterobacteriaceae resistant to β-lactams, etc.) makes it necessary to investigate alternative treatment options for UTIs, such as antiadhesive agents, probiotics or vaccines (De Nunzio et al., 2021; Maisto et al., 2023). In the last few years, natural bioactive molecules, especially phenolic compounds (Table 4.1), have been identified and selected for the prevention and treatment of UTIs. Some examples will be developed in this section.

Proanthocyanidins

Proanthocyanidins (PAC), a form of condensed tannin, are oligomers or polymers of monomeric flavan-3-ols. Due to their high variability, they are classified into several types. In this sense, only PAC A have demonstrated an effect against bacterial adhesion and motility of uropathogenic strains in the urothelial cells of the urinary tract, in addition to having high antioxidant activity. This is why cranberries, one of the most important natural sources, have been used as UTI prophylaxis for a long time (Jagannathan & Viswanathan, 2018).

In in vitro studies, Nicolosi et al. studied the antiadhesion activity of procyanidin A2 in pathogenic strains such as *uropathogenic E. coli (UPEC)* and *P. mirabilis* using a human bladder cancer cell line (HT-1376). They showed that concentrations of 50 $\mu g\ mL^{-1}$ significantly reduced the adhesion of both bacteria to the cell surface. Both bacteria have mannose-resistant (MR) adhesins as a virulence factor, composed mainly of the *Fla A protein*, encoded by the gene of the same name, whose expression is inhibited by PAC (Nicolosi et al., 2014). In addition, Llano et al. evaluated the antiadhesive activity of pure cranberry PACs and their metabolites produced by the intestinal microbiota (such as simple phenols: benzoic acid, phenylacetic acid, and phenylpropionic acids) against *UPEC* in a bladder cell culture model. Intestinal metabolites showed dose-dependent antiadhesive activity against UPEC, reducing the number of bacteria that adhered to bladder epithelial cells at all concentrations tested (100–500 uM). Therefore the authors suggest that the in vitro antiadhesion effect of cranberry PACs against *UPEC* could be mainly due to its intestinal metabolites (De Llano et al., 2015). All these studies show that,

Table 4.1 Summary beneficial effects of plant extracts/phenolic compounds in urinary diseases.

Disorder	Extract/phenolic compound	Assay	Concentration	Action mechanism	Results	Refs.
BPH	ECGC	BPH-1 cells	2.39–89.12 μM	Decrease the activation of FAK, levels of paxillin, RhoA, Cdc42 and PAK1	Inhibition of cell proliferation	Tepedelen et al. (2017)
		BPH rats	50–100 mg/kg/d	Increased the prostatic expression of IGFBP-3 and PPARs	Alleviated the disordered levels of glucose, TC and TG, as well as attenuated the prostate enlargement	Chen and Song (2016)
	Green Tea	Human models	500–1000 mg/d	Improved urine flow, C-reactive protein, postvoid residual volume	Improves AUAS symptoms	Katz et al. (2014)
	Resveratrol	BPH-1 cells	0, 10, 20, 30, 40, and 50 μM	Increased phosphorylation of p38 MAPK and repression of FOXO3a, leading to the accumulation of reactive oxygen species (ROS).	Inhibition of cell proliferation	Li et al. (2019)
		Obesity-induced mouse BPH	100 mg/kg/d	Increased proapoptotic protein Bax and decreased Bcl-2 and Bcl-XL, decreased expression of COX-2 and iNOS	Apoptosis BPH cells	Calmasini et al. (2018)
		Human	1000 mg/d	Reduced serum levels of the androgen precursor androstenedione	No changes in prostate size were found.	Kjaer et al. (2015)

	Red clover isoflavones	Rats	12 mg/kg/d	Increased estrogen receptor-β expression.	Inhibit cell proliferation, and to the adhesion protein E-cadherin	Slater et al. (2002)
	Soy isoflavones	BPH induced rats	-	Increased nitric oxide (NO) and nitric oxide synthase expression	Inhibit prostatic proliferation	Yang et al. (2009)
		humans	40 mg/d	Improved urine flow, postvoid residual volume	Improves AUAS symptoms	Wong et al. (2012)
UTI	PAC A2	*E. coli uropathogenic and P. mirabilis*	50 μg/ml	Inhibit flagella mobility	Reduced bacterial adhesion	Nicolosi et al. (2014)
	PAC from Blueberry	*E. coli uropathogenic*	100–500 μM	Inhibit flagella mobility	Reduced bacterial adhesion	De Llano et al. (2015)
	Cranberry	*E. Coli*	12.5 μg/mL	COX-2, Inhibition, Cytokines reduction	Reduction proinflammatory mediators	Huang et al. (2009)
	PAC	Human	0–72 mg/d	Inhibit flagella mobility	Reduced bacterial adhesion	Howell et al. (2010)
	ECGC	*E. coli*	2.5 mg/mL	Increased cytokines, Reduced TFN-α	Antibacterial activity	Cho et al. (2007)
		Humans	500 mg/d	Decreased bacterial concentration	Reduced Cystitis symptoms	Kheirabadi et al. (2019)
	Quercetin	*E. coli*	0.125–1 mg/mL	Inhibit FirmH protein	Reduced bacterial adhesion	Jaiswal et al. (2018)
		P. aeruginosa	0.125 mg/mL	Reduced biofilm formation	Reduced MIC antimicrobics	Vipin et al. (2020)

(*Continued*)

Table 4.1 (Continued)

Disorder	Extract/phenolic compound	Assay	Concentration	Action mechanism	Results	Refs.
UL	Apigenin	Human	0.005–0.02 g/kg	Antioxidant, inhibition signals TGF-β	reduces epithelial injury, protection against lithiasis	Azimi et al. (2021)
	Vitexin	Rat model	20 mg/kg/d	Inhibiting pyroptosis activation, apoptosis, epithelial-mesenchymal transition (EMT), and macrophage infiltration	Reduced crystal deposition	Ding et al. (2021)
	Epicatechin and procyanidin	Rat model	100 mg/kg/d	Inhibitory activities on gene expression of COX-2 and IL-1β.	Anti-urolithiasis effect	Youn et al. (2017)
	Anthocyanins from *Rubus idaeus*	Rat model	100–200 mg/kg/d	Reduced excretion of calcium, phosphorus and oxalate in urine	Reduced CaOx deposits	Ghalayini et al. (2011)
	Resveratrol	Rat model	10 mg/Kg/d	Promoted intracellular LC3II upregulation, p62 downregulation, and autophagosome formation and induced the expression of transcription factor EB,	Regulated oxalate-induced renal inflammation, oxidative injury, and CaOx crystal deposition	Wu et al. (2021)

AUAss, Urologic Association symptom score; *BPH*, benign prostatic hyperplasia; *Cdc42*, cell division cycle 42; *ECGC*, Epicalgatechin galate; *FAK*, focal adhesion kinase; *PAC*, proanthocyanidins; *PAK 1*, p21 protein-activated kinase 1; *RhoA*, Ras homolog gene family, member A; *TNF-α*, tumor necrotic factor alpha; *UL*, urolithiasis; Urologic Association symptom score; *UTI*, urinary tract infections.

although PACs do not have antimicrobial activity *per se*, they do hinder the first step in the production of UTIs and cell adhesion, and therefore their consumption can help prevent UTIs (Jagannathan & Viswanathan, 2018). Similarly, the effect of PAC on bacterial motility was evaluated, observing that a concentration of 60 μg mL − 1 reduces the mobility of *P. aeruginosa*, reducing its infective capacity (O'May & Tufenkji, 2011).

Other studies evaluated the antioxidant effect of PACs as modulators of proinflammatory mediators induced by infections caused by *E. coli*. To do this, they used cranberry methanol extract (CR-ME) on two strains of *E. coli* (ATCC 700336 and ATCC 25922). Although the extract did not show any direct antibacterial effect against the strains, it did have an effect on the cellular mediators of inflammation by significantly inhibiting the activity of COX-2, with a maximum inhibitory concentration (IC50) of 12.8 μg/ mL. Based on these results, the authors also investigated the effects of CR-ME on the expression of proinflammatory mediators such as NF-κB in the T lymphocyte cell line and on the release of cytokines (IL-1β, IL-6, IL-8) in human peripheral blood mononuclear cells. They found that NF-κB transcriptional activation was significantly inhibited ($p < 0.01$) with an IC50 of 12.8 μg/mL, while cytokine production was significantly reduced with 10 μg/mL CR-ME extract. These data suggest that PACs could reduce the expression of inflammatory mediators stimulated by *UPEC* and, consequently, improve symptoms associated with UTIs (Huang et al., 2009).

Additionally, different in vivo studies have been carried out in humans to assess the most effective concentration of PAC against bacterial adhesion. Howell et al. observed that PACs have a dose and time-dependent effect. The study was optimized with the consumption of PAC in the form of standard cranberry powder and there was a linear decrease in *E. coli* adhesion with increasing PAC dosage. The maximum antiadhesive activity was observed 6 hours after ingestion, the period of elimination of PAC in the urine. However, only above 36 mg of PAC, the residual anti-adhesion effect was observed and validated by mannose resistance hemagglutination assay. Furthermore, the effect was maximal with a PAC dose of 72 mg per day (Howell et al., 2010).

Epigallocatechin

Green tea catechins have demonstrated antimicrobial activity against a wide variety of microorganisms. Epigallocatechin (EGC) is the most abundant in green tea and the one that is excreted in the highest concentration in the urine, which is why it may show a potential effect for the treatment of UTIs. In fact, through in vitro studies, Cho et al. demonstrated the inhibitory activity of these catechins against *E. coli*. The mechanism of action includes an increase in the production of cytokines (IL-10 and IL-12), the reduction of the expression of the TFN-α gene, damage to the cell membrane of the bacteria and bactericidal and antitoxin effects, reducing, therefore, the infectivity (Cho et al., 2007).

In vivo studies have also been carried out using rats as a chronic bacterial prostatitis (CBP) model. An example is that carried out by Lee et al., which showed a synergistic effect between catechins and ciprofloxacin, reducing bacterial growth and prostate inflammation (Y.S. Lee et al., 2005). It has also been shown that the combination of catechins with traditional antibiotics can produce a synergistic effect, improving the minimum inhibitory concentration (MIC) of some antibiotics such as chloramphenicol, amoxicillin, azithromycin, and ciprofloxacin (Passat, 2012). In another study, Kheirabadi et al. conducted a clinical trial in women with uncomplicated cystitis. They showed, in women treated with green tea (four 500 mg capsules/day) together with antibiotic medication, a statistically significant decrease in the prevalence of symptoms related to cystitis along with an improvement in the results of urine tests after 3 days of treatment (Kheirabadi et al., 2019).

Quercetin

Quercetin is a flavonoid found in onions, apples, berries, tea, red wine, and kale. It has antibacterial and immunostimulant action. It stands out as a powerful inhibitor of DNA gyrase, the enzyme responsible for bacterial DNA replication (Lan et al., 2022). In addition, it has antiadhesive properties by inhibiting the FirmH protein related to the attachment of *E. coli* to the uroepithelium (Jaiswal et al., 2018). Likewise, its ability to degrade the extracellular polymeric substance of the matrix of polymicrobial biofilms on urinary catheters has been observed (Hamzah et al., 2020). In in vitro studies, Vipin et al. combined quercetin with different antibiotics for the treatment of *P. aeruginosa*, observing a significant reduction in the MIC of levofloxacin, ceftriaxone, gentamicin, tobramycin, and amikacin (Vipin et al., 2020). Quercetin has demonstrated mutagenic activity in bacterial models that were subsequently confirmed in eukaryotic models at relatively high concentrations (10 mg per incubated culture); however, this activity has not been confirmed in animal models with rats or mice (Harwood et al., 2007).

The evaluation of quercetin *in vivo* in humans was studied in a clinical trial carried out by Torella et al. in 145 postmenopausal women with recurrent UTIs treated with an oral supplement of quercetin, curcumin, and hyaluronic acid accompanied or not with local estrogens. In patients with the combined treatment, recurrences were reduced, almost double that of the other treatment groups (Torella et al., 2016).

Urinary lithiasis

The term urolithiasis (UL) refers to the formation of stones in any part of the urinary tract. This phenomenon constitutes a public health problem as its incidence has increased considerably in various regions of the world, with rates ranging between 7% and−13% in North America, 5% and 9% in Europe, 5% and 9%, and 1% and 5% in Asia (Sorokin et al., 2017). Both environmental—diet, physical exercise—and metabolic factors—metabolic alterations, urinary composition and volume, presence of UTI, and genetic predisposition—intervene in the predisposition to the formation of UL (Sansores-España et al., 2008; Siener, 1917).

UL arises because of the imbalance between the solubility and precipitation of salts in the urinary system, where stone formation inhibitors, such as citrate, magnesium, and Tamm-Horsfall protein (THP), among others, are altered. All of this, together with low urinary volume, hypercalciuria, hyperoxaluria, hyperuricosuria, or alterations in urinary pH, promote stone formation (Ratkalkar & Kleinman, 2011).

At a molecular level, there is an increase in OS due to an increase in ROS and a decrease in protective antioxidant systems (superoxide dismutase (SOD), glutathione peroxidase (GPx), catalase, etc.). All this favors damage to the cell membrane, the adhesion of crystals to the renal tubules and, therefore the formation of **"Randall's plaque,"** the basis for the formation of intrarenal stones (Sansores-España et al., 2008). There are different types of urinary stones, with calcium stones being the most prevalent, comprising approximately 80%. Among them, the calcium oxalate (CaOx) stone stands out as the most frequent, although to a lesser extent, struvite or magnesium ammonium phosphate, uric acid, and cystine stones can be found (Alelign & Petros, 2018; Zeng et al., 2019).

Prevention is the most efficient method of treatment for this disease, based on the implementation of pharmacological treatment to prevent the formation or growth of stones (Han et al., 2015). However, these treatments are not free of adverse effects. For example, hydrochlorothiazide used to treat calcium stones can cause hypotension and hypokalemia, while potassium citrate, used to alkalinize urine in patients with uric acid stones, can cause hyperkalemia (Martínez et al., 2021).

Medicinal plants have been used over the years to treat UL, although a lack of knowledge about their bioactive compounds and effects limits their widespread use (Ahmed et al., 2018). Currently, the demonstrated potential of phenolic compounds as antioxidants and anti-inflammatory agents could be beneficial for preventing UL by inhibiting the overproduction of ROS and stimulating protective antioxidant systems, thereby protecting the renal cellular epithelium from oxidative damage. In addition, some are diuretics and inhibit the angiotensin-converting enzyme (ACE): The renin-angiotensin-aldosterone system (RAAS) activates NADPH oxidase in kidney cells, which produces ROS. Inhibition of ACE-I, by decreasing the production of ROS, significantly reduces the deposition of CaOx crystals and renal inflammation. Furthermore, the free hydroxyl groups of phenolic compounds chelate the zinc present in the active center of ACE, inhibiting it (Jucá et al., 2020). This is why using certain phenolic compounds (Table 4.1), such as flavonoids, phenolic acids, or stilbenes, among others, can be useful for the prevention and treatment of this urinary condition.

Flavonoids

Flavonoids such as catechin, epicatechin (present in green tea), and diosmin (present in the peel of some citrus fruits) have shown beneficial effects on the prevention and treatment of UL (Shen et al., 2022). They have diuretic, antioxidant, antiapoptotic, anti-inflammatory, and antibacterial activity (Jucá et al., 2020). Flavonoids are capable of reducing ROS, inducing gene expression of antioxidant enzymes, and chelation of transition metals, thereby reducing membrane lipid peroxidation and renal cell injury. They have been shown to reduce ACE. All of this prevents the accumulation of CaOx (Zeng et al., 2019). Besides, they are anti-inflammatory, reducing the damage caused by hyperoxaluria (Ahmed et al., 2018; Guzel et al., 2021). However, in vitro studies indicate that it can promote the aggregation and invasion of CaOx crystals, presenting a dual effect (Chaiyarit et al., 2024).

Apigenin, present as an active component in fruits such as grapefruit and orange, or vegetables such as bell pepper or cabbage, attenuates kidney damage by reducing urinary levels of stone-producing components, antioxidant activities and inhibiting TGF-β signaling (Azimi et al., 2021).

Vitexin, derived from medicinal plants such as *Ficus deltoidea*, bamboo, and dried hawthorn leaves (*Crataegus pinnatifida*), demonstrated a protective activity in the formation of UL in a study carried out in mice with nephrolithiasis by inhibiting pyroptosis, apoptosis, epithelial-mesenchymal transition (EMT), and macrophage infiltration (Ding et al., 2021).

In in vivo studies in animal models, the anti-inflammatory role of herbs rich in flavonoids, such as *Orthosiphon stamineus*, Java tea, or *Desmodium styracifolium* (Chao et al., 2020; Zhou et al., 2018), and herbs rich in catechins and epicatechins, such as Quercus gilva Bume (Youn et al., 2017), has been identified.

Another mechanism by which these medicinal herbs have a prophylactic role on UL is due to their ability to inhibit or reduce the excretion of oxalate in urine. Among the polyphenols identified with this function are phlorotannins from the brown algae *Sargassum* or chlorogenic acid from the *shrub Viburnum opulus L.* (Sujatha et al., 2015).

Treatment with anthocyanins (from raspberries) in mice with CaOx nephrolithiasis reduced the excretion of calcium, phosphorus, and oxalate in urine and presented different antioxidant effects (Ghalayini et al., 2011). The administration of bioflavonoids obtained from citrus fruits to prevent and cure UL induced by hyperoxaluria in mice was able to regulate CaOx deposition in addition to regulating lithogenic factors such as THP and the proinflammatory cytokine NF-κB (Badrinathan et al., 2015).

Phenolic acids, coumarins, and stilbenes

Coumarins, present in the *Ammi visnaga L.* plant, have been shown to have anti-inflammatory effects, protect renal epithelial cells from oxalate and calcium crystals, and increase urinary pH, thus preventing the formation of oxalate crystals (Khalil et al., 2020).

Other compounds, such as gallic acid (Zhou et al., 2022), chlorogenic acid (Hoseinynejad et al., 2022), or RSV (Hong et al., 2013; Oksay et al., 2017; Wu et al., 2021), have been shown in vivo to prevent OS and the formation of Randall's plaque (Oksay et al., 2017), as well as enhance antioxidant enzymes (S.H. Hong et al., 2013), leading to a decrease in hyperoxaluria and CaOx crystals in urine. Finally, caffeic acid is capable of regulating stone formation modulators such as THP and producing a reduction in CaOx deposits in in vivo models (Yasir et al., 2018).

Prostate cancer

PCa is the most common non-cutaneous tumor in men in the world (Bergengren et al., 2023). Incidence rates vary significantly globally, being highest in regions such as Oceania, North America, and Europe and lowest in Africa and Asia. Risk factors for PCa include older age, obesity, physical activity, and dietary habits. Diet plays a predominant role in the differences observed in the ethnic and global incidence of this disease (Rawla, 2019).

The treatment of the disease varies depending on its stage. When the disease is organ-confined, surgical treatment or radiotherapy has adequate results, with side effects such as urinary incontinence or erectile dysfunction, which can reduce the patient's QoL (EAU Guidelines on Prostate Cancer, 2024). In advanced or metastatic disease cases, current treatments are based on androgen deprivation therapy associated with new molecules that inhibit androgen signaling: enzalutamide, darolutamide, and apalutamide or molecules that alter androgen formation, such as abiraterone. These drugs usually cause symptoms such as hot flashes, weakness, or bone loss, among others. However, in certain cases where the disease burden is high, chemotherapy therapy such as Docetaxel is also accepted, with typical side effects of chemotherapy such as weakness, nausea, hair loss, etc. (EAU Guidelines on Prostate Cancer, 2024).

In this scenario, polyphenols emerge as an element of interest in PCa research (Table 4.2), the subject of numerous studies that seek to understand their impact on preventing and treating this disease.

Epigallocatechin-3-gallate

EGCG has an effect on AR in androgen-dependent cell lines by inhibiting their expression and transcription. In vitro studies demonstrated the ability of EGCG to arrest the cell cycle of androgen-dependent and independent cells, favoring their apoptosis. In addition, it produces an increase in the tumor suppressor protein p53, favoring the apoptosis of neoplastic cells (Chuu et al., 2009; Lee et al., 2012). Likewise, EGCG can suppress

Table 4.2 Summary of beneficial effects of phenolic compounds in urinary track cancers.

Cancer	Bioactive compound	Experimental model	Concentration	Results	Refs
Prostate	EGCG	Androgen-dependent prostate cancer cells	50 μM	Suppression of androgen receptor signaling	Lee et al. (2012)
		LNCaP and DU145 cells	10–80 μg/mL	Apoptosis cancer cells	Gupta et al. (2000)
		Human	600 mg of tea /d	Chemopreventive effects	Jian et al. (2004)
	Curcumin	PC3 cell line	5 μg/mL	Apoptosis cancer cells	Rivera et al. (2017)
		PC3 and LNCaP cells	0–20 μM	Induction of apoptosis cancer cells	Srivastava et al. (2007)
	Resveratrol	DU145	0.99 nM	Inhibit cancer cell growth	Kampa et al. (2000)
Bladder	EGCG	NBT-II bladder tumor cell line	10–40 μM	Inhibition of cancer cells growth	Chen et al. (2004)
		SW780 cells	25–100 μM	Reduced cell migration and growth	Luo et al. (2017)
	Curcumin	J82, TCCSUP and T24 cells	10 μM	Reduce cell migration and increase cancer cell apoptosis	Wang et al. (2018)
	Resveratrol	T24 cells	0–50 μM	Increase cancer cell apoptosis reduced cell migration	Bai et al. (2017)
	Ellagic acid	T24, UM-UC-3, 5637 and HT-1376 cells	12 mM	Reduced cancer cell growth	Ceci et al. (2016)
	Honokiol	UBC cell	0–4.8 μg/mL	Reduced cellular invasion	Shen et al. (2017)
Kidney	Curcumin	ACHN cancer cells	5–80 μM	Increase radiosensitivity	Li et al. (2017)
	Resveratrol	Rats	60 mg/kg/d	Improves sensitivity of renal cancer cells to chemotherapeutics	Tian et al. (2020)

ECGC, Epigalocatechin galate; *AR*, androgenic receptor; *LNCaP*, lymph node carcinoma of the prostate cells; DU145, human prostate cancer cell; *UBC*, urinary bladder cancer.

the growth of PCa tumor cells in a dose-dependent manner through different mechanisms of action by inducing the expression of Bax, procaspase 6 and 9, and PARP, as well as inhibiting NF-kB, causing apoptosis of tumor cells. However, in vivo studies have shown that these effects are more effective in the early stages of the disease (Gupta et al., 2000; Siddiqui et al., 2011).

Furthermore, EGCG acts on the ability of PCa cells to invade and metastasize. In this sense, it has been observed how EGCG presented effects on the invasiveness of different human PCa cell lines through the reactivation of TIMP-3. This protein plays an important role in suppressing invasion and cancer progression (Deb et al., 2019).

In in vivo studies in humans, they determined that consumption of green tea, rich in EGCG, decreased the risk of developing PCa by up to 72% (95% CI: 0.17, 0.47) (Jian et al., 2004). Moreover, in other clinical trials, this chemopreventive effect was evaluated on patients diagnosed with high-grade intraepithelial prostate neoplasia, demonstrating that the consumption of green tea catechins led to a reduction in the diagnosis of PCa of almost 80% ($p < 0.01$), maintaining this effect after two years of follow-up (Brausi et al., 2008). These results revealed the preventive effect of EGCG on the development of PCa.

Curcumin

Curcumin is capable of acting on human PCa cells through different pathways, presenting activity at different times of the disease. On the one hand, in vitro, studies demonstrated how curcumin-induced apoptosis through the production of ROS, which increased chronic stress of the endoplasmic reticulum in PC3 cells, leading to apoptosis of these cells (Rivera et al., 2017; Rodriguez-Garcia et al., 2017). On the other hand, curcumin inhibited the AR by modulating the Wnt/β-catenin signaling pathway (Hong et al., 2015), negatively regulating AR and AR-related cofactors activator protein-1 (AP-1), NF-κB, and cyclic AMP response element binding protein (CREB) (CBP) (Nakamura et al., 2002). Furthermore, curcumin presented the ability to modify methylation patterns in PCa cells, which may influence the expression of key proteins related to cellular function and cancer development (Epigenetic CpG Demethylation of the Promoter & Reactivation of the Expression of Neurog1 by Curcumin in Prostate LNCaP Cells, 2023). Finally, it has been demonstrated through both in vitro and in vivo studies how curcumin affects the regulation of the cell cycle, influencing the viability and proliferation of cancer cells (Srivastava et al., 2007; Teiten et al., 2011).

Other studies have demonstrated the effects of curcumin on cancer invasiveness, reducing the secretion of metallopeptidase protein (MMP) 2 and 9 (Hong et al., 2006). Besides, it reduced the adhesion and motility of cells overexpressing protein kinase C (PKC) and inhibited CC motif chemokine ligand 2 (CCL2) by reducing the motility and invasion of cancer cells (Herman et al., 2009).

For all these reasons, curcumin also showed an important role in the prophylaxis of PCa since, in in vitro and in vivo studies, this compound has managed to reduce tumor formation (Narayanan et al., 2009), and in cases where the tumor was established, manages to reduce cell proliferation and increase the percentage of apoptotic cells (Shankar & Srivastava, 2007; Yu et al., 2014).

Resveratrol

RSV has activity against PCa cells by reducing AR, regulating signaling pathways, cell cycle arrest, increasing apoptosis, and reducing their metastatic potential (Rauf et al., 2018).

In in vitro studies, it has been observed that RSV has a synergistic effect on reducing the quantity and activity of AR, which are essential in the proliferation of PCa and the development of androgen resistance (Kai & Levenson, 2011).

In an in vitro study using androgen-dependent LNCaP cells and C4−2 cells, RSV was shown to increase PTEN expression by inhibiting AR. Furthermore, independently of AR, it inhibits the phosphorylation of EGFR and HER2, which suppresses AKT signaling (Wang et al., 2010). RSV has also been found to inhibit AKT/miR-21 signaling pathways, resulting in the inhibition of ERK-1/2, reduction of estrogen levels, and IGF-1 (Sheth et al., 2012). Related to the cell cycle, RSV was shown to inhibit the growth of PCa DU145 cells by modulating NO production (Kampa et al., 2000).

RSV also showed the ability to inhibit the MTA1/NuRD complex and restore p53 activity, which would cause apoptosis in PCa cells, in addition to the ability to inhibit cell survival by decreasing the levels of survivin (a protein that facilitates the evasion of apoptosis and aberrant mitotic progression of tumor cells) and sensitize cancer cells to cell death induced by TNF-related apoptosis-inducing ligand (TRAIL) (Fulda & Debatin, 2004).

Bladder cancer

Bladder cancer is the most common tumor of the urinary tract. Histologically, urothelial carcinoma, derived from the urothelium, is the most common type. Among the main risk factors are smoking, responsible for approximately 50% of tumors, followed by exposure to aromatic amines, polycyclic aromatic hydrocarbons, and chlorinated hydrocarbons.

At the time of diagnosis, approximately 70% of tumors are found in the most superficial layers of the bladder, being defined as non-muscle invasive bladder cancer. Its prognosis is favorable, with an overall 5-year survival of 80%. However, these tumors are characterized by their capacity for recurrence, which entails high costs in their follow-up and treatment. In cases in which the disease invades the bladder, survival drops to 50% at five years, and the treatment in these cases involves surgery with greater comorbidity, such as cystectomy, in some cases combining chemotherapy and radiotherapy (Non-muscle-Invasive Bladder Cancer – Introduction – Uroweb, 2023).

In recent decades, the preventive effect and use as an alternative treatment of phenolic compounds from plant matrices have been investigated to reduce the impact of this disease on the population (Table 4.2).

Epigallocatechin-3-gallate

The different effects against bladder cancer of EGCG in green tea have been evaluated through in vitro studies, demonstrating effects at various levels, from inhibition of cell growth to anti-migratory and proapoptotic capacity (J.J. Chen et al., 2004; Luo et al., 2017). Thus EGCG has the ability to inhibit growth and arrest the cell cycle in the NBT-II bladder tumor cell line in a time and dose-dependent manner (J.J. Chen et al., 2004). On the other hand, green tea extract rich in EGCG also has a proapoptotic function through the inactivation of the PI3K/AKT pathway and a negative regulation of the NF-kB pathway (Luo et al., 2017). Finally, EGCG has anti-migratory capacity, in isolation, through the negative regulation of MMP-9 (Luo et al., 2017) or in conjunction with a mixture of lysine, proline, arginine, ascorbic acid, decreasing the production of MMP-2 and MMP-9 in the T24 bladder cancer cell line (Roomi et al., 2006).

Curcumin

Curcumin has been shown to have beneficial properties in suppressing the development of bladder cancer since it has been shown in vitro that curcumin has antiproliferative effects in a dose-dependent manner. This activity is attributed to its ability to inhibit the PL3K/AKT/mTOR AKT pathway and promote c-myc expression (Binqiang Tian et al., 2008). Additionally, curcumin inhibits β-catenin, which is overexpressed in bladder cancer cells and causes the induction of EMT in tumor cells (Shi, Wang, et al., 2017). Curcumin also has effects on tumor cell migration and invasiveness: it inhibits vascular endothelial growth factor (VEGF) and reduces MMP levels (Ashrafizadeh et al., 2020), in addition to suppressing MMP-2 and 9 (Shi, Zhang, et al., 2017). Curcumin also increases HIF-1α and reduces the expression of miRNA-764, an oncogenic factor that promotes cell invasion and increases apoptosis (Wang, Tan, et al., 2018).

Resveratrol

The antiproliferative role of RSV has been described by inhibiting the phosphorylation of AKT, leading to the inhibition of miR-21 (Y. Bai et al., 2010). Likewise, it manages to arrest the cell cycle through the activation of p21 and p38 (Yang et al., 2017) and is capable of causing an increase in ROS that causes increased cell permeability, DNA fragmentation, and cell apoptosis (Stocco et al., 2012). In vivo studies confirm the antiproliferative role of RSV, identifying as underlying mechanisms the inhibition of STAT3 signaling (Mo-Li Wu et al., 2014) and the decrease of VEGF and FGF-2 factors (Y. Bai et al., 2010). In other studies, RSV is also indicated as a protective factor against metastasis by decreasing the phosphorylation of JNK1/2 and ERK-1/2, resulting in the inhibition of MMP-2 and 9 (Yu Bai et al., 2017).

Ellagic acid

Ellagic acid has shown a preventive effect on bladder cancer by inhibiting the enzyme N-acetyl transferase (NAT), which prevents the acetylation of aromatic amines and the formation of carcinogenic DNA adducts (Lin et al., 2001). It also has an antiproliferative effect on tumor cells by stopping the cell cycle in the G0/G1 phase, promoting apoptosis by increasing the levels of p53 and p21, reducing the levels of CDK2, and increasing the activity of the caspase 3 (Li et al., 2005). Ellagic acid metabolites also induce apoptosis by arresting the cell cycle in the G2/M phase and inhibiting the PL3K/AKT and MAPK pathway (Liberal et al., 2017).

In vivo studies have shown that ellagic acid acts as an antiproliferative agent by reducing angiogenesis and tumor invasion. This is achieved by decreasing the expression of the VEGFR-2 receptor and programmed cell death ligand 1 (PD-L1) (Ceci et al., 2016). Similarly, pomegranate peel extract rich in ellagic acid inhibited cell proliferation by increasing the p53/miR-34 pathway (Chang et al., 2018; Zhou et al., 2015).

Honokiol

Honokiol is a compound isolated from *Magnolia officinalis*, coming from the hydrolysis of the lignins of this species. This compound has shown the ability to slow the migration and invasion of bladder cancer cells. Its effectiveness is related to the dose used, and it manages to inhibit EMT by inducing the production of E-cadherin and suppressing that of N-cadherin. Furthermore, it significantly decreases the expression of genes associated with cell invasion, such as steroid receptor coactivator-3 (SRC-3), MMP-2, and Twist1 (Shen et al., 2017).

Renal cancer

Renal cell carcinoma constitutes approximately 3% of all cancers, with a growing global incidence, estimated at 400,000 new cases per year, and a global mortality rate approaching 175,000 deaths annually. The incidence is estimated to continue to increase in the coming years (Cirillo et al., 2024). It is most prevalent in Western countries, with North America being the most common, followed by Western Europe and Australia/New Zealand. Furthermore, a predominance is observed in men. Its established risk factors include advanced age, obesity, smoking, and hypertension (Capitanio et al., 2019).

Systemic treatment of renal cancer has undergone changes in recent years, and single-agent VEGF-targeted therapy such as sunitinib has been replaced by a combination therapy based on immune checkpoint drugs (immunotherapy) together with an inhibitor of tyrosine kinase (TKIs). Some examples are nivolumab/cabozantinib, pembrolizumab/axitinib, or pembrolizumab/levantinib. The adverse effects of immunotherapy are immune-related, so they can affect any organ, although they are usually reversible and controllable with corticosteroids. TKIs can cause asthenia, arterial hypertension, mucositis, hand-foot syndrome, diarrhea, hematological toxicity, and endocrine alterations (EAU Guidelines on RCC, 2023).

Due to these adverse effects, various studies have been carried out in recent years that reveal the potential use of phenolic compounds as interesting alternatives for preventing and treating this disease.

Curcumin

Curcumin presents antiproliferative effects against renal cancer cells through negative regulation of the PI3K-AKT-mTOR pathway (Zhan et al., 2021). It induces apoptosis and cell cycle arrest by suppressing NF-kB signaling and VEGF and COX pathways, which improves cell radiosensitivity (Gang Li et al., 2017). Moreover, it reduces the levels of MMP-2 and 9, which facilitates apoptosis and reduces the invasion of renal tumor cells (Gao et al., 2022). Finally, curcumin administration slows tumor progression by inhibiting the ERK5/AP-1 axis and reversing EMT through upregulation of E-cadherin and through downregulation of vimentin, N-cadherin, and TWIST (Zhang et al., 2020).

Resveratrol

RSV has shown effects that promote apoptosis and suppress the proliferation, migration, and invasion of renal cancer cells by inhibiting the NLRP3 pathway involved in inflammation and the immune response in both in vitro studies and in live models (Xixi Tian et al., 2020). It has also been described as improving the sensitivity of renal cancer cells to chemotherapy drugs such as paclitaxel through the inhibition of the Pl3K/AKT pathway (Ying-Jie et al., 2019). Finally, it has been described how the administration of RSV through nanoparticles has a dose-dependent effect on kidney cells, causing a decrease in cell viability, a reduction in the migration and invasion capacity through the reduction of MMP-2 and elevation of TIMP-2 and a decrease in ERK-1/2 expression (Zhu et al., 2022).

Conclusions

Throughout this chapter, the published evidence on the use of phenolic compounds in the most prevalent urinary pathologies has been compiled.

In the etiopathogenesis of BPH, the role of chronic inflammation has gained great importance in recent years, and, among other processes, phenolic compounds such as EGCG, RSV, or isoflavones seem to act on this in in vitro studies. Although there are promising studies that demonstrate an improvement in urinary symptoms in patients with BPH treated with isoflavones, studies with a larger sample size are required to recommend their use.

Antibiotic resistance is a global public health problem, so research into alternative treatments is essential. In this sense, PAC derived from cranberry have been the most studied, both in vitro and in vivo, demonstrating a certain decrease in the incidence of recurrent UTIs. Although the level of evidence on the effectiveness of PAC in UTIs is low, they are currently recommended in patients with recurrent UTIs in urological clinical practice guidelines.

Inflammatory factors intervene in the formation of UL, which generates an imbalance between inhibitors and urinary crystal formers. Phenolic compounds such as flavonoids, coumarins, and stilbenes have been investigated, and it was observed that these molecules have anti-inflammatory and antioxidant activity that prevents the formation of UL.

Regarding oncological diseases, such as PCa, bladder cancer, and renal cancer, the main phenolic compounds studied have been EGCG, curcumin, and RSV. In PCa, ECGC, curcumin, and RSV have shown regulation of the AR and induction of apoptosis in tumor cells. In human studies, consumption of green tea, rich in EGCG, was shown to reduce the risk of developing PCa. In bladder and renal cancer, these phenolic compounds have been shown, in vitro, to inhibit cell growth and migratory capacity and stimulate cell apoptosis, but no in vivo studies have been published that support the use of these compounds in oncological patients.

In view of the results collected, using phenolic compounds as prophylaxis against different urinary conditions seems to be an interesting alternative since, thanks to their antioxidant, anti-inflammatory, or diuretic potential, among others, it seems to improve the prognosis and treatment of some of these urinary disorders. However, there is still a long way to go before research can verify the effectiveness of its use. Despite the promising results, most of these studies have been in vitro, with few in vivo models and even fewer studies in humans.

Although there is scientific evidence about the potential beneficial effect of phenolic compounds in urinary disorders, there is still a lack of in vitro and, to a greater extent, in vivo information that reveals the mechanism of action on different urological diseases and that demonstrates the need for its use.

References

Ahmed, S., Hasan, M. M., Khan, H., Mahmood, Z. A., & Patel, S. (2018). The mechanistic insight of polyphenols in calcium oxalate urolithiasis mitigation. *Biomedicine and Pharmacotherapy*, *106*, 1292–1299. Available from https://doi.org/10.1016/j.biopha.2018.07.080, https://www.journals.elsevier.com/biomedicine-and-pharmacotherapy.

Alelign, T., & Petros, B. (2018). Kidney stone disease: An update on current concepts. *Advances in Urology*, *2018*, 1–12. Available from https://doi.org/10.1155/2018/3068365.

Ashrafizadeh, M., Yaribeygi, H., & Sahebkar, A. (2020). Therapeutic effects of curcumin against bladder cancer: A review of possible molecular pathways. *Anti-cancer Agents in Medicinal Chemistry*, *20*(6), 667–677. Available from https://doi.org/10.2174/1871520620666200203143803.

Azimi, A., Eidi, A., Mortazavi, P., & Rohani, A. H. (2021). Protective effect of apigenin on ethylene glycol-induced urolithiasis via attenuating oxidative stress and inflammatory parameters in adult male Wistar rats. *Life Sciences*, *279*119641. Available from https://doi.org/10.1016/j.lfs.2021.119641.

Badrinathan, S., Shiju, M. T., Arya, R., Rajesh, G. N., & Viswanathan, P. (2015). Citrus bioflavonoids ameliorate hyperoxaluria induced renal injury and calcium oxalate crystal deposition in wistar rats. *Advanced Pharmaceutical Bulletin*, *5*(3), 419–427. Available from https://doi.org/10.15171/apb.2015.057, http://apb.tbzmed.ac.ir/Portals/0/Archive/Vol5-2015/57-Viswanathan.pdf.

Bai, Y., Mao, Q. Q., Qin, J., Zheng, X. Y., Wang, Y. B., Yang, K., Shen, H. F., & Xie, L. P. (2010). Resveratrol induces apoptosis and cell cycle arrest of human T24 bladder cancer cells in vitro and inhibits tumor growth in vivo. *Cancer Science*, *101*(2), 488–493. Available from https://doi.org/10.1111/j.1349-7006.2009.01415.x.

Bai, Y., Yang, H., Zhang, G., Hu, L., Lei, Y., Qin, Y., Yang, Y., Wang, Q., Li, R., & Mao, Q. (2017). Inhibitory effects of resveratrol on the adhesion, migration and invasion of human bladder cancer cells. *Molecular Medicine Reports*, *15*(2), 885–889. Available from https://doi.org/10.3892/mmr.2016.6051.

Bergengren, O., Pekala, K. R., Matsoukas, K., Fainberg, J., Mungovan, S. F., Bratt, O., Bray, F., Brawley, O., Luckenbaugh, A. N., Mucci, L., Morgan, T. M., & Carlsson, S. V. (2023). 2022 update on prostate cancer epidemiology and risk factors—A systematic review. *European Urology*, *84*(2), 191–206. Available from https://doi.org/10.1016/j.eururo.2023.04.021, http://www.europeanurology.com/.

Brausi, M., Rizzi, F., & Bettuzzi, S. (2008). Chemoprevention of human prostate cancer by green tea catechins: two years later. A follow-up update. *European urology*, *54*(2), 472–473.

Calmasini, F. B., de Oliveira, M. G., Alexandre, E. C., Silva, F. H., Tavares, E. B. G., André, D. M., Zapparoli, A., & Antunes, E. (2018). Obesity-induced mouse benign prostatic hyperplasia (BPH) is improved by treatment with resveratrol: Implication of oxidative stress, insulin sensitivity and neuronal growth factor. *Journal of Nutritional Biochemistry*, *55*, 53–58. Available from https://doi.org/10.1016/j.jnutbio.2017.12.009, http://www.elsevier.com/locate/jnutbio.

Capitanio, U., Bensalah, K., Bex, A., Boorjian, S. A., Bray, F., Coleman, J., Gore, J. L., Sun, M., Wood, C., & Russo, P. (2019). Epidemiology of renal cell carcinoma. *European Urology*, *75*(1), 74–84. Available from https://doi.org/10.1016/j.eururo.2018.08.036.

Capitanio, U., & Montorsi, F. (2016). Renal cancer. *The Lancet*, *387*(10021), 894–906. Available from https://doi.org/10.1016/s0140-6736(15)00046-x.

Ceci, C., Tentori, L., Atzori, M., Lacal, P., Bonanno, E., Scimeca, M., Cicconi, R., Mattei, M., de Martino, M., Vespasiani, G., Miano, R., & Graziani, G. (2016). Ellagic acid inhibits bladder cancer invasiveness and in vivo tumor growth. *Nutrients*, *8*(11), 744. Available from https://doi.org/10.3390/nu8110744.

Chaiyarit, S., Phuangkham, S., & Thongboonkerd, V. (2024). Quercetin inhibits calcium oxalate crystallization and growth but promotes crystal aggregation and invasion. *Current Research in Food Science*, *8*100650. Available from https://doi.org/10.1016/j.crfs.2023.100650.

Chang, C. P., Chan, Y. Y., Li, C. F., Chien, L. H., Lee, S. T., & Wu, T. F. (2018). Deciphering the molecular mechanism underlying the inhibitory efficacy of taiwanese local pomegranate peels against urinary bladder urothelial carcinoma. *Nutrients*, *10*(5). Available from https://doi.org/10.3390/nu10050543, http://www.mdpi.com/2072-6643/10/5/543/pdf.

Chao, Y., Gao, S., Li, N., Zhao, H., Qian, Y., Zha, H., Chen, W., & Dong, X. (2020). Lipidomics reveals the therapeutic effects of EtOAc extract of Orthosiphon stamineus Benth. on nephrolithiasis. *Frontiers in Pharmacology*, *11*. Available from https://doi.org/10.3389/fphar.2020.01299.

Chen, J., & Song, H. (2016). Protective potential of epigallocatechin-3-gallate against benign prostatic hyperplasia in metabolic syndrome rats. *Environmental Toxicology and Pharmacology*, *45*, 315–320. Available from https://doi.org/10.1016/j.etap.2016.06.015.

Chen, J. J., Ye, Z.-Q., & Koo, M. W. L. (2004). Growth inhibition and cell cycle arrest effects of epigallocatechin gallate in the NBT-II bladder tumour cell line. *BJU International*, *93*(7), 1082–1086. Available from https://doi.org/10.1111/j.1464-410x.2004.04785.x.

Chen, Z., Miao, L., Gao, X., Wang, G., & Xu, Y. (2015). Effect of obesity and hyperglycemia on benign prostatic hyperplasia in elderly patients with newly diagnosed type 2 diabetes. E-Century Publishing Corporation. *China International Journal of Clinical and Experimental Medicine*, *8*(7), 11289–11294. Available from http://www.ijcem.com/files/ijcem0007877.pdf.

Cho, Y. S., Schiller, N. L., Kahng, H. Y., & Oh, K. H. (2007). Cellular responses and proteomic analysis of *Escherichia coli* exposed to green tea polyphenols. *Current Microbiology*, *55*(6), 501–506. Available from https://doi.org/10.1007/s00284-007-9021-8.

Chung, K. S., Cheon, S. Y., & An, H. J. (2015). Effects of resveratrol on benign prostatic hyperplasia by the regulation of inflammatory and apoptotic proteins. American Chemical Society, South Korea. *Journal of Natural Products*, *78*(4), 689–694. Available from https://doi.org/10.1021/np500810c, http://pubs.acs.org/journal/jnprdf.

Chuu, C. P., Chen, R. Y., Kokontis, J. M., Hiipakka, R. A., & Liao, S. (2009). Suppression of androgen receptor signaling and prostate specific antigen expression by (-)-epigallocatechin-3-gallate in different progression stages of LNCaP prostate cancer cells. *Cancer Letters*, *275*(1), 86–92. Available from https://doi.org/10.1016/j.canlet.2008.10.001.

Cirillo, L., Innocenti, S., & Becherucci, F. (2024). Global epidemiology of kidney cancer. *Nephrology Dialysis Transplantation*. Available from https://doi.org/10.1093/ndt/gfae036.

Daniyal, M., Siddiqui, Z. A., Akram, M., Asif, H. M., Sultana, S., & Khan, A. (2014). Epidemiology, etiology, diagnosis and treatment of prostate cancer. Asian Pacific Organization for Cancer Prevention, Pakistan. *Asian Pacific Journal of Cancer Prevention*, *15*(22), 9575–9578. Available from https://doi.org/10.7314/APJCP.2014.15.22.9575, http://www.apocpcontrol.org/paper_file/issue_abs/Volume15_No22/9575-9578/208.17/20Muhammad/20Daniyal/20[MINI-REVIEW].pdf.

De Nunzio, C., Bartoletti, R., Tubaro, A., et al. (2021). Role of D-Mannose in the Prevention of Recurrent Uncomplicated Cystitis: State of the Art and Future Perspectives. *Antibiotics (Basel)*, *10*, 373.

Deb, G., Shankar, E., Thakur, V. S., Ponsky, L. E., Bodner, D. R., Fu, P., & Gupta, S. (2019). Green tea–induced epigenetic reactivation of tissue inhibitor of matrix metalloproteinase-3 suppresses prostate cancer progression through histone-modifying enzymes. John Wiley and Sons Inc., United States. *Molecular Carcinogenesis*, *58*(7), 1194–1207. Available from https://doi.org/10.1002/mc.23003, http://onlinelibrary.wiley.com/journal/10.1002/(ISSN)1098-2744.

Ding, T., Zhao, T., Li, Y., Liu, Z., Ding, J., Ji, B., Wang, Y., & Guo, Z. (2021). Vitexin exerts protective effects against calcium oxalate crystal-induced kidney pyroptosis in vivo and in vitro. *Phytomedicine: International Journal of Phytotherapy and Phytopharmacology*, *86*153562. Available from https://doi.org/10.1016/j.phymed.2021.153562.

¿Qué es la Urología? Diccionario Médico (2024) - Clínica U. Navarra. https://www.cun.es, https://www.cun.es/diccionario-medico/terminos/urologia (accessed 9 March 2024).

EAU Guidelines on Urological Infections – THE GUIDELINE – Uroweb. (2024). European Association of Urology.

EAU Guidelines on the Management of Non-neurogenic Male LUTS - INTRODUCTION - Uroweb. *Uroweb - European Association of Urology*, 2023. https://uroweb.org/guidelines/management-of-non-neurogenic-male-luts (accessed 21 June 2023).

EAU Guidelines on Prostate Cancer - TREATMENT - Uroweb. *Uroweb - European Association of Urology*, 2024. https://uroweb.org/guidelines/prostate-cancer/chapter/treatment (accessed 3 January 2024).

EAU Guidelines on RCC - DISEASE MANAGEMENT - Uroweb. *Uroweb - European Association of Urology*, 2023. https://uroweb.org/guidelines/renal-cell-carcinoma/chapter/disease-management (accessed 16 June 2023).

Epigenetic CpG Demethylation of the Promoter and Reactivation of the Expression of Neurog1 by Curcumin in Prostate LNCaP Cells, (2023). PubMed.

Flores-Mireles, A. L., Walker, J. N., Caparon, M., & Hultgren, S. J. (2015). Urinary tract infections: Epidemiology, mechanisms of infection and treatment options. *Nature Reviews. Microbiology*, *13*(5), 269–284. Available from https://doi.org/10.1038/nrmicro3432, http://www.nature.com/nrmicro/index.html.

Fulda, S., & Debatin, K. M. (2004). Sensitization for tumor necrosis factor-related apoptosis-inducing ligand-induced apoptosis by the chemopreventive agent resveratrol. *Cancer Research*, *64*(1), 337–346. Available from https://doi.org/10.1158/0008-5472.CAN-03-1656.

Gao, R., Ma, Y., Wang, Y., Xu, H., & Rajakani, K. (2022). Curcumin inhibits the proliferation of renal cancer 786-O cells through MTOR signaling pathway and its mechanism. *Journal of Healthcare Engineering*, *2022*, 1–7. Available from https://doi.org/10.1155/2022/1842389.

Ghalayini, I. F., Al-Ghazo, M. A., & Harfeil, M. N. A. (2011). Prophylaxis and therapeutic effects of Raspberry (Rubus idaeus) on renal stone formation in Balb/c mice. *International Braz J Urol*, *37*(2), 259–266. Available from https://doi.org/10.1590/S1677-55382011000200013, http://www.scielo.br/pdf/ibju/v37n2/a13v37n02.pdf.

González de Llano, D., Moreno-Arribas, M. V., & Bartolomé, B. (2020). Cranberry polyphenols and prevention against urinary tract infections: relevant considerations. *Molecules*, *25*(15), 3523. Available from https://doi.org/10.3390/molecules25153523.

Gupta, S., Ahmad, N., Nieminen, A. L., & Mukhtar, H. (2000). Growth inhibition, cell-cycle dysregulation, and induction of apoptosis by green tea constituent (-)-epigallocatechin-3-gallate in androgen-sensitive and androgen-insensitive human prostate carcinoma cells. *Toxicology and Applied Pharmacology*, *164*(1), 82–90. Available from https://doi.org/10.1006/taap.1999.8885.

Guzel, A., Yunusoglu, S., Calapoglu, M., et al. (2021). Protective Effects of Quercetin on Oxidative Stress-Induced Tubular Epithelial Damage in the Experimental Rat Hyperoxaluria Model. *Medicina (Kaunas)*, *57*, 566.

Hamzah, H., Hertiani, T., Pratiwi, S. U. T., & Nuryastuti, T. (2020). Efficacy of quercetin against polymicrobial biofilm on catheters. *Research Journal of Pharmacy and Technology*, *13* (11), 5277–5282. Available from https://doi.org/10.5958/0974-360X.2020.00923.3, http://www.rjptonline.org/.

Han, H., Segal, A. M., Seifter, J. L., & Dwyer, J. T. (2015). Nutritional management of kidney stones (Nephrolithiasis). *Clinical Nutrition Research*, *4*(3), 137. Available from https://doi.org/10.7762/cnr.2015.4.3.137.

Harwood, M., Danielewska-Nikiel, B., Borzelleca, J. F., Flamm, G. W., Williams, G. M., & Lines, T. C. (2007). A critical review of the data related to the safety of quercetin and lack of evidence of in vivo toxicity, including lack of genotoxic/carcinogenic properties. *Food and Chemical Toxicology*, *45*(11), 2179–2205. Available from https://doi.org/10.1016/j.fct.2007.05.015.

Herman, J. G., Stadelman, H. L., & Roselli, C. E. (2009). Curcumin blocks CCL2-induced adhesion, motility and invasion, in part, through down-regulation of CCL2 expression and proteolytic activity. *International Journal of Oncology*, *34*(5), 1319–1327. Available from https://doi.org/10.3892/ijo_00000259, http://www.spandidos-publications.com/serveFile/ijo_34_5_1319_PDF.pdf?type = article&article_id = ijo_34_5_1319&item = PDF.

Hong, J. H., Ahn, K. S., Bae, E., Jeon, S. S., & Choi, H. Y. (2006). The effects of curcumin on the invasiveness of prostate cancer in vitro and in vivo. *Prostate Cancer and Prostatic Diseases*, *9*(2), 147–152. Available from https://doi.org/10.1038/sj.pcan.4500856.

Hong, J. H., Lee, G., & Choi, H. Y. (2015). Effect of curcumin on the interaction between androgen receptor and Wnt/β-catenin in LNCaP xenografts. Korean Urological Association, South Korea. *Korean Journal of Urology*, *56*(9), 656–665. Available from https://doi.org/10.4111/kju.2015.56.9.656, http://kjurology.org/Synapse/Data/PDFData/0020KJU/kju-56-656.pdf.

Hong, S. H., Lee, H. J., Sohn, E. J., Ko, H. S., Shim, B. S., Ahn, K. S., & Kim, S. H. (2013). Anti-nephrolithic potential of resveratrol via inhibition of ROS, MCP-1, hyaluronan and osteopontin in vitro and in vivo. *Pharmacological Reports*, *65*(4), 970–979. Available from https://doi.org/10.1016/s1734-1140(13)71078-8, http://www.if-pan.krakow.pl/pjp/pdf/2013/4_970.pdf.

Hoseinynejad, K., Mard, S. A., Mansouri, Z., Lamoochi, Z., & Kazemzadeh, R. (2022). Efficacy of chlorogenic acid against ethylene glycol-induced renal stone model: The role of NFKB-RUNX2-AP1-OSTERIX signaling pathway. *Tissue and Cell*, *79*101960. Available from https://doi.org/10.1016/j.tice.2022.101960.

Howell, A. B., Botto, H., Combescure, C., Blanc-Potard, A. B., Gausa, L., Matsumoto, T., Tenke, P., Sotto, A., & Lavigne, J. P. (2010). Dosage effect on uropathogenic *Escherichia coli* anti-adhesion activity in urine following consumption of cranberry powder standardized for proanthocyanidin content: A multicentric randomized double blind study. *BMC Infectious Diseases*, *10*. Available from https://doi.org/10.1186/1471-2334-10-94, http://www.biomedcentral.com/1471-2334/10/94.

Huang, Y., Nikolic, D., Pendland, S., Doyle, B. J., Locklear, T. D., & Mahady, G. B. (2009). Effects of cranberry extracts and ursolic acid derivatives on P-fimbriated *Escherichia coli*, COX-2 activity, pro-inflammatory cytokine release and the NF-κβ transcriptional response in vitro. *Pharmaceutical Biology*, *47*(1), 18–25. Available from https://doi.org/10.1080/13880200802397996.

Jagannathan, V., & Viswanathan, P. (2018). Proanthocyanidins—Will they effectively restrain conspicuous bacterial strains devolving on urinary tract infection? *Journal of Basic Microbiology*, *58*(7), 567–578. Available from https://doi.org/10.1002/jobm.201800131.

Jaiswal, S. K., Sharma, N. K., Bharti, S. K., Krishnan, S., Kumar, A., Prakash, O., Kumar, P., Kumar, A., & Gupta, A. K. (2018). Phytochemicals as uropathognic *Escherichia coli* fimh antagonist: In vitro and in silico approach. *Current Molecular Medicine*, *18*(9), 640–653. Available from https://doi.org/10.2174/1566524019666190104104507, http://www.eurekaselect.com/168752/article.

Jang, H., Bae, W. J., Yuk, S. M., Han, D. S., Ha, U. S., Hwang, S. Y., Yoon, S. H., Kim, S. W., & Han, C. H. (2014). Seoritae extract reduces prostate weight and suppresses prostate cell proliferation in a rat model of benign prostate hyperplasia. *Evidence-based Complementary and Alternative Medicine*, *2014*. Available from https://doi.org/10.1155/2014/475876.

Jian, L., Xie, L. P., Lee, A. H., & Binns, C. W. (2004). Protective effect of green tea against prostate cancer: A case-control study in southeast China. *International Journal of Cancer*, *108*(1), 130–135. Available from https://doi.org/10.1002/ijc.11550.

Jucá, M. M., Cysne Filho, F. M. S., de Almeida, J. C., Mesquita, D. D. S., Barriga, J. R. D. M., Dias, K. C. F., Barbosa, T. M., Vasconcelos, L. C., Leal, L. K. A. M., Ribeiro, J. E., & Vasconcelos, S. M. M. (2020). Flavonoids: Biological activities and therapeutic potential. *Natural Product Research*, *34*(5), 692–705. Available from https://doi.org/10.1080/14786419.2018.1493588, http://www.tandf.co.uk/journals/titles/14786419.asp.

Kai, L., & Levenson, A. S. (2011). Combination of resveratrol and antiandrogen flutamide has synergistic effect on androgen receptor inhibition in prostate cancer cells. *Anticancer Research*, *31*(10), 3323–3330.

Kampa, M., Hatzoglou, A., Notas, G., Damianaki, A., Bakogeorgou, E., Gemetzi, C., Kouroumalis, E., Martin, P. M., & Castanas, E. (2000). Wine antioxidant polyphenols inhibit the proliferation of human prostate cancer cell lines. *Nutrition and Cancer*, *37* (2), 223–233. Available from https://doi.org/10.1207/S15327914NC372_16.

Katz, A., Efros, M., Kaminetsky, J., Herrlinger, K., Chirouzes, D., & Ceddia, M. (2014). A green and black tea extract benefits urological health in men with lower urinary tract symptoms. *Therapeutic Advances in Urology*, *6*(3), 89–96. Available from https://doi.org/10.1177/1756287214526924.

Khalil, N., Bishr, M., Desouky, S., & Salama, O. (2020). A Visnaga L., a potential medicinal plant: A review. *Molecules*, *25*(2), 301. Available from https://doi.org/10.3390/molecules25020301.

Kheirabadi, Z., Mehrabani, M., Sarafzadeh, F., Dabaghzadeh, F., & Ahmadinia, N. (2019). Green tea as an adjunctive therapy for treatment of acute uncomplicated cystitis in women: A randomized clinical trial. *Complementary Therapies in Clinical Practice*, *34*, 13–16. Available from https://doi.org/10.1016/j.ctcp.2018.10.018.

Kjaer, T. N., Ornstrup, M. J., Poulsen, M. M., Jørgensen, J. O. L., Hougaard, D. M., Cohen, A. S., Neghabat, S., Richelsen, B., & Pedersen, S. B. (2015). Resveratrol reduces the levels of circulating androgen precursors but has no effect on, testosterone, dihydrotestosterone, PSA levels or prostate volume. A 4-month randomised trial in middle-aged men. *The Prostate*, *75*(12), 1255–1263. Available from https://doi.org/10.1002/pros.23006.

Lan, T., Nguyen, A., & Bhattacharya, D. (2022). Antimicrobial activity of quercetin: An approach to its mechanistic principle. *Molecules*, *27*(8), 2494. Available from https://doi.org/10.3390/molecules27082494.

Lee, Y. H., Kwak, J., Choi, H. K., Choi, K. C., Kim, S., Lee, J., Jun, W., Park, H. J., & Yoon, H. G. (2012). EGCG suppresses prostate cancer cell growth modulating acetylation of androgen receptor by anti-histone acetyltransferase activity. *International Journal of Molecular Medicine*, *30*(1), 69–74. Available from https://doi.org/10.3892/ijmm.2012.966, http://www.spandidos-publications.com/serveFile/ijmm_30_1_69_PDF.pdf?type = article&article_id = ijmm_30_1_69&item = PDF.

Lee, Y. S., Han, C. H., Kang, S. H., Lee, S. J., Kim, S. W., Shin, O. R., Sim, Y. C., Lee, S. J., & Cho, Y. H. (2005). Synergistic effect between catechin and ciprofloxacin on chronic bacterial prostatitis rat model. *International Journal of Urology*, *12*(4), 383–389. Available from https://doi.org/10.1111/j.1442-2042.2005.01052.x.

Lephart, E. D. (2013). Severe & moderate BPH symptoms in mid-aged men improve with isoflavonoid-equol treatment: pilot intervention study. *Open Journal of Urology*, *03*(01), 21–27. Available from https://doi.org/10.4236/oju.2013.31004.

Lephart, E. D. (2014). Review: Anti-oxidant and anti-aging properties of equol in prostate health (BPH). *Open Journal of Endocrine and Metabolic Diseases*, *04*(01), 1–12. Available from https://doi.org/10.4236/ojemd.2014.41001.

Leyva-Jiménez, F. J., Lozano-Sánchez, J., Cádiz-Gurrea, M. de la L., Arráez-Román, D., & Segura-Carretero, A. (2019). Functional ingredients based on nutritional phenolics. A case study against inflammation: Lippia Genus. *Nutrients*, *11*(7), 1646. Available from https://doi.org/10.3390/nu11071646.

Li, C., Hu, W. L., Lu, M. X., & Xiao, G. F. (2019). Resveratrol induces apoptosis of benign prostatic hyperplasia epithelial cell line (BPH-1) through p38 MAPK-FOXO3a pathway. *BMC Complementary and Alternative Medicine*, *19*(1). Available from https://doi.org/10.1186/s12906-019-2648-8, http://www.biomedcentral.com/bmccomplementalternmed/.

Li, G., Wang, Z., Chong, T., Yang, J., Li, H., & Chen, H. (2017). Curcumin enhances the radiosensitivity of renal cancer cells by suppressing NF-κB signaling pathway. *Biomedicine & Pharmacotherapy*, *94*, 974–981. Available from https://doi.org/10.1016/j.biopha.2017.07.148.

Li, T. M., Chen, G. W., Su, C. C., Lin, J. G., Yeh, C. C., Cheng, K. C., & Chung, J. G. (2005). Ellagic acid induced p53/p21 expression, G1 arrest and apoptosis in human bladder cancer T24 cells. *Anticancer Research*, *25*(2 A), 971–979.

Liberal, J., Carmo, A., Gomes, C., Cruz, M. T., & Batista, M. T. (2017). Urolithins impair cell proliferation, arrest the cell cycle and induce apoptosis in UMUC3 bladder cancer cells. *Investigational New Drugs*, *35*(6), 671–681. Available from https://doi.org/10.1007/s10637-017-0483-7.

Lin, S., Hung, C., Tyan, Y., Yang, C., Hsia, T., Yang, M., & Chung, J. (2001). Ellagica acid inhibits arylamine N-acetyltransferase activity and DNA adduct formation in human bladder tumor cell lines (T24 and TSGH 8301). *Urological Research*, *29*(6), 371–376. Available from https://doi.org/10.1007/s002400100213.

De Llano, D. G., Esteban-Fernández, A., Sánchez-Patán, F., Martín-Álvarez, P. J., Moreno-Arribas, M. V., & Bartolomé, B. (2015). Anti-adhesive activity of cranberry phenolic compounds and their microbial-derived metabolites against uropathogenic *Escherichia coli* in bladder epithelial cell cultures. MDPI AG, Spain. *International Journal of Molecular Sciences*, *16*(6), 12119–12130. Available from https://doi.org/10.3390/ijms160612119, http://www.mdpi.com/1422-0067/16/6/12076/pdf.

Luo, K. W., Chen, W., Lung, W. Y., Wei, X. Y., Cheng, B. H., Cai, Z. M., & Huang, W. R. (2017). EGCG inhibited bladder cancer SW780 cell proliferation and migration both in vitro and in vivo via down-regulation of NF-κB and MMP-9. *Journal of Nutritional Biochemistry*, *41*, 56–64. Available from https://doi.org/10.1016/j.jnutbio.2016.12.004, https://www.sciencedirect.com/science/journal/09552863.

Rivera, M., Ramos, Y., Rodríguez-Valentín, M., López-Acevedo, S., Cubano, L. A., Zou, J., Zhang, Q., Wang, G., Boukli, N. M., & Ahmad, A. (2017). Targeting multiple pro-apoptotic signaling pathways with curcumin in prostate cancer cells. *PLoS One*, *12*(6), e0179587. Available from https://doi.org/10.1371/journal.pone.0179587.

Maisto, M., Iannuzzo, F., Novellino, E., Schiano, E., Piccolo, V., & Tenore, G. C. (2023). Natural polyphenols for prevention and treatment of urinary tract infections.

International Journal of Molecular Sciences, *24*(4), 3277. Available from https://doi.org/10.3390/ijms24043277.

Martínez, J. M. L., Sierra del Rio, A., & Galvez, M. P. L. (2021). Tratamiento farmacológico de la litiasis renal. *Spain Archivos Espanoles de Urologia*, *74*(1), 63–70. Available from https://aeurologia.com/journals.php#.

Minciullo, P. L., Inferrera, A., Navarra, M., Calapai, G., Magno, C., & Gangemi, S. (2015). Oxidative stress in benign prostatic hyperplasia: A systematic review. *Urologia Internationalis*, *94*(3), 249–254. Available from https://doi.org/10.1159/000366210, http://www.karger.com/journals/uin/uin_jh.htm.

Morales-Martínez, A., Melgarejo-Segura, M. T., & Arrabal-Polo, M. A. (2021). Epidemiología de la utiasis urinaria en el mundo y en España. Iniestares, S.A.,. *Spain Archivos Espanoles de Urologia*, *74*(1), 4–14. Available from https://aeurologia.com/journals.php#.

Nakamura, K., Yasunaga, Y., Segawa, T., Ko, D., Moul, J. W., Srivastava, S., & Rhim, J. S. (2002). Curcumin down-regulates AR gene expression and activation in prostate cancer cell lines. *International Journal of Oncology*, *21*(4), 825–830. Available from 10.3892/ijo.21.4.825.

Narayanan, N. K., Nargi, D., Randolph, C., & Narayanan, B. A. (2009). Liposome encapsulation of curcumin and resveratrol in combination reduces prostate cancer incidence in PTEN knockout mice. *International Journal of Cancer*, *125*(1), 1–8. Available from https://doi.org/10.1002/ijc.24336, http://www3.interscience.wiley.com/cgi-bin/fulltext/121684296/PDFSTART.

Nicolosi, D., Tempera, G., Genovese, C., & Furneri, P. (2014). Anti-adhesion activity of a2-type proanthocyanidins (a cranberry major component) on uropathogenic *E. coli* and *P. mirabilis* Strains. *Antibiotics*, *3*(2), 143–154. Available from https://doi.org/10.3390/antibiotics3020143.

Non-muscle-Invasive Bladder Cancer – Introduction – Uroweb. (2023). European Association of Urology.

O'May, C., & Tufenkji, N. (2011). The swarming motility of *Pseudomonas aeruginosa* is blocked by cranberry proanthocyanidins and other tannin-containing materials. *Applied and Environmental Microbiology*, 77(9), 3061–3067. Available from https://doi.org/10.1128/aem.02677-10.

Oksay, T., Yunusoğlu, S., Calapoğlu, M., Candan, İ. A., Onaran, İ., Ergün, O., & Özorak, A. (2017). Protective impact of resveratrol in experimental rat model of hyperoxaluria. *International Urology and Nephrology*, *49*(5), 769–775. Available from https://doi.org/10.1007/s11255-017-1534-x.

Ortega-Lozano, A. J., Hernández-Cruz, E. Y., Gómez-Sierra, T., & Pedraza-Chaverri, J. (2023). Antimicrobial activity of spices popularly used in mexico against urinary tract infections. *Antibiotics*, *12*(2), 325. Available from https://doi.org/10.3390/antibiotics12020325.

Passat, D. N. (2012). Interactions of black and green tea water extracts with antibiotics activity in local urinary isolated *Escherichia coli*. *Journal of Al-Nahrain University Science*, *15*(3), 134–142. Available from https://doi.org/10.22401/jnus.15.3.19.

Ratkalkar, V. N., & Kleinman, J. G. (2011). Mechanisms of stone formation. *Clinical Reviews in Bone and Mineral Metabolism*, *9*(3-4), 187–197. Available from https://doi.org/10.1007/s12018-011-9104-8.

Rauf, A., Imran, M., Butt, M. S., Nadeem, M., Peters, D. G., & Mubarak, M. S. (2018). Resveratrol as an anti-cancer agent: A review. *Critical Reviews in Food Science and Nutrition*, *58*(9), 1428–1447. Available from https://doi.org/10.1080/10408398.2016.1263597, http://www.tandf.co.uk/journals/titles/10408398.asp.

Rawla, P. (2019). Epidemiology of prostate cancer. *World Journal of Oncology*, *10*(2), 63–89. Available from https://doi.org/10.14740/wjon1191.

Refaie, M. M. M., Rifaai, R. A., & Zenhom, N. M. (2018). Role of PPAR -α agonist fenofibrate in the treatment of induced benign prostatic hyperplasia with dysplastic changes in rats. *Fundamental & Clinical Pharmacology*, *32*(6), 617–626. Available from https://doi.org/10.1111/fcp.12384.

Rodriguez-Garcia, A., Hevia, D., Mayo, J. C., Gonzalez-Menendez, P., Coppo, L., Lu, J., Holmgren, A., & Sainz, R. M. (2017). Thioredoxin 1 modulates apoptosis induced by bioactive compounds in prostate cancer cells. *Redox Biology*, *12*, 634–647. Available from https://doi.org/10.1016/j.redox.2017.03.025, http://www.journals.elsevier.com/redox-biology.

Roomi, M. W., Ivanov, V., Kalinovsky, T., Niedzwiecki, A., & Rath, M. (2006). Antitumor effect of ascorbic acid, lysine, proline, arginine, and green tea extract on bladder cancer cell line T-24. *International Journal of Urology*, *13*(4), 415–419. Available from https://doi.org/10.1111/j.1442-2042.2006.01309.x.

Sheth, S., Jajoo, S., Kaur, T., Mukherjea, D., Sheehan, K., Rybak, L. P., Ramkumar, V., & Kyprianou, N. (2012). Resveratrol reduces prostate cancer growth and metastasis by inhibiting the Akt/MicroRNA-21 Pathway. *PLoS One*, 7(12), e51655. Available from https://doi.org/10.1371/journal.pone.0051655.

Sansores-España, D., Pech-Aguilar, A. G., & Cua-Pech, K. G. (2008). Plants used in mexican traditional medicine for the management of urolithiasis: a review of preclinical evidence, bioactive compounds, and molecular mechanisms. *Molecules (Basel, Switzerland)*, *27*, 2008.

Sausville, J., & Naslund, M. (2010). Benign prostatic hyperplasia and prostate cancer: An overview for primary care physicians. *International Journal of Clinical Practice*, *64*(13), 1740–1745. Available from https://doi.org/10.1111/j.1742-1241.2010.02534.x.

Scholes, D., Hooton, T. M., Roberts, P. L., Stapleton, A. E., Gupta, K., & Stamm, W. E. (2000). Risk factors for recurrent urinary tract infection in young women. *The Journal of Infectious Diseases*, *182*(4), 1177–1182. Available from https://doi.org/10.1086/315827.

Shankar, S., & Srivastava, R. K. (2007). Involvement of Bcl-2 family members, phosphatidylinositol 3′-kinase/AKT and mitochondrial p53 in curcumin (diferulolylmethane)-induced apoptosis in prostate cancer. *International Journal of Oncology*, *30*(4), 905–918. Available from https://doi.org/10.3892/ijo.30.4.905, http://www.spandidos-publications.com/serveFile/ijo_30_4_905_PDF.pdf?type = article&article_id = ijo_30_4_905 &item = PDF.

Shen, L., Zhang, F., Huang, R., Yan, J., & Shen, B. (2017). Honokiol inhibits bladder cancer cell invasion through repressing SRC-3 expression and epithelial-mesenchymal transition. *Oncology Letters*, *14*(4), 4294–4300. Available from https://doi.org/10.3892/ol.2017.6665.

Shen, N., Wang, T., Gan, Q., Liu, S., Wang, L., & Jin, B. (2022). Plant flavonoids: Classification, distribution, biosynthesis, and antioxidant activity. *Food Chemistry*, *383*132531. Available from https://doi.org/10.1016/j.foodchem.2022.132531.

Shi, J., Wang, Y., Jia, Z., Gao, Y., Zhao, C., & Yao, Y. (2017). Curcumin inhibits bladder cancer progression via regulation of β-catenin expression. *Tumor Biology*, *39*(7). Available from https://doi.org/10.1177/1010428317702548, 101042831770254.

Shi, J., Zhang, X., Shi, T., & Li, H. (2017). Antitumor effects of curcumin in human bladder cancer in vitro. *Oncology Letters*, *14*(1), 1157–1161. Available from https://doi.org/10.3892/ol.2017.6205.

Siddiqui, I. A., Asim, M., Hafeez, B. B., Adhami, V. M., Tarapore, R. S., & Mukhtar, H. (2011). Green tea polyphenol EGCG blunts androgen receptor function in prostate cancer. *FASEB Journal*, *25*(4), 1198–1207. Available from https://doi.org/10.1096/fj.10-167924, http://www.fasebj.org/content/25/4/1198.full.pdf + html.

Siener, R. (1917). Nutrition and kidney stone disease. *Nutrients*, *13*(1917).

Silverio, F. D., Gentile, V., Matteis, A. D., Mariotti, G., Giuseppe, V., Luigi, P. A., & Sciarra, A. (2003). Distribution of inflammation, pre-malignant lesions, incidental carcinoma in histologically confirmed benign prostatic hyperplasia: A retrospective analysis. *European Urology*, *43*(2), 164–175. Available from https://doi.org/10.1016/s0302-2838(02)00548-1.

Slater, M., Brown, D., & Husband, A. (2002). In the prostatic epithelium, dietary isoflavones from red clover significantly increase estrogen receptor β and E-cadherin expression but decrease transforming growth factor β1. *Prostate Cancer and Prostatic Diseases*, *5*(1), 16–21. Available from https://doi.org/10.1038/sj.pcan.4500546.

Sorokin, I., Mamoulakis, C., Miyazawa, K., Rodgers, A., Talati, J., & Lotan, Y. (2017). Epidemiology of stone disease across the world. *World Journal of Urology*, *35*(9), 1301–1320. Available from https://doi.org/10.1007/s00345-017-2008-6.

Sreenivasulu, K., Nandeesha, H., Dorairajan, L. N., Rajappa, M., Vinayagam, V., & Cherupanakkal, C. (2018). Gene expression of insulin receptor, insulin-like growth factor increases and insulin-like growth factor-binding protein-3 reduces with increase in prostate size in benign prostatic hyperplasia. *The Aging Male: the Official Journal of the International Society for the Study of the Aging Male*, *21*(2), 138–144. Available from https://doi.org/10.1080/13685538.2017.1401994.

Srivastava, R. K., Chen, Q., Siddiqui, I., Sarva, K., & Shankar, S. (2007). Linkage of curcumin-induced cell cycle arrest and apoptosis by cyclin-dependent kinase inhibitor p21/WAF1/CIP1. *Cell Cycle (Georgetown, Tex.)*, *6*(23), 2953–2961. Available from https://doi.org/10.4161/cc.6.23.4951, http://www.landesbioscience.com/journals/cc/article/SrivastavaCC6-23.pdf.

Stewart, K. L., & Lephart, E. D. (2023). Overview of BPH: Symptom relief with dietary polyphenols, vitamins and phytochemicals by nutraceutical supplements with implications to the prostate microbiome. *International Journal of Molecular Sciences*, *24*(6). Available from https://doi.org/10.3390/ijms24065486, http://www.mdpi.com/journal/ijms.

Stocco, B., Toledo, K., Salvador, M., Paulo, M., Koyama, N., & Toloi, M. R. T. (2012). Dose-dependent effect of Resveratrol on bladder cancer cells: Chemoprevention and oxidative stress. *Maturitas*, *72*(1), 72–78. Available from https://doi.org/10.1016/j.maturitas.2012.02.004.

Sujatha, D., Singh, K., Vohra, M., Vijay Kumar, K., & Sunitha, S. (2015). Antilithiatic activity of phlorotannin rich extract of *Sarghassum wightii* on calcium oxalate urolithiais – In vitro and in vivo evaluation. *International Braz J Urol*, *41*(3), 511–520. Available from https://doi.org/10.1590/S1677-5538.IBJU.2014.0357, http://www.scielo.br/pdf/ibju/v41n3/1677-5538-ibju-41-3-0511.pdf.

Teiten, M. H., Gaascht, F., Cronauer, M., Henry, E., Dicato, M., & Diederich, M. (2011). Anti-proliferative potential of curcumin in androgen-dependent prostate cancer cells occurs through modulation of the Wingless signaling pathway. *International Journal of Oncology*, *38*(3), 603–611. Available from https://doi.org/10.3892/ijo.2011.905Luxembourg, http://www.spandidos-publications.com/serveFile/ijo_38_3_603_PDF.pdf?type = article&article_id = ijo_38_3_603&item = PDF.

Tepedelen, B. E., Soya, E., & Korkmaz, M. (2017). Epigallocatechin-3-gallate reduces the proliferation of benign prostatic hyperplasia cells via regulation of focal adhesions. *Life Sciences*, *191*, 74–81. Available from https://doi.org/10.1016/j.lfs.2017.10.016, http://www.elsevier.com/locate/lifescie.

Thompson, I. M., Goodman, P. J., Tangen, C. M., Lucia, M. S., Miller, G. J., Ford, L. G., Lieber, M. M., Cespedes, R. D., Atkins, J. N., Lippman, S. M., Carlin, S. M., Ryan, A., Szczepanek, C. M., Crowley, J. J., & Coltman, C. A. (2003). The influence of finasteride on the development of prostate cancer. *New England Journal of Medicine*, *349*(3), 215–224. Available from https://doi.org/10.1056/NEJMoa030660.

Tian, B., Wang, Z., Zhao, Y., Wang, D., Li, Y., Ma, L., Li, X., Li, J., Xiao, N., Tian, J., & Rodriguez, R. (2008). Effects of curcumin on bladder cancer cells and development of urothelial tumors in a rat bladder carcinogenesis model. *Cancer Letters*, *264*(2), 299–308. Available from https://doi.org/10.1016/j.canlet.2008.01.041.

Tian, X., Zhang, S., Zhang, Q., Kang, L., Ma, C., Feng, L., Li, S., Li, J., Yang, L., Liu, J., Qi, Z., & Shen, Y. (2020). Resveratrol inhibits tumor progression by down-regulation of NLRP3 in renal cell carcinoma. *The Journal of Nutritional Biochemistry*, *85*108489. Available from https://doi.org/10.1016/j.jnutbio.2020.108489.

Torella, M., Del Deo, F., Grimaldi, A., Iervolino, S. A., Pezzella, M., Tammaro, C., Gallo, P., Rappa, C., De Franciscis, P., & Colacurci, N. (2016). Efficacy of an orally administered combination of hyaluronic acid, chondroitin sulfate, curcumin and quercetin for the prevention of recurrent urinary tract infections in postmenopausal women. *European Journal of Obstetrics & Gynecology and Reproductive Biology*, *207*, 125–128. Available from https://doi.org/10.1016/j.ejogrb.2016.10.018.

Vestergaard, M., & Ingmer, H. (2019). Antibacterial and antifungal properties of resveratrol. *International Journal of Antimicrobial Agents*, *53*(6), 716–723. Available from https://doi.org/10.1016/j.ijantimicag.2019.02.015.

Vipin, C., Saptami, K., Fida, F., Mujeeburahiman, M., Rao, S. S., Athmika, A. B., Arun, P. D., Rekha, Y. W., & Lin. (2020). Potential synergistic activity of quercetin with antibiotics against multidrug-resistant clinical strains of *Pseudomonas aeruginosa*. *PLoS One*, *15* (11), e0241304. Available from https://doi.org/10.1371/journal.pone.0241304.

Wang, G., Zhao, D., Spring, D. J., & Depinho, R. A. (2018). Genetics and biology of prostate cancer. *Genes and Development*, *32*(17-18), 1105–1140. Available from https://doi.org/10.1101/gad.315739.118, http://genesdev.cshlp.org/content/32/17-18/1105.full.pdf.

Wang, K., Tan, S. L., Lu, Q., Xu, R., Cao, J., Wu, S. Q., Wang, Y. H., Zhao, X. K., & Zhong, Z. H. (2018). Curcumin suppresses microRNA-7641-mediated regulation of p16 expression in bladder cancer. *American Journal of Chinese Medicine*, *46*(6), 1357–1368. Available from https://doi.org/10.1142/S0192415X18500714, http://www.worldscinet.com/ajcm/ajcm.shtml.

Wang, Y., Romigh, T., He, X., Orloff, M. S., Silverman, R. H., Heston, W. D., & Eng, C. (2010). Resveratrol regulates the PTEN/AKT pathway through androgen receptor-dependent and -independent mechanisms in prostate cancer cell lines. *Human Molecular Genetics*, *19*(22), 4319–4329. Available from https://doi.org/10.1093/hmg/ddq354.

Weber, K. S., Setchell, K. D., Stocco, D. M., & Lephart, E. D. (2001). Dietary soy-phytoestrogens decrease testosterone levels and prostate weight without altering LH, prostate 5alpha-reductase or testicular steroidogenic acute regulatory peptide levels in adult male Sprague-Dawley rats. *Journal of Endocrinology*, *170*(3), 591–599. Available from https://doi.org/10.1677/joe.0.1700591.

Wong, W. C. W., Wong, E. L. Y., Li, H., You, J. H., Ho, S., Woo, J., & Hui, E. (2012). Isoflavones in treating watchful waiting benign prostate hyperplasia: A double-blinded, randomized controlled trial. *Journal of Alternative and Complementary Medicine*, *18*(1), 54–60. Available from https://doi.org/10.1089/acm.2010.0077.

Wu, M. L., Li, H., Yu, L. J., Chen, X. Y., Kong, Q. Y., Song, X., Shu, X. H., Liu, J., & Lebedeva, I. V. (2014). Short-term resveratrol exposure causes in vitro and in vivo growth inhibition and apoptosis of bladder cancer cells. *PLoS One*, *9*(2), e89806. Available from https://doi.org/10.1371/journal.pone.0089806.

Wu, Y., Xun, Y., Zhang, J., Hu, H., Qin, B., Wang, T., Wang, S., Li, C., & Lu, Y. (2021). Resveratrol attenuates oxalate-induced renal oxidative injury and calcium oxalate crystal deposition by regulating TFEB-induced autophagy pathway. *Frontiers in Cell and Developmental Biology*, *9*. Available from https://doi.org/10.3389/fcell.2021.638759, https://www.frontiersin.org/journals/cell-and-developmental-biology#.

Yang, A., Ren, G., Tang, L., & Jiang, W. (2009). Effects of soy bean isoflavone on inhibition of benign prostatic hyperplasia and the expressions of NO and NOS of rats. *Wei Sheng Yan Jiu = Journal of Hygiene Research*, *38*(2), 172–174.

Yang, Y., Li, C., Li, H., Wu, M., Ren, C., Zhen, Y., Ma, X., Diao, Y., Ma, X., Deng, S., Liu, J., & Shu, X. (2017). Differential sensitivities of bladder cancer cell lines to resveratol are unrelated to its metabolic profile. *Oncotarget*, *8*(25), 40289–40304. Available from https://doi.org/10.18632/oncotarget.15041, http://www.impactjournals.com/oncotarget/index.php?journal = oncotarget&page = article&op = download&path%5B%5D = 15041&path%5B%5D = 48072.

Yasir, F., Wahab, A. T., & Choudhary, M. I. (2018). Protective effect of dietary polyphenol caffeic acid on ethylene glycol-induced kidney stones in rats. *Urolithiasis*, *46*(2), 157–166. Available from https://doi.org/10.1007/s00240-017-0982-1, http://link.springer.com/journal/240.

Ying-Jie, K., Ling-Wei, C., Min, Z., Jia-Ping, G., Ying, W., Dan, Z., & Sen, Z. (2019). Resveratrol enhances chemosensitivity of renal cell carcinoma to paclitaxel. *Frontiers in Bioscience – Landmark*, *24*(8), 1452–1461. Available from https://doi.org/10.2741/4790, https://www.bioscience.org/fbs/getfile.php?FileName = /2019/v24/af/4790/4790.pdf.

Youn, S. H., Kwon, J. H., Yin, J., Tam, L. T., Ahn, H. S., Myung, S. C., & Lee, M. W. (2017). Anti-inflammatory and anti-urolithiasis effects of polyphenolic compounds from Quercus gilva Blume. *Molecules (Basel, Switzerland)*, *22*(7), 1121. Available from https://doi.org/10.3390/molecules22071121.

Yu, X. L., Jing, T., Zhao, H., Li, P. J., Xu, W. H., & Shang, F. F. (2014). Curcumin inhibits expression of inhibitor of DNA binding 1 in PC3 cells and xenografts. *Asian Pacific Journal of Cancer Prevention*, *15*(3), 1465–1470. Available from https://doi.org/10.7314/APJCP.2014.15.3.1465, http://www.apocpcontrol.org/paper_file/issue_abs/Volume15_No3/1465-1470%2012.24%20Xiaoling%20YU.pdf.

Zeng, X., Xi, Y., & Jiang, W. (2019). Protective roles of flavonoids and flavonoid-rich plant extracts against urolithiasis: A review. *Critical Reviews in Food Science and Nutrition*, *59*(13), 2125–2135. Available from https://doi.org/10.1080/10408398.2018.1439880.

Zhan, B., Zhang, Z., Piao, C., Dong, X., Du, Y., Kong, C., & Jiang, Y. (2021). The sigma-2 receptor/TMEM97 agonist PB28 suppresses cell proliferation and invasion by regulating the PI3K-AKT-mTOR signalling pathway in renal cancer. *Journal of Cellular and Molecular Medicine*, *25*(24), 11244–11256. Available from https://doi.org/10.1111/jcmm.17047.

Zhang, T., Zhao, L., Zhang, T., Wu, W., Liu, J., Wang, X., Wan, Y., Geng, H., Sun, X., Qian, W., & Yu, D. (2020). Curcumin negatively regulates cigarette smoke-induced renal cell carcinoma epithelial– mesenchymal transition through the erk5/ap-1 pathway. Dove Medical Press Ltd, China. *OncoTargets and Therapy*, *13*, 9689–9700. Available from https://doi.org/10.2147/OTT.S265847, https://www.dovepress.com/getfile.php?fileID = 61938.

Zhou, B., Yi, H., Tan, J., Wu, Y., Liu, G., & Qiu, Z. (2015). Anti-proliferative effects of polyphenols from pomegranate rind (Punica granatum L.) on EJ bladder cancer cells via regulation of p53/miR-34a axis. *Phytotherapy Research*, *29*(3), 415–422. Available from https://doi.org/10.1002/ptr.5267.

Zhou, D., Wu, Y., Yan, H., Shen, T., Li, S., Gong, J., Li, G., Mai, H., Wang, D., & Tan, X. (2022). Gallic acid ameliorates calcium oxalate crystal-induced renal injury via upregulation of Nrf2/HO-1 in the mouse model of stone formation. *Phytomedicine: International Journal of Phytotherapy and Phytopharmacology*, *106*154429. Available from https://doi.org/10.1016/j.phymed.2022.154429.

Zhou, J., Jin, J., Li, X., Zhao, Z., Zhang, L., Wang, Q., Li, J., Zhang, Q., & Xiang, S. (2018). Total flavonoids of Desmodium styracifolium attenuates the formation of hydroxy-l-proline-induced calcium oxalate urolithiasis in rats. *Urolithiasis*, *46*(3), 231–241. Available from https://doi.org/10.1007/s00240-017-0985-y.

Zhu, X. W., Dong, F. M., Liu, J., & Li, M. S. (2022). Resveratrol nanoparticles suppresses migration and invasion of renal cell carcinoma cells by inhibiting matrix metalloproteinase 2 expression and extracellular signal-regulated kinase pathway. *Journal of Biomedical Nanotechnology*, *18*(4), 1001–1008. Available from https://doi.org/10.1166/jbn.2022.3310.

CHAPTER FIVE

Sport: Improving exercise performance with polyphenol extracts

Francisco J. Osuna-Prieto[1,2,3], Juan J. Martin-Olmedo[1,4], Carlos Hernández-Martínez[1], Jonatan R. Ruiz[1,5,6] and Lucas Jurado-Fasoli[1,4]

[1]Department of Physical Education and Sports, Faculty of Sports Science, Sport and Health University Research Institute (iMUDS), University of Granada, Granada, Spain
[2]Hospital Universitari Joan XXIII of Tarragona, Pere Virgili Health Research Institute (IISPV), Tarragona, Spain
[3]CIBER of Diabetes and Associated Metabolic Diseases (CIBERDEM)-Institute of Health Carlos III (ISCIII), Madrid, Spain
[4]Department of Physiology, Faculty of Medicine, University of Granada, Granada, Andalucía, Spain
[5]CIBER of Obesity and Nutrition Physiopathology (CIBEROBN), Institute of Health Carlos III, Granada, Spain
[6]Biosanitary Research Institute, Ibs.Granada, Granada, Spain

Introduction

Throughout history, polyphenols have played a crucial role in promoting health and preventing diseases due to their antioxidant and anti-inflammatory properties (Pandey & Rizvi, 2009). Dietary polyphenols are naturally occurring nonnutritive compounds in plants—especially fruits, vegetables, nuts, seeds, and leaves—that are characterized by the presence of multiple phenolic rings within their structure. Supplements containing polyphenols are commonly employed by athletes to mitigate exercise-induced oxidative stress and inflammation, which can contribute to muscle fatigue and tissue damage, thereby hampering physical performance (Myburgh, 2014). Today, there is a growing level of scientific evidence supporting their use to enhance exercise performance and prevent and/or mitigate exercise-induced muscle damage and inflammation (Bojarczuk & Dzitkowska-Zabielska, 2023; Carey et al., 2021; Somerville et al., 2017). Although polyphenols possess antioxidant properties that could enhance exercise performance, certain polyphenols may disrupt specific adaptive signaling pathways, thereby impairing overall performance (Gomez-Cabrera et al., 2015). For this reason, it is crucial to identify which polyphenols enhance athletic performance considering

Bioactive Polyphenols for Health and Pathology Treatment.
DOI: https://doi.org/10.1016/B978-0-443-13821-8.00005-6

the type of exercise—i.e., endurance or resistance exercise - along with optimum supplementation strategies.

This chapter provides an updated, critical review of the impact of polyphenols on endurance and resistance exercise performance in humans.

Physiological responses to exercise: balancing redox and inflammatory pathways for optimal adaptations

In simple terms, oxidative stress is a condition where there is an imbalance between the production of reactive species or free radicals and the body's antioxidant system. This imbalance can damage cellular components such as lipids, proteins, and deoxyribonucleic acid (DNA) (Pizzino et al., 2017). Paradoxically, although exercise yields numerous health benefits, the physiological response to acute exercise triggers an adaptative, augmented generation of reactive oxygen species, and the initiation of inflammatory pathways (Mari Carmen Gomez-Cabrera et al., 2021; Powers & Jackson, 2008). On the other hand, long-term exercise training promotes adaptations that enhance the body's natural antioxidant defense system—i.e., higher expression/activity of superoxide dismutase, catalase, and glutathione peroxidase (Simioni et al., 2018)—and exert anti-inflammatory effects by regulating the production of cytokines (Bowtell & Kelly, 2019).

However, certain conditions such as exercise bouts of high intensity and/or duration or engaging in excessive or unbalanced exercise training programs without adequate recovery may result in increased oxidative stress and inflammation, thereby reducing exercise performance (Powers et al., 2020; Woods et al., n.d.). In this regard, supplementation with dietary polyphenols resembles a promising strategy for mitigating the adverse effects stemming from strenuous or high-frequency exercise sessions, as these antioxidants can help to prevent and/or mitigate oxidative stress and inflammation (Bojarczuk & Dzitkowska-Zabielska, 2023; Carey et al., 2021; Somerville et al., 2017).

How polyphenols improve exercise performance: mechanistic insights

Polyphenols potentially enhance exercise performance through various mechanisms (Fig. 5.1). These mechanisms can unfold individually or

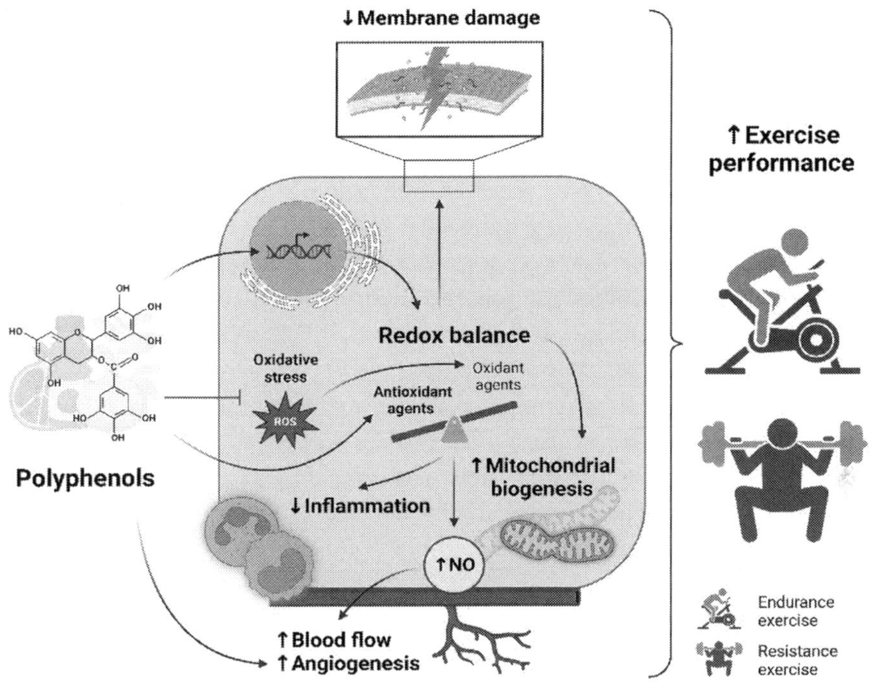

Figure 5.1 Proposed mechanisms of action of polyphenols impacting exercise performance in humans. Polyphenol intake may reduce pro-inflammatory signaling pathways, decrease cellular membrane damage, improve blood flow and angiogenesis, and increase mitochondrial biogenesis by reducing oxidative stress and maintaining redox balance. These mechanisms may lead to improvements in exercise performance following polyphenol ingestion. *NO*, Nitric oxide; *ROS*, reactive oxygen species.

synergistically, depending on the specific polyphenol, offering a dynamic spectrum of benefits that contribute to improved physical performance.

Polyphenols reduce oxidative stress and maintain the redox balance

During exercise, polyphenols emerge as key contributors to counteracting cellular damage induced by oxidative stress (Kruk et al., 2022). Polyphenols also activate the factors involved in the endogenous antioxidant system, such as the Keap1-Nrf2-ARE pathway (Kelch-like ECH-associated protein 1—nuclear factor erythroid 2-related factor 2—antioxidant response element), a cellular signaling mechanism promoting the transcription of genes involved in antioxidant processes and oxidative stress (Bellezza et al., 2018; Eynon et al., 2010; Huang et al., 2015).

Polyphenols decrease inflammation

The anti-inflammatory properties of polyphenols are primarily driven by their ability to modulate key molecular pathways involved in the inflammatory response. For instance, some flavonoids modulate the production of key enzymes involved in the production of inflammatory molecules such as cyclooxygenase and lipoxygenase (Malireddy et al., 2012; Yahfoufi et al., 2018; Yoon & Baek, 2005). Polyphenols also regulate other inflammatory key signaling pathways, such as the nuclear factor-κB pathway, reducing the expression of pro-inflammatory genes and suppressing the inflammatory response. Furthermore, they can modulate immune responses by controlling inflammatory cytokine production by immune cells (Yahfoufi et al., 2018).

Polyphenols improve mitochondrial biogenesis and function

Polyphenols may enhance exercise performance by promoting mitochondrial biogenesis through two different pathways. First, they activate stress-related cell signaling pathways, leading to increased expression of genes encoding cytoprotective proteins like nuclear factor erythroid 2-related factor 2 [Nrf2 (Eynon et al., 2010; Huang et al., 2015)]. Nrf2 is also a transcription factor that plays a crucial role in mitochondrial biogenesis. Indeed, variations in the Nrf2 gene have been linked to endurance performance (Somerville et al., 2017). Secondly, specific polyphenols like quercetin, resveratrol, or curcumin may increase the expression of genes like sirtuin 1 and peroxisome proliferator-activated receptor γ co-activator 1 α, orchestrating a coordinated response that promotes the transcription of crucial genes for mitochondrial function and biogenesis (Chodari et al., 2021; Malaguti et al., 2013; Mthembu et al., 2021).

Polyphenols improve blood flood and angiogenesis

There is evidence that various polyphenols, such as anthocyanins and flavonols, could enhance flow-mediated dilatation and endothelial function in humans by increasing endothelial nitric oxide synthase expression either directly or indirectly (Ghosh & Scheepens, 2009; Serreli & Deiana, 2023). In addition, there is evidence supporting an angiogenic effect of polyphenols, which could improve exercise performance by increasing blood flow supply to skeletal muscle (Diniz et al., 2017). However, most of these investigations are conducted in preclinical models, and further research is needed to clarify these mechanisms in humans.

Polyphenols protect cell membrane integrity

Polyphenols like flavonols might potentially help to protect cell membranes from exercise-induced damage. The exercise-induced oxidative stress can lead to lipid peroxidation, a process where free radicals damage cell membranes (Su et al., 2019). Thus polyphenols, by neutralizing free radicals, could mitigate lipid peroxidation thereby preserving the structural integrity of cell membranes (Caruana et al., 2012; Rudrapal et al., 2022).

Other potential mechanisms

High exercise intensity can lead to gastrointestinal distress, including nausea, cramping, and inflammation. Interestingly, some polyphenols might exhibit promising attributes in contributing to the prevention and/or mitigation of these gastrointestinal complications (Ribichini et al., 2023). It has been proposed that polyphenols may contribute to maintaining the integrity of the gut barrier preventing leakage of toxins and harmful bacteria into the bloodstream, which can trigger the "leaky gut" of the athlete and induce gastrointestinal complications (Sandoval-Ramírez et al., 2021). Lastly, recent evidence suggests that polyphenols improve exercise performance through improvements in cognition by regulating enzymes that play a significant role in regulating neurotransmitters, such as monoamine oxidase inhibition (Lomiwes et al., 2019).

Polyphenols supplementation to improve endurance and resistance exercise performance

We have extracted information from randomized, double-blind, placebo-controlled trials where dietary polyphenols were consumed from food or supplements. By limiting our analysis to studies with the most robust experimental designs, we aim to provide a more reliable synthesis of evidence, minimizing the impact of potential bias (Schulz et al., 2010).

Polyphenols and endurance exercise performance

Understanding the effects of polyphenols on endurance performance is an active area of research. However, despite an abundance of studies scrutinizing the impact of dietary polyphenols on endurance performance, the findings reveal a compilation of mixed evidence (Best et al., 2022; Cook et al., 2015; Howatson et al., 2022; Keane et al., 2018; Levers et al., 2016; Morgan et al., 2019; Nieman et al., 2010; Sharp et al., 2012; Willems et al., 2015). The characteristics and results of the selected studies evaluating the effect of polyphenol supplementation on endurance exercise performance can be found in Table 5.1.

Several studies have investigated the effects of acute anthocyanin supplementation [e.g., Montmorency tart cherry concentrates, New Zealand blackcurrant (NZBC)] on endurance time trial (TT) (Braakhuis et al., 2014; Montanari et al., 2023; Morgan et al., 2019; Moss et al., 2023; Willems et al., 2015), and time-to-exhaustion (TTE) performance tests (Keane et al., 2018). Yet it has been proposed that anthocyanins could improve endurance performance by increasing peripheral vasodilatation (Matsumoto et al., 2005) and increasing lactate clearance (Gladden, 2004), most of the studies reported a lack of ergogenic effects (Braakhuis et al., 2014; Cook et al., 2015; Keane et al., 2018; Moss et al., 2023). For instance, the ingestion of NZBC extract (315 mg anthocyanins) 2 h before exercise improved 5-km running TT performance in men (Montanari et al., 2023), but not in women (300 mg anthocyanins) (Moss et al., 2023). Similarly, in trained cyclists, 315 mg of NZBC extract did not enhance 16.1-km TT performance despite increases in power output (PO) and speed (Cook et al., 2015). In line with this, neither anthocyanin supplementation with 300 mg nor with $\approx$550 mg before an exercise test improved TTE performance in men (Braakhuis et al., 2014; Keane et al., 2018). Considering these results, the current evidence for supporting acute anthocyanin supplementation as an effective strategy to improve endurance performance remains inconclusive.

Regarding tannin supplementation, ingesting 1000 mg of pomegranate extract (tannin content not reported) 30 min before running until exhaustion increased TTE performance at 90% and 100%, but not at 110% of peak velocity in young adults (Trexler et al., 2014). Since it has been shown that tannins promote angiogenesis, these findings might be explained by increased oxygen and nutrient delivery to highly active metabolic tissues during exercise (Roelofs et al., 2017). Nevertheless, further

Table 5.1 Effects of polyphenol supplementation on endurance exercise performance.

Polyphenol type	Participants	Type, dosage, and timing	Exercise protocol	Performance	Refs.
Flavonoids, anthocyanins	27 (9 women) endurance-trained runners or triathletes (age, 21.8 ± 3.9 years)	Anthocyanins, from CherryPURE (Freeze Dried Tart Cherry Powder) 66 mg anthocyanins per d for 7 d before, the d of, and 2 d after exercise protocol	1/2-marathon (21.1 km)	↓ Finish time ↓ Split time ↔ Projected and actual race pace	Levers et al. (2016)
	9 highly trained male water polo athletes (age, 18.6 ± 1.4 years)	Anthocyanins, from 90 mL tart cherry juice concentrate (Montmorency, Prunus Cerasus by CherryActive, UK) ≈ 820 mg for 6 consecutive d before the testing d	Water Polo Intermittent Shuttle Test	↔ Water Polo test	McCormick et al. (2016)
	10 highly trained male cyclists (age, 28 ± 7 years)	Anthocyanins, from 60 mL of commercially available Montmorency tart cherry concentrate (CherryActive, UK) ≈ 550 mg of anthocyanins per d 1.5 h before tests	Cycling TTE and 60 s all-out sprint	↔ TTE performance ↑ Peak power over the first 20 s and total work completed during the all-out sprint (total kJ)	Keane et al. (2018)

(*Continued*)

Table 5.1 (Continued)

Polyphenol type	Participants	Type, dosage, and timing	Exercise protocol	Performance	Refs.
	8 highly trained male cyclists (age, 19.7 ± 1.6 years)	Anthocyanins, from Montmorency tart cherry powder capsules (powdered freeze-dried Montmorency cherry, CherryActive, MC) 256.8 mg/d of anthocyanin consumed for 7 d (6 d before the testing d; on the testing d, 3 pills 1 h before tests)	15-km cycling TT	↑ TT performance	Morgan et al. (2019)
	34 trained male cyclists (age, 38 ± 7 years)	Anthocyanins, from New Zealand Blackcurrant (NZBC) extract (CurraNZ, New Zealand) 315 mg anthocyanins 2 h before exercise	Two 16.1 km cycling TT	↔ TT performance ↑ TT performance in the slower group ↑ PO and speed	Montanari et al. (2023)
	13 male rugby union players (age, 21 ± 2 years).	Anthocyanins, from a concentrated New Zealand Blackcurrant (NZBC) extract (CurraNZ, New Zealand) 210 mg of anthocyanins every d for 7 consecutive d before breakfast	RSA, Illinois Agility, and Seated Medicine Ball Throw Tests	↑ 35m-sprints (RSA test) ↔ Illinois Agility and Seated Medicine Ball Throw Tests Best performance	Braakhuis et al. (2014)

	17 trained male runners (age, 26 ± 5 years).	Anthocyanins, from New Zealand Blackcurrant (NZBC) extract (CurraNZ, New Zealand) 315 mg anthocyanins 2 h before exercise	5-km running TT	↑ 5-km TT performance (i.e., ↓ Finish time)	Moss et al. (2023)
	15 healthy males (age, 24.4 ± 3.6 years)	Anthocyanins, from 300 mg of blackcurrant extract (CurraNZ New Zealand Blackcurrant Extract, UK) 105 mg anthocyanin daily from d -1 and test d before incremental exercise, and 315 mg anthocyanin was administered before TTE trials	Treadmill running incremental exercise and TTE	↔ Performance during running incremental exercise testing ↔ TTE	Pastellidou et al. (2021)
	13 (5 women) trained triathletes (age, 38 ± 8 years)	Anthocyanins, from 6 g Sujon New Zealand blackcurrant powder containing 138.6 mg anthocyanins daily for 7 d	Two incremental cycling protocols	↑ PO at onset blood lactate accumulation ↔ VO_{2max}	Willems et al. (2015)

(*Continued*)

Table 5.1 (Continued)

Polyphenol type	Participants	Type, dosage, and timing	Exercise protocol	Performance	Refs.
	23 healthy women (age, 31 ± 8 years)	Anthocyanins, from a fruit drink concentrate with blackcurrant juice powder (Tasman Extracts, Nelson, New Zealand) 300 mg of anthocyanins daily	Incremental running test and 5-km TT test	↔ Incremental running test ↔ TT performance	Braakhuis et al. (2014)
	30 male recreational runners (33 ± 7 years)	Anthocyanins, from Haskap berry powder 150 mg for 6 d before, 1 h before VO_{2peak} test and 1 h before TT test		VO_{2peak} test, TTE and 5-km TT running tests	Howatson et al. (2022)
	12 recreationally active women (age, 19.9 ± 1.1 years)	Anthocyanins, from New Zealand Blackcurrant (NZBC) extract (ViBERi) 1.6 mg/kg of anthocyanins consumed daily for 7 d	RSA test	↔ Mean sprint time from baseline ↔ Fastest sprint time from baseline	Best et al. (2022)

	15 healthy men (age, 38 ± 13 years)	Anthocyanins, from a concentrated New Zealand Blackcurrant (NZBC) extract (CurraNZ, Auckland, New Zealand) 105 mg of anthocyanins daily for 7 consecutive d	16.1 km cycling TT	↑ 16.1 km TT performance	Cook et al. (2015)
Flavonoids, flavanols	30 recreationally active healthy young men (Quercitin age, 23.1 ± 2.4; Placebo age, 22.1 ± 1.8 years)	Flavonols, from a sports hydration beverage (946 mL) prepared by The Coca-Cola Company 1000 mg/d of quercetin for 7–16 d	VO_{2peak} test and 10-min cycling TT	↔ VO_{2peak} ↔ 10-min time trial TT distance	Cureton et al. (2009)
	12 (5 women) healthy young adults (age, 22.9 ± 2.4 years)	Flavonols, from a food-grade quercetin powder (Nutravail Technologies, VA) 500 mg/d of quercitin for 7 d	VO_{2max} test and TTE at 75% VO_{2max}	↑ VO_{2max} ↑ TTE performance	Davis et al. (2010)

(*Continued*)

Table 5.1 (Continued)

Polyphenol type	Participants	Type, dosage, and timing	Exercise protocol	Performance	Refs.
	26 young men (aged, 18–28 years)	Flavonols, from a quercetin beverage (a sugarless PowerAde base by Coca-Cola) 1000 mg of quercetin consumed daily for 2 weeks before tests	12-min TT on 15% graded treadmills after 60 min of moderate exercise at 60% VO_{2max}	↑ 12-min TT distance	Nieman et al. (2010)
	17 physically active and moderately fit healthy men (age, 22 ± 3 years)	Flavonols, from bars (Quercegen Pharma, USA) 1000 mg of quercetin per d for 8.5 d	Two 200 kJ cycling TTs	↔ VO_{2peak} ↔ TT performance	Sharp et al. (2012)
Non-flavonoids, tannins	19 (9 women) highly active young adults (age, 22.2 ± 2.2 years)	Tannins, from a standardized pomegranate extract (NITRO2GRANIT, CA) 1000 mg of pomegranate extract 30 min before exercise	Three treadmill runs to exhaustion at 90, 100, and 110% peak velocity previously determined with VO2max treadmill test	↑ TTE performance at 90% and 100% peak velocity ↓TTE performance at 110% peak velocity	Trexler et al. (2014)

research is needed to determine whether acute tannin supplementation could be an effective strategy to improve endurance performance in sports engaging submaximal intensities.

Chronic supplementation protocols—i.e., those lasting ≥ 3 days—are believed to be the minimum period for polyphenols to elicit significant changes in gene and protein expression responsible for increasing mitochondrial biogenesis and function, angiogenesis, and endogenous antioxidant system activity. Various studies have found ergogenic effects on endurance exercise after chronic administration of anthocyanins at both low and high doses. Doses ranging from 50 to 300 mg/d anthocyanins for ≥ 7 d improved cycling (e.g., 15-km TT (Cook et al., 2015; Morgan et al., 2019) and 16.1-km TT (Montanari et al., 2023)) and running performance (e.g., TTE running performance (Howatson et al., 2022), 21.1-km running performance (Levers et al., 2016), and running repeated sprint tests (Burnett & Willems, 2022). Nevertheless, two studies (820.5 mg and 100 mg/d for 6 d and 7 d, respectively) reported no significant performance improvement in the intermittent running (McCormick et al., 2016) and repeated sprint ability tests (Best et al., 2022). The ergogenic properties of chronic anthocyanin supplementation seem to be explained by increased fat utilization during exercise (Strauss et al., 2018), which may preserve glycogen for later use (Yeo et al., 2011). Besides, anthocyanins have been linked to increased vasodilation, resulting in improved oxygen and nutrient delivery to the skeletal muscle (Matsumoto et al., 2005). Thus the current level of evidence suggests that the use of chronic anthocyanin supplementation might be an effective strategy to increase endurance performance in middle-distance cycling and running events.

Other studies have investigated the chronic effect of flavonol—i.e., quercetin—supplementation on endurance performance (Cureton et al., 2009; Davis et al., 2010; Nieman et al., 2010; Sharp et al., 2012). Davis et al. (2010) showed higher VO_{2max} and TTE performance after consuming 500 mg/d of quercetin for 7 d in healthy young participants. Similarly, small but significant running performance improvements were observed by Nieman et al. (2010) during a 12-min running TT after consuming 1000 mg/d of quercetin for 2 weeks in healthy untrained males. The authors suggested that quercetin might have a more pronounced effect in untrained individuals due to their lower mitochondrial count/function, potentially leading to a lesser or negligible impact on trained individuals (Cureton et al., 2009). However, other flavonol supplementation trials (Cureton et al., 2009; Sharp et al., 2012) with analogous exercise protocols but different supplementation periods (1000 mg/d for 7–16 d; 1000 mg/d for 8.5 d, respectively) reported no differences vs. placebo in untrained individuals. In this case, the authors reported that higher doses and longer supplementation periods might be needed to elicit mitochondrial adaptations (Sharp et al., 2012). Thereby, the results from studies using chronic quercetin supplementation, yet promising, are still inconsistent. Additional studies are needed to corroborate whether higher doses and/or longer supplementation periods are needed to achieve significant ergogenic effects during endurance exercise events.

Overall, the current level of evidence supports the use of chronic anthocyanin supplementation strategies for improving exercise performance, especially during middle-distance running and cycling events. Although acute tannin supplementation shows potential for enhancing endurance performance in submaximal and short-to-moderate duration activities, the absence of studies replicating this finding precludes drawing firm conclusions. Whether higher doses and/or longer supplementation periods with flavonols are necessary to achieve significant ergogenic effects during endurance exercise events warrants further research. Finally, acute anthocyanin and flavonol supplementation, regardless of the population or exercise protocols, did not consistently result in significant performance improvements, thus lacking evidence to support their use as ergogenic aids.

Polyphenols and resistance exercise performance

Enhancing resistance exercise performance and facilitating recovery, particularly following unaccustomed, high-intensity eccentric muscle contractions, are common challenges faced by athletes. Therefore it is not surprising the growing interest in the research and application of polyphenol supplementation in resistance exercise nutrition (Carey et al., 2021). The characteristics and results of the selected studies evaluating the effect of polyphenol supplementation on resistance exercise performance can be found in Table 5.2.

Only two studies have investigated the acute effects of polyphenol supplementation on resistance exercise performance (Patrizio et al., 2018; Roelofs et al., 2017). For instance, the supplementation of quercetin (1 g, 3 h before resistance exercise) increased neuromuscular performance without affecting perceived exertion in a resistance exercise session in recreationally active, healthy young men, possibly by increasing neuromuscular efficiency at a central level (Patrizio et al., 2018). Similarly, pomegranate

Effects of polyphenol supplementation on resistance exercise performance in humans.

Polyphenol type	Participants	Type, dosage, and timing	Exercise protocol	Performance	Refs.
Flavonoids, anthocyanins	10 resistance-trained, healthy young men (age, 27.8 ± 1.6 years)	Anthocyanins, from CherryActive (Montmorency Cherry Juice Concentrate) 30 mL/d (≈275 mg anthocyanins/d) for 7 d before, the day of, and 2 d after lower-body exercise	10 sets of 10 repetitions of single-leg knee extensions at a beginning 80% 1RM	↑ MVC of the knee extensors 24 and 48 h after exercise	Bowtell et al. (2011)
	17 recreationally active women (age, 22.2 ± 3.3 years)	Anthocyanins, from a tart cherry extract powder 500 mg twice a day (total dose: 1000 mg/d) for 4 d before, the day of, and 4 d after lower-body exercise	8 sets of 10 maximal isokinetic (60°/s) concentric/eccentric repetitions of knee extensions	↔ Peak torque, time-to-peak torque, peak power, and mean power	Ortega et al. (2023)
Flavonoids, flavonols	10 recreationally active, healthy young men (age, 22.1 ± 1.8 years)	Quercetin 1000 mg 3 h before exercise sessions	3 sets of 8 repetitions on seated row, chest press, knee extension, knee flexion, leg press, and calf raise at 80% 1RM	↑ Volume load ↑ Knee extension torque after exercise session ↑ Maximal voluntary isometric contraction after exercise session ↑ Rate of torque development and neuromuscular efficiency during maximal voluntary isometric contraction after exercise session	Patrizio et al. (2018)

(*Continued*)

Table 5.2 (Continued)

Polyphenol type	Participants	Type, dosage, and timing	Exercise protocol	Performance	Refs.
	16 healthy young men (age, 25.9 ± 3.3 years)	Quercetin 500 mg twice a day (total dose: 1000 mg Q/d) for 14 d before upper-body exercise	10 sets of 10 maximal isokinetic (45°/s) eccentric contractions of the elbow flexors	↑ Elbow flexion torque during a maximal voluntary isometric contraction immediately after, and 24, 72, and 96 h after eccentric exercise ↑ Mean fibers conduction velocity during a maximal voluntary isometric contraction immediately after, and 24 and 96 h after eccentric exercise ↑ F-V relationship and mean fibers conduction velocity during the recovery phase	Bazzucchi et al. (2020)
Flavonoids, flavanols	35 untrained healthy young men (group GTP age, 21.5 ± 0.3 years; group PLA age, 21.2 ± 0.3 years)	Catechins, from green tea extract (*Camellia sinensis*) 640 mg PP/d during the training program (4 weeks)	4 weeks of a supervised training program (3 sessions/week) 3 circuits of 15 repetitions on bench press, back squat, power clean, behind-the-neck press, biceps curls, triceps extensions, trunk flexions, and lunges at 60% 1RM per session	↔ 1RM and repetitions to exhaustion on back squat and bench press at 60% 1RM after the training program	Jówko et al. (2011)

	11 (6 women) trained healthy young adults (age, 26.4 ± 4.8 years)	Epicatechin 98 mg twice a day (total dose: 196 mg/d) for 2 d before, and 60–90 minutes before an experimental session	15.5 CrossFit Open Workout: rowing on a rowing machine, followed by barbell thrusters completed for four rounds to finish as fast as possible	↔ Total time, and split time to complete the workout	Schwarz et al. (2020)
	17 untrained, healthy young men (age, 22.3 ± 0.3 years)	Catechins, from matcha green tea powder 1.5 g twice a day (total dose: 3 g/d; 333 mg PP/d; 189 mg EGCG/d) during the training program (8 weeks)	8 weeks of a supervised training program (2 sessions/week) 3 sets of 10 repetitions on chest press, fly, back extension, seated rowing, leg press, leg extension, leg curl, and sit-up at 10RM per session	↔ Body mass and composition (body fat percentage and muscle mass) after the training intervention ↔ Knee extension, grip strength, and 10RM chest press after the training intervention ↑ 10RM leg press after the training intervention	Shigeta et al. (2023)
	19 untrained, healthy young men (age, 20.9 ± 1.3 years)	Catechins, from matcha green tea powder 1.5 g twice a day (total dose: 3 g; 333 mg PP/d; 189 mg EGCG/d) during the training program (12 weeks)	12 weeks of a supervised training program (2 sessions/week) 3 sets of 10 repetitions on chest press, fly, back extension, seated rowing, leg press, leg extension, leg curl, and sit-up at 10RM per session	↔ Body mass and muscle mass after the training intervention ↓ Body fat percentage after the training intervention	Shigeta et al. (2023)

(*Continued*)

Table 5.2 (Continued)

Polyphenol type	Participants	Type, dosage, and timing	Exercise protocol	Performance	Refs.
Non-flavonoids, phenolic acids	10 healthy young men (age, 28.5 ± 3.4 years)	Curcumin, from Theracurmin (*Curcuma longa* extract with colloidal nanoparticles) 180 mg CUR/d for 7 d before upper-body exercise	30 maximal isokinetic (120°/s) eccentric contractions of the elbow flexors	↔ MVC torque of the elbow flexors	Tanabe et al. (2019)
	10 healthy young men (age, 29.0 ± 3.9 years)	Curcumin, from Theracurmin (*Curcuma longa* extract with colloidal nanoparticles) 180 mg CUR/d for 7 d after upper-body exercise	30 maximal isokinetic (120°/s) eccentric contractions of the elbow flexors	↑ MVC torque of the elbow flexors 3–7 d after eccentric exercise	Tanabe et al. (2019)
	17 healthy active men (age, 33.8 ± 5.4 years)	Curcumin, from *Curcuma longa* extract 2.5 CUR g twice a day (total dose: 5 g CUR/d) for 2 d before, the day of, and 2 d after lower-body exercise	5 sets of 10 repetitions at 120% 1RM + 2 sets of 10 repetitions at 100% 1RM of eccentric single-leg press	↑ Single-leg jump performance 0, 24, and 48 h after eccentric exercise	Nicol et al. (2015)
Non-flavonoids, tannins	16 recreationally active men (age, 24.2 ± 1.4 years)	Ellagitannins and anthocyanins, from POM Wonderful (pomegranate juice) 480 mL (total dose: 960 mL/d; ≈ 1242 mg ellagitannins/d) for 4 d before, the day of, and 4 d after upper-body exercise twice a day, with an additional 480 mL (≈ 621 mg ellagitannins) immediately after exercise session on day 5	2 sets of 20 maximal isokinetic (0.7 rad/s) eccentric contractions of the elbow flexors	↑ Isometric elbow flexion strength 48 and 72 h after eccentric exercise	Trombold et al. (2010)

17 resistance-trained, healthy young men (age, 21.9 ± 2.4 years)	Ellagitannins and anthocyanins, from POM Wonderful (pomegranate juice) 250 mL twice a day (total dose: 500 mL/d; ≈ 192 mg anthocyanins/d; ≈ 61 mg ellagic acid/d) for 7 d before, the day of, and 7 d after exercise, with an additional 250 mL (≈ 96 mg anthocyanins/d; ≈ 30 mg ellagic acid/d) immediately after exercise session on day 8	3 sets of 20 maximal isokinetic (0.7 rad s-1) eccentric contractions of the elbow flexors + 6 sets of 10 eccentric contractions of the knee extensors at 110% 1RM	↔ Isometric strength during elbow flexion ↔ Isometric strength during knee extension	Trombold et al. (2011)
19 recreationally active, healthy young men and women (age, 22.1 ± 1.9 years)	Ellagitannins and anthocyanins, from pomegranate extract powder 1000 mg 30 min before experimental sessions	Repeated sprint ability test, and one set of repetitions to fatigue on bench press and leg press at 80% 1RM, respectively	↑ Average and peak power in the fifth sprint of the repeated sprint ability test ↔ Repetitions to fatigue in bench and leg press	Roelofs et al. (2017)

extract supplementation 30 min before exercise improved sprint capacity, but not lower- and upper-body strength-endurance outcomes in recreationally active, healthy young men and women (Roelofs et al., 2017). Sprint capacity improvements might be linked to increased blood flow, while the absence of effects on strength endurance might be explained by the low-exhaustive protocol employed to assess it (Roelofs et al., 2017). Altogether, acute polyphenols supplementation—i.e., quercetin and tannins—may increase resistance exercise performance when performing high-intensity resistance exercise bouts.

On the other hand, a significant amount of studies have investigated the chronic effect of polyphenols supplementation—i.e., ≥ 3 d supplementation period—on resistance exercise performance (Bazzucchi et al., 2020; Bazzucchi et al., 2019; Bowtell et al., 2011; Jówko et al., 2011; Nicol et al., 2015; Ortega et al., 2023; Schwarz et al., 2020; Shigeta et al., 2023; Tanabe et al., 2019; Trombold et al., 2010, 2011). Moreover, strength performance is usually employed for assessing recovery after inducing muscle damage following resistance exercise. For instance, the ingestion of 5 g of curcumin from a *Curcuma longa* extract (2 d before, on the day of, and 2 d after inducing lower-body muscle damage) in healthy active men improved single-leg jump performance immediately after, 24, and 48 h after the inducing muscle damage session (Nicol et al., 2015). Similarly, Montmorency cherry juice concentrate supplementation (≈ 275 mg anthocyanins/d for 15 d) enhanced muscle strength recovery at 24 and 48 h following a muscle damage-inducing resistance exercise session in resistance-trained men (Bowtell et al., 2011). In light of these results, the use of chronic polyphenol supplementation—i.e., curcumin and anthocyanins—could be an effective strategy to boost recovery from muscle-damaging bouts of resistance exercise.

However, not all studies employing polyphenol supplementation have demonstrated ergogenic resistance exercise effects. In this sense, pomegranate juice consumption (≈ 192 mg anthocyanins/d for 15 d) led to significant reductions in muscle soreness after resistance eccentric exercise, without increasing upper- or lower-body muscle strength (Trombold et al., 2011). Nevertheless, it is recognized that trained athletes exhibit greater resilience to muscle damage and appear to recover more rapidly than non-trained individuals (Clarkson et al., 1992; McHugh, 2003). In this context, ingesting pomegranate juice (≈ 1242 mg ellagitannins/d for 9 d) increased isometric elbow flexion strength at 48 and 72 h after an upper-body eccentric exercise in recreationally active but resistance-untrained men

(Trombold et al., 2010). Altogether, these findings suggest that higher doses and/or longer periods of polyphenol supplementation might be needed to improve recovery from resistance exercise in untrained individuals.

One concern regarding assessments of recovery from muscle damage through muscle strength measurements is whether results from such tests directly translate to long-term or in-competition performance. Green tea extract supplementation (640 mg polyphenols/d) during a 4-week resistance training program did not enhance muscle strength performance in untrained men (Jówko et al., 2011). It is plausible that the relatively short duration of the training program may have been insufficient to reveal significant differences in muscle adaptations and performance between conditions. However, in a study that implemented more prolonged resistance training programs (8 and 12 weeks, respectively), green tea extract supplementation (333 mg polyphenols/d; 189 mg epigallocatechin-3-galate/d) improved lower-body muscle endurance strength in resistance-untrained men, although it did not improve overall muscle strength performance (Shigeta et al., 2023). Additional research is warranted to ascertain which polyphenol supplementation improves long-term resistance performance across these diverse contexts.

The current level of evidence suggests that supplementation with quercetin and tannins may acutely improve resistance exercise performance (Patrizio et al., 2018; Roelofs et al., 2017). Curcumin, anthocyanins, and tannins chronic ingestion may expedite recovery in situations characterized by a high degree of muscle damage after resistance exercise (Bowtell et al., 2011; Nicol et al., 2015; Trombold et al., 2010), as often occurs in specific phases of the competitive season, and in those athletes with low polyphenols intake in their habitual dietary pattern (Close et al., 2022). Moreover, untrained individuals may require higher doses and/or prolonged periods of polyphenol supplementation to expedite recovery from resistance exercise and enhance performance (Clarkson et al., 1992; McHugh, 2003; Trombold et al., 2010).

Limitations

Some considerations and limitations restrain the interpretation of these heterogeneous results, such as the absence of dietary standardization (Nicol et al., 2015; Roelofs et al., 2017; Schwarz et al., 2020; Shigeta et al., 2023; Tanabe et al., 2019; Trombold et al., 2010, 2011), the restriction of polyphenol-rich foods in the habitual dietary pattern (Jówko et al., 2011; Patrizio et al., 2018), and the wide range of individual's fitness levels. Noteworthy, the scarcity of research conducted on women (Ortega et al., 2023; Schwarz et al., 2020) hinders the possibility of evaluating potential sex differences in the neuromuscular function and recovery response to polyphenol supplementation in men vs. women. Additional research is needed to ascertain the most suitable type, dose, and timing of polyphenol supplementation across diverse scenarios. In addition, some polyphenol-rich ingredients contain various bioactive compounds—i.e., pomegranate extract is also a highly concentrated source of dietary nitrate—which could hamper the isolation of the ergogenic effect of each polyphenol type. These interactions should be carefully taken into account by the researchers, as there might exist synergistic or detrimental interactions between bioactive compounds present in the polyphenol extract.

Future perspectives and challenges

Noteworthy, some authors (Best et al., 2022) have reported the effects of polyphenols on exercise performance might be sex-specific—i.e., no effect in women—highlighting the need for studies that include both men and women to investigate potential sex-specific interactions and elucidate the underlying mechanisms.

Identifying potential synergies and interferences with other supplements is essential to providing evidence-based polyphenol supplementation guides. For instance, some polyphenol food sources (e.g., pomegranate) that contain other bioactive compounds (e.g., nitrates) that have also demonstrated performance benefits that could synergistically exert their effects through the same physiological pathway (e.g., reducing the oxygen cost of exercise through increased mitochondrial efficiency) (Trexler et al., 2014).

In addition, future studies should also investigate the long-term potential consequences of health-related outcomes in athletes. For example, polyphenols are inhibitors of iron absorption, and their supplementation, specifically in athletes at risk, might prevent iron regulation, thereby increasing the risk of anemia (Peeling & McKay, 2023). Future research endeavors should prioritize investigating the long-term effects of polyphenol supplementation in non-related performance outcomes considering potential interactions with other dietary ingredients.

Conclusions

In conclusion, polyphenol supplementation with anthocyanins, quercetin, and curcumin seems to be an effective nutritional strategy for improving endurance and resistance exercise performance in humans. Nevertheless, further research is needed to explore the potential benefits of other types of polyphenols, determine optimal dosages, establish supplementation protocols, and elucidate the underlying mechanisms driving these effects.

References

Bazzucchi, I., Patrizio, F., Ceci, R., Duranti, G., Sabatini, S., Sgrò, P., Luigi, L. D., & Sacchetti, M. (2020). Quercetin supplementation improves neuromuscular function recovery from muscle damage. *Nutrients*, *12*(9), 1–12. Available from https://doi.org/10.3390/nu12092850, https://www.mdpi.com/2072-6643/12/9/2850/pdf.

Bazzucchi, I., Patrizio, F., Ceci, R., Duranti, G., Sgrò, P., Sabatini, S., Di Luigi, L., Sacchetti, M., & Felici, F. (2019). The effects of quercetin supplementation on eccentric exercise-induced muscle damage. *Nutrients*, *11*(1), 205. Available from https://doi.org/10.3390/nu11010205.

Bellezza, I., Giambanco, I., Minelli, A., & Donato, R. (2018). Nrf2-Keap1 signaling in oxidative and reductive stress. *Biochimica et Biophysica Acta (BBA) — Molecular Cell Research*, *1865*(5), 721–733. Available from https://doi.org/10.1016/j.bbamcr.2018.02.010.

Best, R., Metekingi, C., Longhurst, G., & Maulder, P. S. (2022). New Zealand blackcurrant extract supplementation does not improve repeated sprint ability. *The Journal of Sport and Exercise Science*, *6*(1), 1–8. Available from https://doi.org/10.36905/jses.2022.01.01.

Bojarczuk, A., & Dzitkowska-Zabielska, M. (2023). Polyphenol supplementation and antioxidant status in athletes: A narrative review. *Nutrients*, *15*(1), 158. Available from https://doi.org/10.3390/nu15010158.

Bowtell, J. L., Sumners, D. P., Dyer, A., Fox, P., & Mileva, K. N. (2011). Montmorency cherry juice reduces muscle damage caused by intensive strength exercise. *Medicine and Science in Sports and Exercise*, *43*(8), 1544–1551. Available from https://doi.org/10.1249/MSS.0b013e31820e5adc.

Bowtell, J., & Kelly, V. (2019). Fruit-derived polyphenol supplementation for athlete recovery and performance. *Sports Medicine*, *49*(S1), 3–23. Available from https://doi.org/10.1007/s40279-018-0998-x.

Braakhuis, A. J., Hopkins, W. G., & Lowe, T. E. (2014). Effects of dietary antioxidants on training and performance in female runners. *European Journal of Sport Science*, *14*(2), 160–168. Available from https://doi.org/10.1080/17461391.2013.785597.

Burnett, P. J. M., & Willems, M. E. T. (2022). Effects of New Zealand blackcurrant extract on sequential performance testing in male rugby union players. *Sports (Basel, Switzerland)*, *10*(10), 152. Available from https://doi.org/10.3390/sports10100152.

Carey, C. C., Lucey, A., & Doyle, L. (2021). Flavonoid containing polyphenol consumption and recovery from exercise-induced muscle damage: A systematic review and meta-analysis. *Sports Medicine*, *51*(6), 1293–1316. Available from https://doi.org/10.1007/s40279-021-01440-x, http://rd.springer.com/journal/40279.

Caruana, M., Neuner, J., Högen, T., Schmidt, F., Kamp, F., Scerri, C., Giese, A., & Vassallo, N. (2012). Polyphenolic compounds are novel protective agents against lipid membrane damage by α-synuclein aggregates in vitro. *Biochimica et Biophysica Acta (BBA) – Biomembranes*, *1818*(11), 2502–2510. Available from https://doi.org/10.1016/j.bbamem.2012.05.019.

Chodari, L., Aytemir, M. D., Vahedi, P., Alipour, M., Vahed, S. Z., Khatibi, S. M. H., Ahmadian, E., Ardalan, M., & Eftekhari, A. (2021). Targeting mitochondrial biogenesis with polyphenol compounds. *Oxidative Medicine and Cellular Longevity*, *2021*. Available from https://doi.org/10.1155/2021/4946711, http://www.hindawi.com/journals/oximed/.

Clarkson, P. M., Nosaka, K., & Braun, B. (1992). Muscle function after exercise-induced muscle damage and rapid adaptation. *Medicine and Science in Sports and Exercise*, *24*(5), 512–520. Available from https://doi.org/10.1249/00005768-199205000-00004.

Close, G. L., Kasper, A. M., Walsh, N. P., & Maughan, R. J. (2022). "Food first but not always food only": Recommendations for using dietary supplements in sport. *International Journal of Sport Nutrition and Exercise Metabolism*, *32*(8), 371–386. Available from https://doi.org/10.1123/ijsnem.2021-0335, https://journals.humankinetics.com/view/journals/ijsnem/32/5/article-p371.xml.

Cook, M. D., Myers, S. D., Blacker, S. D., & Willems, M. E. T. (2015). New Zealand blackcurrant extract improves cycling performance and fat oxidation in cyclists. *European Journal of Applied Physiology*, *115*(11), 2357–2365. Available from https://doi.org/10.1007/s00421-015-3215-8, link.springer.de/link/service/journals/00421/index.htm.

Cureton, K. J., Tomporowski, P. D., Singhal, A., Pasley, J. D., Bigelman, K. A., Lambourne, K., Trilk, J. L., McCully, K. K., Arnaud, M. J., & Zhao, Q. (2009). Dietary quercetin supplementation is not ergogenic in untrained men. *Journal of Applied Physiology*, *107*(4), 1095–1104. Available from https://doi.org/10.1152/japplphysiol.00234.2009UnitedStates, http://jap.physiology.org/cgi/reprint/107/4/1095.

Davis, J. M., Carlstedt, C. J., Chen, S., Carmichael, M. D., & Murphy, E. A. (2010). The dietary flavonoid quercetin increases VO2max and endurance capacity. *International Journal of Sport Nutrition and Exercise Metabolism*, *20*(1), 56–62. Available from https://doi.org/10.1123/ijsnem.20.1.56, http://hk.humankinetics.com/eJournalMedia/pdfs/17824.pdf.

Diniz, C., Suliburska, J., & Ferreira, I. M. P. L. V. O. (2017). New insights into the anti-angiogenic and proangiogenic properties of dietary polyphenols. *Molecular Nutrition & Food Research*, *61*(6). Available from https://doi.org/10.1002/mnfr.201600912.

Eynon, N., Alves, A. J., Sagiv, M., Yamin, C., Sagiv, M., & Meckel, Y. (2010). Interaction between SNPs in the NRF2 gene and elite endurance performance. *Physiological Genomics*, *41*(1), 78–81. Available from https://doi.org/10.1152/physiolgenomics.00199.2009, http://physiolgenomics.physiology.org/cgi/reprint/41/1/78.

Ghosh, D., & Scheepens, A. (2009). Vascular action of polyphenols. *Molecular Nutrition & Food Research*, *53*(3), 322–331. Available from https://doi.org/10.1002/mnfr.200800182.

Gladden, L. B. (2004). Lactate metabolism: a new paradigm for the third millennium. *The Journal of Physiology*, *558*(1), 5–30. Available from https://doi.org/10.1113/jphysiol.2003.058701.

Gomez-Cabrera, M. C., Carretero, A., Millan-Domingo, F., Garcia-Dominguez, E., Correas, A. G., Olaso-Gonzalez, G., & Viña, J. (2021). Redox-related biomarkers in physical exercise. *Redox Biology*, *42*, 101956. Available from https://doi.org/10.1016/j.redox.2021.101956.

Gomez-Cabrera, M. C., Salvador-Pascual, A., Cabo, H., Ferrando, B., & Vina, J. (2015). Redox modulation of mitochondriogenesis in exercise. Does antioxidant supplementation blunt the benefits of exercise training? *Free Radical Biology and Medicine*, *86*, 37–46. Available from https://doi.org/10.1016/j.freeradbiomed.2015.04.006, http://www.elsevier.com/locate/freeradbiomed.

Howatson, G., Snaith, G. C., Kimble, R., Cowper, G., & Keane, K. M. (2022). Improved endurance running performance following haskap berry (Lonicera caerulea L.) ingestion. *Nutrients*, *14*(4). Available from https://doi.org/10.3390/nu14040780, https://www.mdpi.com/2072-6643/14/4/780/pdf.

Huang, Y., Li, W., Su, Z. Y., & Kong, A. N. T. (2015). The complexity of the Nrf2 pathway: Beyond the antioxidant response. *Journal of Nutritional Biochemistry*, *26*(12), 1401–1413. Available from https://doi.org/10.1016/j.jnutbio.2015.08.001, http://www.elsevier.com/locate/jnutbio.

Jówko, E., Sacharuk, J., Balasińska, B., Ostaszewski, P., Charmas, M., & Charmas, R. (2011). Green tea extract supplementation gives protection against exercise-induced oxidative damage in healthy men. *Nutrition Research*, *31*(11), 813–821. Available from https://doi.org/10.1016/j.nutres.2011.09.020.

Keane, K. M., Bailey, S. J., Vanhatalo, A., Jones, A. M., & Howatson, G. (2018). Effects of montmorency tart cherry (L. Prunus Cerasus) consumption on nitric oxide biomarkers and exercise performance. *Scandinavian Journal of Medicine and Science in Sports*, *28*(7), 1746–1756. Available from https://doi.org/10.1111/sms.13088, http://onlinelibrary.wiley.com/journal/10.1111/(ISSN)1600-0838.

Kruk, J., Aboul-Enein, B. H., Duchnik, E., & Marchlewicz, M. (2022). Antioxidative properties of phenolic compounds and their effect on oxidative stress induced by severe physical exercise. *Journal of Physiological Sciences*, *72*(1). Available from https://doi.org/10.1186/s12576-022-00845-1, http://www.springer.com/biomed/human + physiology/journal/12576.

Levers, K., Dalton, R., Galvan, E., O'Connor, A., Goodenough, C., Simbo, S., Mertens-Talcott, S. U., Rasmussen, C., Greenwood, M., Riechman, S., Crouse, S., & Kreider, R. B. (2016). Effects of powdered Montmorency tart cherry supplementation on acute endurance exercise performance in aerobically trained individuals. *Journal of the International Society of Sports Nutrition*, *13*(1). Available from https://doi.org/10.1186/s12970-016-0133-z, http://www.jissn.com/.

Lomiwes, D., Ha, B., Ngametua, N., Burr, N. S., Cooney, J. M., Trower, T. M., Sawyer, G., Hedderley, D., Hurst, R. D., & Hurst, S. M. (2019). Timed consumption of a New Zealand blackcurrant juice support positive affective responses during a self-motivated moderate walking exercise in healthy sedentary adults. *Journal of the International Society of Sports Nutrition*, *16*(1). Available from https://doi.org/10.1186/s12970-019-0300-0, http://www.jissn.com/.

Malaguti, M., Angeloni, C., & Hrelia, S. (2013). Polyphenols in exercise performance and prevention of exercise-induced muscle damage. *Oxidative Medicine and Cellular Longevity*, *2013*, 1–9. Available from https://doi.org/10.1155/2013/825928.

Malireddy, S., Kotha, S. R., Secor, J. D., Gurney, T. O., Abbott, J. L., Maulik, G., Maddipati, K. R., & Parinandi, N. L. (2012). Phytochemical antioxidants modulate mammalian cellular epigenome: Implications in health and disease. *Antioxidants and Redox Signaling*, *17*(2), 327–339. Available from https://doi.org/10.1089/ars.2012.4600.

Matsumoto, H., Takenami, E., Iwasaki-Kurashige, K., Osada, T., Katsumura, T., & Hamaoka, T. (2005). Effects of blackcurrant anthocyanin intake on peripheral muscle circulation during typing work in humans. *European Journal of Applied Physiology*, *94*(1-2), 36–45. Available from https://doi.org/10.1007/s00421-004-1279-y.

McCormick, R., Peeling, P., Binnie, M., Dawson, B., & Sim, M. (2016). Effect of tart cherry juice on recovery and next day performance in well-trained Water Polo players. *Journal of the International Society of Sports Nutrition, 13*(1). Available from https://doi.org/10.1186/s12970-016-0151-x.

McHugh, M. P. (2003). Recent advances in the understanding of the repeated bout effect: The protective effect against muscle damage from a single bout of eccentric exercise. *Scandinavian Journal of Medicine and Science in Sports, 13*(2), 88−97. Available from https://doi.org/10.1034/j.1600-0838.2003.02477.x.

Montanari, S., Blacker, S. D., & Willems, M. E. T. (2023). Acute effects of New Zealand blackcurrant extract on cycling time-trial are performance dependent in endurance-trained cyclists: A home-based study. *Sports, 11*(5), 93. Available from https://doi.org/10.3390/sports11050093.

Morgan, P. T., Barton, M. J., & Bowtell, J. L. (2019). Montmorency cherry supplementation improves 15-km cycling time-trial performance. *European Journal of Applied Physiology, 119*(3), 675−684. Available from https://doi.org/10.1007/s00421-018-04058-6, http://link.springer.de/link/service/journals/00421/index.htm.

Moss, S. L., Brindley, E., Enright, K., Highton, J., & Bott, R. (2023). The effects of an acute dose of New Zealand blackcurrant extract on 5-km running performance. *International Journal of Sport Nutrition and Exercise Metabolism, 33*(6), 323−330. Available from https://doi.org/10.1123/ijsnem.2023-0029, https://journals.humankinetics.com/view/journals/ijsnem/33/6/article-p323.xml.

Mthembu, S. X. H., Dludla, P. V., Ziqubu, K., Nyambuya, T. M., Kappo, A. P., Madoroba, E., Nyawo, T. A., Nkambule, B. B., Silvestri, S., Muller, C. J. F., & Mazibuko-Mbeje, S. E. (2021). The potential role of polyphenols in modulating mitochondrial bioenergetics within the skeletal muscle: A systematic review of preclinical models. *Molecules, 26*(9). Available from https://doi.org/10.3390/molecules26092791, https://www.mdpi.com/1420-3049/26/9/2791/pdf.

Myburgh, K. H. (2014). Polyphenol supplementation: Benefits for exercise performance or oxidative stress? *Sports Medicine, 44*(1), S57. Available from https://doi.org/10.1007/s40279-014-0151-4, http://rd.springer.com/journal/40279.

Nicol, L. M., Rowlands, D. S., Fazakerly, R., & Kellett, J. (2015). Curcumin supplementation likely attenuates delayed onset muscle soreness (DOMS). *European Journal of Applied Physiology, 115*(8), 1769−1777. Available from https://doi.org/10.1007/s00421-015-3152-6, http://link.springer.de/link/service/journals/00421/index.htm.

Nieman, D. C., Williams, A. S., Shanely, R. A., Jin, F., McAnulty, S. R., Triplett, N. T., Austin, M. D., & Henson, D. A. (2010). Quercetin's influence on exercise performance and muscle mitochondrial biogenesis. *Medicine and Science in Sports and Exercise, 42*(2), 338−345. Available from https://doi.org/10.1249/MSS.0b013e3181b18fa3.

Ortega, D. G., Coburn, J. W., Galpin, A. J., & Costa, P. B. (2023). Effects of a tart cherry supplement on recovery from exhaustive exercise. *Journal of Functional Morphology and Kinesiology, 8*(3). Available from https://doi.org/10.3390/jfmk8030121, http://www.mdpi.com/journal/jfmk.

Pandey, K. B., & Rizvi, S. I. (2009). Plant polyphenols as dietary antioxidants in human health and disease. *Oxidative Medicine and Cellular Longevity, 2*(5), 270−278. Available from https://doi.org/10.4161/oxim.2.5.9498India, http://www.landesbioscience.com/journals/oximed/article/PandeyOMCL2-5.pdf.

Pastellidou, E., Gillespie, E., McGrotty, A., Spence, J., McCloskey, W., Johnston, L., Wilson, J., & Kemi, O. J. (2021). Blackcurrant extract does not affect the speed−duration relationship during high-intensity running. *European Journal of Sport Science, 21*(4), 552−561. Available from https://doi.org/10.1080/17461391.2020.1771428, http://www.tandf.co.uk/journals/journal.asp?issn = 1746-1391&linktype = 1.

Patrizio, F., Ditroilo, M., Felici, F., Duranti, G., De Vito, G., Sabatini, S., Sacchetti, M., & Bazzucchi, I. (2018). The acute effect of Quercetin on muscle performance following a single resistance training session. *European Journal of Applied Physiology*, *118* (5), 1021–1031. Available from https://doi.org/10.1007/s00421-018-3834-y.

Peeling, P., & McKay, A. (2023). Iron regulation and absorption in athletes: Contemporary thinking and recommendations. *Current Opinion in Clinical Nutrition & Metabolic Care*, *26* (6), 551–556. Available from https://doi.org/10.1097/mco.0000000000000966.

Pizzino, G., Irrera, N., Cucinotta, M., Pallio, G., Mannino, F., Arcoraci, V., Squadrito, F., Altavilla, D., & Bitto, A. (2017). Oxidative stress: Harms and benefits for human health. *Oxidative Medicine and Cellular Longevity*, *2017*, 1–13. Available from https://doi.org/10.1155/2017/8416763.

Powers, S. K., Deminice, R., Ozdemir, M., Yoshihara, T., Bomkamp, M. P., & Hyatt, H. (2020). Exercise-induced oxidative stress: Friend or foe? *Journal of Sport and Health Science*, *9*(5), 415–425. Available from https://doi.org/10.1016/j.jshs.2020.04.001, http://www.elsevier.com/wps/find/journaldescription.cws_home/728044/description.

Powers, S. K., & Jackson, M. J. (2008). Exercise-induced oxidative stress: Cellular mechanisms and impact on muscle force production. *Physiological Reviews*, *88*(4), 1243–1276. Available from https://doi.org/10.1152/physrev.00031.2007, http://physrev.physiology.org/cgi/reprint/88/4/1243, UnitedStates.

Ribichini, E., Scalese, G., Cesarini, A., Mocci, C., Pallotta, N., Severi, C., & Corazziari, E. S. (2023). Exercise-induced gastrointestinal symptoms in endurance sports: A review of pathophysiology, symptoms, and nutritional management. *Dietetics*, *2*(3), 289–307. Available from https://doi.org/10.3390/dietetics2030021.

Roelofs, E. J., Smith-Ryan, A. E., Trexler, E. T., Hirsch, K., & Mock, M. G. (2017). Effects of pomegranate extract on blood flow and vessel diameter after high-intensity exercise in young, healthy adults. *European Journal of Sport Science*, *17*, 317–325. Available from https://doi.org/10.1080/17461391.2016.1230892.

Rudrapal, M., Khairnar, S. J., Khan, J., Dukhyil, A. B., Ansari, M. A., Alomary, M. N., Alshabrmi, F. M., Palai, S., Deb, P. K., & Devi, R. (2022). Dietary polyphenols and their role in oxidative stress-induced human diseases: Insights into protective effects, antioxidant potentials and mechanism(s) of action. *Frontiers in Pharmacology*, *13*. Available from https://doi.org/10.3389/fphar.2022.806470, http://www.frontiersin.org/Pharmacology.

Sandoval-Ramírez, B. A., Catalán, Ú., Pedret, A., Valls, R. M., Motilva, M. J., Rubió, L., & Solà, R. (2021). Exploring the effects of phenolic compounds to reduce intestinal damage and improve the intestinal barrier integrity: A systematic review of in vivo animal studies. *Clinical Nutrition*, *40*(4), 1719–1732. Available from https://doi.org/10.1016/j.clnu.2020.09.027, http://www.elsevier-international.com/journals/clnu/.

Schulz, K. F., Altman, D. G., Moher, D., & Consort. (2010). Statement: Updated guidelines for reporting parallel group randomised trials. *BMC Med*, *8*, 1–9, 2010.

Schwarz, N. A., Theodore, A. P., Funderburg, B. R., Waldhelm, A., McKinley-Barnard, S. K., & Hudson, G. M. (2020). Acute (-)-epicatechin consumption: Effects on local vasodilation following resistance exercise and high-intensity exercise performance. *Sports*, *8*(2). Available from https://doi.org/10.3390/sports8020022, https://www.mdpi.com/2075-4663/8/2/22/pdf.

Serreli, G., & Deiana, M. (2023). Role of dietary polyphenols in the activity and expression of nitric oxide synthases: A review. *Antioxidants*, *12*(1), 147. Available from https://doi.org/10.3390/antiox12010147.

Sharp, M. A., Hendrickson, N. R., Staab, J. S., Mcclung, H. L., Nindl, B. C., & Michniak-Kohn, B. B. (2012). Effects of short-term quercetin supplementation on soldier performance. *Journal of Strength and Conditioning Research*, *26*(2), S53. Available from https://doi.org/10.1519/JSC.0b013e31825cf22d.

Shigeta, M., Aoi, W., Morita, C., Soga, K., Inoue, R., Fukushima, Y., Kobayashi, Y., & Kuwahata, M. (2023). Matcha green tea beverage moderates fatigue and supports

resistance training-induced adaptation. *Nutrition Journal*, *22*(1). Available from https://doi.org/10.1186/s12937-023-00859-4.
Simioni, C., Zauli, G., Martelli, A. M., Vitale, M., Sacchetti, G., Gonelli, A., & Neri, L. M. (2018). Oxidative stress: Role of physical exercise and antioxidant nutraceuticals in adulthood and aging. *Oncotarget*, *9*(24), 17181–17198. Available from https://doi.org/10.18632/oncotarget.24729, http://www.oncotarget.com/index.php?journal = oncotarget&page = article&op = download&path%5B%5D = 24729&path%5B%5D = 77567.
Somerville, V., Bringans, C., & Braakhuis, A. (2017). Polyphenols and performance: A systematic review and meta-analysis. *Sports Medicine*, *47*(8), 1589–1599. Available from https://doi.org/10.1007/s40279-017-0675-5.
Strauss, J. A., Willems, M. E. T., & Shepherd, S. O. (2018). New Zealand blackcurrant extract enhances fat oxidation during prolonged cycling in endurance-trained females. *European Journal of Applied Physiology*, *118*(6), 1265–1272. Available from https://doi.org/10.1007/s00421-018-3858-3, http://link.springer.de/link/service/journals/00421/index.htm.
Su, L. J., Zhang, J. H., Gomez, H., Murugan, R., Hong, X., Xu, D., Jiang, F., & Peng, Z. Y. (2019). Reactive oxygen species-induced lipid peroxidation in apoptosis, autophagy, and ferroptosis. *Oxidative Medicine and Cellular Longevity*, *2019*. Available from https://doi.org/10.1155/2019/5080843, http://www.hindawi.com/journals/oximed/.
Tanabe, Y., Chino, K., Ohnishi, T., Ozawa, H., Sagayama, H., Maeda, S., & Takahashi, H. (2019). Effects of oral curcumin ingested before or after eccentric exercise on markers of muscle damage and inflammation. *Scandinavian Journal of Medicine & Science in Sports*, *29*(4), 524–534. Available from https://doi.org/10.1111/sms.13373.
Trexler, E. T., Smith-Ryan, A. E., Melvin, M. N., Roelofs, E. J., & Wingfield, H. L. (2014). Effects of pomegranate extract on blood flow and running time to exhaustion. *Applied Physiology, Nutrition and Metabolism*, *39*(9), 1038–1042. Available from https://doi.org/10.1139/apnm-2014-0137, http://pubs.nrc-cnrc.gc.ca/rp-ps/volumes.jsp?jcode = apnm&lang = eng&exp34.
Trombold, J. R., Barnes, J. N., Critchley, L., & Coyle, E. F. (2010). Ellagitannin consumption improves strength recovery 2-3 d after eccentric exercise. *Medicine and Science in Sports and Exercise*, *42*(3), 493–498. Available from https://doi.org/10.1249/MSS.0b013e3181b64edd.
Trombold, J. R., Reinfeld, A. S., Casler, J. R., & Coyle, E. F. (2011). The effect of pomegranate juice supplementation on strength and soreness after eccentric exercise. *Journal of Strength and Conditioning Research*, *25*(7), 1782–1788. Available from https://doi.org/10.1519/JSC.0b013e318220d992.
Willems, M. E. T., Myers, S. D., Gault, M. L., & Cook, M. D. (2015). Beneficial physiological effects with blackcurrant intake in endurance athletes. *International Journal of Sport Nutrition and Exercise Metabolism*, *25*(4), 367–374. Available from https://doi.org/10.1123/ijsnem.2014-0233, http://journals.humankinetics.com/AcuCustom/Sitename/Documents/DocumentItem/07%20Willems%20IJSNEM_2014_0233_367-374.pdf.
Woods, J. A., Wilund, K. R., Martin, S. A., & Kistler, B.M. Exercise, inflammation and aging.
Yahfoufi, N., Alsadi, N., Jambi, M., & Matar, C. (2018). The immunomodulatory and anti-inflammatory role of polyphenols. *Nutrients*, *10*(11), 1618. Available from https://doi.org/10.3390/nu10111618.
Yeo, W. K., Carey, A. L., Burke, L., Spriet, L. L., & Hawley, J. A. (2011). Fat adaptation in well-trained athletes: Effects on cell metabolism. *Applied Physiology, Nutrition and Metabolism*, *36*(1), 12–22. Available from https://doi.org/10.1139/H10-089.
Yoon, J. H., & Baek, S. J. (2005). Molecular targets of dietary polyphenols with anti-inflammatory properties. *Yonsei Medical Journal*, *46*(5), 585–596. Available from https://doi.org/10.3349/ymj.2005.46.5.585, http://www.eymj.org/abstracts/viewArticle.asp?year = 2005&month = 10&page = 585.

SECTION 3

Polyphenols against neurodegenerative and cognitive disorders

CHAPTER SIX

Polyphenol effects in pathologies related to cerebral stroke

Marina Romero-Bernal[1], Ángela González-Díaz[1], Ana M Najar[1,2], Carmen del Río[1] and Joan Montaner[1,2]

[1]Institute of Biomedicine of Seville (IBiS), Hospital Universitario Virgen del Rocío, CSIC, Universidad de Sevilla, Seville, Spain

[2]Department of Neurology, Hospital Universitario Virgen Macarena, Seville, Spain

Introduction

Cardiovascular diseases (CVs) are a range of disorders of the heart or blood vessels that encompass numerous afflictions, including coronary heart disease or cerebrovascular disease. Common risk factors are shared by several CVs, which allows the development of health policies that simultaneously target different CVs (Thiriet, 2019). According to the World Health Organisation (WHO), CVs are responsible for more than half of all deaths across Europe, with females accounting for more new cases than males. CVs are also the leading cause of death worldwide and significantly contribute to the cost of health systems globally (Timmis et al., 2022).

Stroke, which is the most common neurovascular disease, is caused by the focal and sustained disruption of blood supply to the brain, resulting in oxygen and nutrient deprivation. Only in 2019, there were 12.2 million new stroke cases and an estimated 143 million of disability-adjusted life years (DALYs) due to stroke. Stroke represented the second cause of death globally accounting for more than 6 million deaths (Collaborators, 2019/2019). Recently, annual stroke cost in the European countries was calculated in €60 billion, with healthcare accounting for €27 billion (Luengo-Fernandez et al., 2020).

Despite stroke mortality has declined in high-income countries during the last five decades concurrent with cardiovascular risk factor control interventions, the number of deaths have increased in many middle-income countries influenced by the aging of populations accompanied by

Bioactive Polyphenols for Health and Pathology Treatment.
DOI: https://doi.org/10.1016/B978-0-443-13821-8.00006-8

their urbanization and globalization (Lackland et al., 2014). It is widely accepted that a significant proportion of strokes can be attributed to unhealthy lifestyle. A multifactorial approach that includes dietary modification can result in a reduction of up to 80% of stroke recurrence. Therefore an optimized diet is a major strategy in controlling modifiable stroke risk factors (Kernan et al., 2014; Kleindorfer, 2021).

Here, we reviewed studies on polyphenols for the treatment of neurovascular diseases. Preclinical studies have reported that different polyphenols can ameliorate neuron loss and reduce infarct volume, and epidemiological studies suggested an inverse relationship between polyphenol intake and cardiovascular risk factors. The wide range of health benefits of polyphenols has drawn the attention of the food industry and many phenolic compounds are commercially available as dietary supplements with different health claims. The direct and indirect neuroprotective effects of polyphenols support their use as prophylactic substances or as therapeutic molecules in different types of neurovascular diseases.

Stroke

The Stroke Council of the American Heart Association (AHA)/ American Stroke Association (ASA) published an updated definition of stroke in 2013. The new stroke definition incorporates clinical and tissue criteria and emphasizes central nervous system (CNS) infarction, defined as brain, spinal cord or retinal cell death attributable to ischemia, based on pathological, neuroimaging or other objective and/or clinical evidence with symptoms persisting ≥ 24 h or until death, and excluding other aetiologies. CNS infarction includes, according to the clinical spectrum, symptomatic (ischemic stroke) and asymptomatic CNS lesion (silent infarction), as well as intracerebral hemorrhage and subarachnoid hemorrhage (Sacco et al., 2013).

Ischemic stroke accounts for approximately 85% (Khaku & Tadi, 2023) of all strokes and is caused by obstruction of blood flow to the brain by a blood clot. The remaining 15% of strokes are of the hemorrhagic type, in which the rupture of a cerebral blood vessel impedes blood flow to a brain region (Unnithan et al., 2023). When a transient episode of neurologic dysfunction occurs, it is known as transient ischemic attack

(TIA). TIAs have been long time considered as any focal cerebral ischemic event with symptoms lasting less than 24 h. However, recent studies using clinical observation and modern brain imaging have shown that the duration and reversibility of cerebral ischemia are variable among patients. Because of this, it is now considered that time should not be the main distinguishing factor between stroke and TIA (Easton et al., 2009; Sacco et al., 2013).

A system for categorization of subtypes of ischemic stroke mainly according to etiology has been developed based on the Trial of Org 10172 in Acute Stroke Treatment (TOAST). The TOAST classification includes five subtypes of ischemic stroke (Adams et al., 1993) (1) large-artery atherosclerosis: related to a stenosis of more than 50% of a large caliber extracranial or intracranial artery or to a stenosis of less than 50% associated with the presence of cardiovascular risk factors; (2) cardioembolism: caused by emboligenic heart disease such as atrial fibrillation or flutter, among others; (3) small-vessel occlusion: infarction smaller than 1.5 cm in the territory of a perforating artery, clinically manifested by a typical lacunar syndrome; (4) stroke of other determined etiology: due to arterial dissections, vasculitis, prothrombotic conditions, etc.; (5) stroke of undetermined etiology (Sobrino García et al., 2013).

During ischemic stroke cerebral blood flow is reduced, which limits the bioavailability of glucose and oxygen to a brain area (Fig. 6.1). This causes a disruption in ATP synthesis and energy deficiency that initiates a cascade of cellular events affecting specially neurons (Campbell, 2019/2019). Energy imbalance leads to mitochondrial dysfunction and oxidative stress-induced damage. Increased production of mitochondria-derived reactive oxygen species (ROS) exacerbates cellular damage (Datta et al., 2020). Consequently, neurons cannot maintain the membrane potential and neuronal signaling is impaired. To compensate for detrimental effects, excessive secretion of the excitatory neurotransmitters glutamate and aspartate occurs, which activates AMPA, NMDA, and kainate receptors, resulting in $Ca2+$ and $Na+$ overload in mitochondria and cytosol, further contributing to impaired ion homeostasis, ROS production, and mitochondrial permeability transition pore (MPTP) opening. This process results in the activation of proteases and the dysfunction of the blood-brain barrier (BBB) resulting in an increased permeability that contributes to local inflammation, causing brain edema (Datta et al., 2020). When cerebral circulation is restored after a period of occlusion, tissue re-oxygenation leads to reperfusion injury resulting in paradoxical extensive

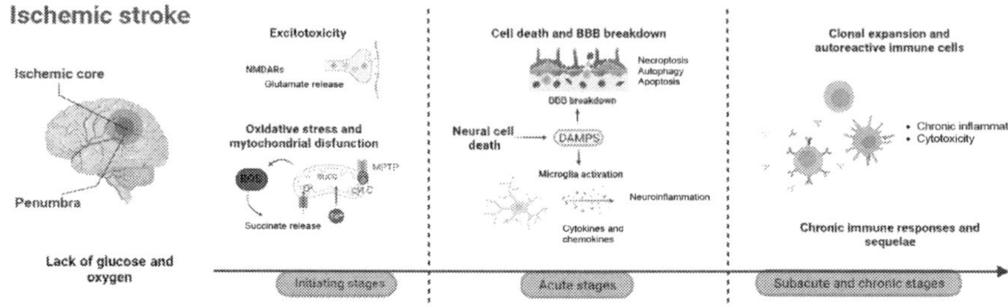

Figure 6.1 Pathophysiological mechanisms involved in ischemic stroke. Ischemic core undergoes irreversible damage mainly caused by necrotic cell death within minutes while the penumbra region is characterized by a reversible loss of homeostasis. During the initial stages, excitotoxicity and oxidative stress contribute to Ca2 + influx in mitochondria, MPTP opening, and succinate release, while changes in neuronal signaling produce excessive excitatory neurotransmitters. Consequently, cell death occurs during the acute stages and the release of damage-associated molecular patterns (DAMPs) promotes BBB breakdown and microglia activation, leading to neuroinflammation. Chronic immune responses produced by autoreactive immune cells contribute to ischemic stroke sequelae.

tissue destruction. Tissue damage is irreversible in the core ischemic region, where necrosis mostly takes place, and neurons promptly die due to hypoperfusion. Within the penumbra, where the cellular homeostasis is altered and mainly apoptosis takes place, tissue damage can be reversible if rapid reperfusion is achieved (Datta et al., 2020).

Cerebroprotective effect of dietary polyphenols

Polyphenol daily intake is estimated to be 1 g/day, but its bioavailability is limited by poor absorption, rapid metabolism, and massive system elimination (Di Lorenzo et al., 2021; Perrone et al., 2020). Unabsorbed polyphenols have been estimated at 90%–95% of the ingested dose. Most of them are present in food in the form of esters, glycosides, or polymers that resist acidic conditions during gastric digestion and cannot be absorbed in their native form. They are orally ingested and once in the gut, they must be hydrolyzed into a wide diversity of metabolites, which can cross the intestinal barrier or undergo first-pass metabolism in the liver and phase II enzymatic conversion, such as methylation, sulfation, and glucuronidation. These modifications enable polyphenol absorption but make them more hydrophilic, and less likely to reach the central nervous system. Polyphenols are finally eliminated by the biliary and urinary systems (Arfaoui, 2021; Manach et al., 2004; Williamson, 2017).

The endothelial transport of polyphenols through the BBB could be influenced by chemicals modifications, ability to establish hydrogen exchanges, or their capacity to bind serum albumin. Transmembrane diffusion of some polyphenol metabolites in vitro is related to their lipophilicity, where less polar is related to better uptake. Nevertheless, it is not clear whether the primary route by which polyphenol metabolites cross the BBB is by simple diffusion or by carrier-mediated transport. The presence of efflux transporters in BBB was confirmed in HBMEC cells and these ABC-type transporters may contribute to overall BBB transport of the polyphenol metabolites (Figueira et al., 2017). Other studies have also highlighted the involvement of Breast Cancer Resistance Protein (BCRP) in limiting polyphenol access to the brain (Youdim et al., 2003). It should be noted that the BBB is a complex cellular structure composed not only by endothelial cells but also by astrocytes and pericytes which can also influence BBB transport, so further studies are needed to study the exact mechanisms by which circulating phenolic metabolites interact with the BBB (Figueira et al., 2017).

Direct action of polyphenols on CNS function related to stroke

Effect of polyphenols on oxidative stress

Several studies pointed out that a dietary regimen abundant in polyphenols, such as the Mediterranean diet, is correlated with a diminished occurrence of vascular disorders, including atherosclerosis and ischemic stroke (Martínez-González et al., 2019). Polyphenols have been documented to possess the capability of modulating different mechanisms associated to ischemia, thereby mitigating the deleterious effects caused by this condition. The way in which polyphenols attenuate the damage induced by ischemia varies depending on the specific pathway (Pacifici et al., 2021).

One of the primary attributes of polyphenols lies in their considerable antioxidant capacity (Nani et al., 2021). This phenomenon confers notable advantages given that oxidative stress is recognized as an important process in the pathological cascade during ischemia (Orellana-Urzúa et al., 2020). Many polyphenols exhibit scavenging activity toward free radicals and can directly target the reduction of ROS, such as rutin (Nassiri-Asl et al., 2020), taxifolin (Turovskaya et al., 2019), vanillic acid (Tai et al., 2012), chrysin (Jian-Bin et al., 2005), naringin (Jinghan Feng et al., 2018), and catechin (Bernatoniene & Kopustinskiene, 2018). The strength of the antioxidant effect depends on their structural properties, including the number of hydroxyl groups (Platzer, 2022). In addition, certain polyphenols such as gallic acid (Sun et al., 2014), caffeic acid (Liang et al., 2015), ferulic acid (Ren et al., 2017), and P-coumaric acid (Guven et al., 2015) have been shown to reduce levels of malondialdehyde (MDA), directly contributing to the mitigation of oxidative stress-induced damage in addition to induce an upregulation of superoxide dismutase (SOD) levels.

Another molecular target for polyphenols is the Nuclear factor erythroid 2-related factor 2 (Nrf-2), a redox-sensitive transcription factor that modulates the cellular defense pathways to promote survival under stress conditions. In this context, certain polyphenols, including quercetin (Li et al., 2021), chlorogenic acid (Fan et al., 2022), kaempferol (Yuan et al., 2021), epicatechin (Leonardo et al., 2013), myricetin (Zhang et al., 2019), phloretin (Yang et al., 2019) and isorhamnetin (Wu et al., 2021) exert their effects by enhancing the activity of this transcription factor and its subsequent downstream signaling cascades. Remarkably, the activation of Nrf-2 exhibits an inverse correlation with inflammation, constituting yet another mechanism underlying its protective effect on cerebral ischemia. This interplay arises from the fact that Nrf2 deletion leads to increased Nuclear factor kappa B (NFκB) activation. Conversely, an upregulation of Nrf-2 expression effectively inhibits the pro-inflammatory response mediated by NFκB (Qin et al., 2022).

It has been shown that NFκB can regulate the expression of genes involved in controlling the transport of glucose (Zha et al., 2015). Cerebral cells, particularly neurons, require a constant supply of glucose to execute their functions optimally. Initially, hypoxia impairs the mTOR pathway, thereby impeding the degradation of HIF-1α. Consequently, the accumulation of HIF-1α facilitates a transient increase in GLUT-3 glucose transporter expression within neurons in an attempt to improve glucose utilization within neurons (Wu et al., 2012). However, due to the lack of oxygen during ischemia, glucose is metabolized anaerobically, leading to the production of lactate and subsequent acidosis (Frerichs et al., 1990). When reperfusion occurs, and oxygen becomes available again, the excessive influx of glucose into the cells can lead to high-glucose-induced inflammation and cytotoxicity (Lindsberg & Roine, 2004). Some polyphenols, such as epigallochatechin gallate and epichatechin gallate have been shown to decrease glucose uptake in vitro, and resveratrol reduced the ischemia-induced upregulation of GLUT-3 (Aguilar et al., 2020).

Polyphenols and brain-derived neurotrophic factor

Brain Derived Neurotrophic Factor (BDNF) is the most abundant neurotrophin in the adult brain. BDNF is known to repair brain damage and thereby plays a crucial role in stroke prognosis, pathogenesis, and rehabilitation (Liu et al., 2020). It has been shown that low levels of circulating BDNF could be associated with increased risk of suffering a stroke or poor recovery (Bejot, 2011; Béjot et al., 2011; Stanne et al., 2016). Besides, in animal models of stroke BDNF reduces cell death and maintains the survival of neurons by upregulating Trk, PI3K/Akt, and MAPK/ERK pathways (Kubo et al., 1995) or by inhibiting apoptosis (Han et al., 2000; Jantas et al., 2009; Zhang et al., 2011). It has also been reported that BDNF release is able to enhance neuronal resistance to glutamate toxicity in cultured neurons (Favaron et al., 1993). BDNF may also contribute to sustained regenerative signaling at synaptic sites, or increase the expression of markers for synaptogenesis and axonal sprouting, contributing to neuronal regeneration after stroke (Bamji et al., 2006; Horch et al., 1999; Lessmann et al., 2003).

In fact, some neuroprotective approaches against stroke to indirectly increase endogenous BDNF production include the administration of antioxidants. Several polyphenols have been reported to have neuroprotective effects through the production of BDNF. Their effects may be due to the prevention of the of CREB and BDNF/TrkB downregulation signaling caused by neurotoxic substances (Ding et al., 2017) or the activation of both PI3k/Akt and MAPK/ERK, involving neuroregeneration and antiapoptotic effects (Hwang et al., 2012; Tsai et al., 2013). By enhancing these two molecular pathways, the natural isoflavonoid puerarin stimulated BDNF expression in astrocytes and suppressed apoptosis when administered after ischemia-reperfusion in rats (Wang et al., 2014). Chlorogenic acid has been shown to mitigate brain damage and reduce nerve injury while increasing BDNF in rats with cerebral ischemia-reperfusion (Liu et al., 2020). Curcumin or resveratrol, among others, showed to reverse BDNF downregulation after nerve injury and exerted neuroprotection in different neurodegenerative disease models (Ali et al., 2015; Jin et al., 2022; Wu et al., 2006). Overall, polyphenol neurotrophic action is complex and further studies on the exact molecular mechanisms of these compounds may help to develop new neuroprotective strategies.

Molecular mechanisms of polyphenol on inflammation

Neuroinflammation after stroke is characterized by activation of both, resident (microglia, astrocytes) and invading (macrophages and leukocytes) inflammatory cells through different pathways (Pacifici et al., 2021; Qin et al., 2022). In the infarct area of mice after experimental stroke, activated microglia take a ramified shape in the border zone and ameboid phenotype in the ischemic core with the ability to synthesize cytokines mediators that attract other inflammatory cells (Jayaraj et al., 2019). This phenotype shifts over the time toward a proliferative phagocytic state (Hernandez et al., 2023). On the other hand, astrocytes acutely respond to ischemia by entering a proinflammatory phenotype characterized by a molecular footprint that persists for days after ischemia (Hernandez et al., 2023). The release of proinflammatory cytokines, such as TNFα and IL-1β and the enhanced oxidative stress contribute to neuronal death and a reduction in neurogenesis (Pacifici et al., 2021). Microglia and astrocytes are shown to have a double role and also promote neuroprotection by phagocyting damaged neurons and producing neurotrophic factors such as BDNF, respectively (Jayaraj et al., 2019).

Polyphenols may exert antiinflammatory effects by their radical scavenging properties and the regulation of cellular and enzymatic activities in inflammatory cells. Their mechanisms are diverse but they predominantly converge in a downregulation of MAPK, NF-κB and arachidonic pathways (Arulselvan et al., 2016; Yahfoufi et al., 2018). Some polyphenols, such as quercetin, resveratrol or procyanidins, or high-polyphenols products, have also been reported to reduce C-reactive protein (CRP) levels in vitro (Kaur et al., 2007) or in vivo (Terra et al., 2009), but also in patients with a clinical condition (Nemzer et al., 2011; Sarkhosh-Khorasani & Hosseinzadeh, 2021). Other polyphenols of the flavonoid subgroup have been shown to reduce the expression of proinflammatory cytokines and chemokines (IL-6, TNFα, IL-1β, IL-2, MCP-1) in macrophages and astrocytes (Comalada et al., 2006; Sharma et al., 2007). Similarly, some studies have established a role for resveratrol in the regulation the polarization of T lymphocytes and microglia from a proinflammatory to an antiinflammatory phenotype, reducing inflammatory damage in vitro and in vivo (Pacifici et al., 2021; Yang et al., 2017). In the same way, epigallocatechin-3-gallate from green tea and a proanthocyanidin extract showed immunomodulatory roles (Wang et al., 2013). Accordingly, several polyphenol-rich extracts showed to downregulate the levels of proinflammatory cytokines such as IL-1β, IL-6 or TNF and upregulate antiinflammatory factors such as IL-10 (Abulizi et al., 2023; Hong, 2020).

Following ischemic injury, arachidonic acid (AA) is liberated by membrane phospholipids upon phospholipase A2 (PLA2) cleavage. Cyclooxygenase (COX) and lipoxygenase can produce inflammatory mediator known as eicosanoids by AA metabolism to produce prostaglandins, thromboxane A2, and hydroxyeicosatetraenoic acids and leukotrienes, respectively (Yahfoufi et al., 2018). Some polyphenols, such as oleuropein, hydroxytyrosol, and tyrosol, among others, have been reported to inhibit inflammation by blocking the production of eicosanoids (Larussa et al., 2017; Rosignoli et al., 2013).

Indirect impact of polyphenols on brain health

Antithrombotic effect of polyphenols

Molecular mechanisms by which dietary polyphenols may reduce thrombotic risk are not fully understood. Polyphenols have been reported to play a role in thrombosis by the inhibition of platelet activation and platelet microparticles release, the reduction of ROS, enhancement of nitric oxide production, and binding to thromboxane A2 receptors (Nignpense, 2019/2019). Some in vitro studies have shown an antithrombotic by decreasing platelet aggregation induced by ADP when incubating whole blood with polyphenolic extracts (B.E. Nignpense et al., 2020; Tresserra-Rimbau et al., 2014) or washed platelets with chlorogenic acid (Bejot, 2011). Similarly, grape seed extracts containing a mixture of polyphenols inhibited thrombin and thrombin receptor activation peptide (TRAP)-induced platelet aggregation, the expression of surface receptors, and the formation of microparticle and superoxide anion radicals (Olas et al., 2012). Some specific polyphenols have also been analyzed for their antithrombotic effect. Hesperetin was reported to inhibit platelet aggregation induced by collagen and AA by interfering with COX-1, an intracellular enzyme responsible for catalyzing the conversion of AA to thromboxane A2, a potent aggregating agent (Jin et al., 2007). Gallic acid has also demonstrated to inhibit platelet aggregation by reducing intracellular Ca2 + concentration (Chang et al., 2012). And resveratrol has shown multiple effects on blood platelet activation (Olas et al., 2012). Interestingly, a comparative study between hydroxytyrosol acetate (HT-AC), a polyphenol in olive oil, and aspirin, an antiplatelet drug, has revealed similar effects in inhibiting platelet aggregation (Sarkhosh-Khorasani & Hosseinzadeh, 2021).

In animal mouse models of hypertension, it has been shown that anthocyanines attenuated thrombin-induced platelet activation and ATP secretion (Yao et al., 2017) and red wine polyphenols, such as proanthocyanins, catechins, flavonols, and tannins, were found to prevent the increase of circulating microparticles, inflammation, oxidative stress, and endothelial dysfunction (López Andrés et al., 2012). Studies in humans have revealed an antiaggregating platelet action of polyphenol-rich extracts, such as cocoa (Flammer et al., 2007) or hydroxyyrosol-rich extract, which produced its effect by reducing serum thromboxane production (Léger et al., 2005).

Despite there is evidence from different studies that dietary polyphenols may improve cardiovascular health by their antiplatelet properties, better studies are needed to further understand the exact mechanisms by which they could exert antithrombotic effects in stroke.

Polyphenols in the regulation of vascular function

Vascular aging is regarded as the primary non-modifiable risk factor for the development of CV. High levels of ROS, among other factors, have been associated with the generation of senescent cells, which secrete pro-inflammatory molecules and create a state of chronic inflammation pertinent to vascular aging (De Silva & Faraci, 2020). As a consequence of age-related vascular inflammation, the functional deterioration of certain proteins, including endothelial Nitric Oxide Synthase (eNOS), promotes endothelial dysfunction paving the way for vascular diseases (Ungvari et al., 2018). Certain polyphenols, such as resveratrol (Leikert et al., 2002), quercetin (Shen et al., 2012), and proanthocyanidins (Feng et al., 2010), can enhance the expression of eNOS. Consequently, the augmented availability of NO by polyphenol treatments may contribute to a significant improvement in vascular function.

NO generated from eNOS maintains endothelial homeostasis and regulates vascular tone. As an essential gas signal molecule, the decrease in NO blood levels diminishes arterial vasodilation and contributes to the development of high blood pressure. It is worth noting that hypertension is the main risk factor implicated in the development of an ischemic stroke (Leong et al., 2015). Synringic acid, ferulic acid, gallic acid, and p-coumaric acid have been demonstrated to function as vasodilators, thereby preserving the optimal level of NO in blood (Huang et al., 2021).

The release of angiotensin II, facilitated by the renin-angiotensin-aldosterone system, also represents physiological mechanisms involved in development of hypertension (Atlas, 2007). In this sense, quercetin has been demonstrated to reduce blood pressure in rats by attenuating the operation of the renin-angiotensin-aldosterone system reducing the transcription of the angiotensin-converting enzyme (ACE) (Marunaka et al., 2017). The same effect on ACE inhibition was reported for resveratrol (Jang et al., 2018), cinnamic acid (Mnafgui et al., 2015), caffeic acid and its derivatives (Bhullar et al., 2014). Moreover, a structure—activity relationship for ACE inhibition has been suggested for flavonoids (Guerrero et al., 2012).

Antioxidant molecules, like polyphenols, have also been suggested to exert beneficial effects on the vascular BBB. This structure is composed of pericytes, astrocytes, and endothelial cells and constitutes a pivotal position separating the brain from the vasculature, playing a fundamental role in maintaining the homeostasis of the central nervous system. Following an ischemic injury, this complex structure starts to deteriorate, leading to a compromised performance of its essential functions (Jiang et al., 2018). A multitude of polyphenols have been the subject of in vitro studies, seeking to determine their therapeutic efficacy in restoring the functional integrity of BBB. Ferulic acid (Huang et al., 2022) and quercetin (Li et al., 2021) exemplify such polyphenols and demonstrate to protect endothelial cells against ischemic injury. Furthermore, certain polyphenols, such as apigenin, may contribute to the preservation of the BBB integrity by stimulating angiogenesis (Tu et al., 2017).

Additionally, it must be noted that the integrity of the BBB is also influenced beyond ischemic damage. Notably, the gut microbiota exerts an indirect, yet significant, impact on the functional propriety of this structure (Braniste et al., 2014).

Polyphenols and gut microbiota in ischemic stroke

Alterations in the composition of the gut microbiota have been associated with fluctuations in BBB permeability, which highlight the complex interplay between gut health and neurological function (Braniste et al., 2014). Moreover, gut microbiota can directly impact thrombosis by regulating factors related to platelet function and hemostasis (Jäckel et al., 2017). In the opposite way, several neurological diseases, including stroke, have been shown to cause a dysbiosis in gut microbiota (Jemimah et al., 2023; Xia et al., 2019). In fact, changes in microbiome composition after experimental stroke can influence the subsequent immune response (Singh et al., 2016).

It is known that most dietary polyphenol intake remains unabsorbed and is therefore metabolized by the gut microbiota to produce active metabolites, which at the same time regulate the composition of the host microbiota (Wang et al., 2022). In this sense, specific polyphenols can alter the composition of the gut microbiota. Cyanidin-3-O-glucoside (Tu et al., 2022), cyanidin, and delphinidin (Cremonini et al., 2019) have been noticed to exert regulatory influences over the constitution of the gut microbiota, in addition to influencing the production of microbial metabolites. Hesperetin supplementation in rats resulted in a significant increase in starch excretion in the feces and decreased the proportion of *Clostridium* subcluster XIVa probably through the stimulation of Short Chain Fatty Acid (SFCA) production (Unno et al., 2015) and persimmon tannin administration by gavage modified the *Firmicutes/Bacteroidetes* ratio and *Bacteroidetes/Proteobacteria* ratio in control and high cholesterol rats (Zhu et al., 2018).

Actually, the beneficial effects of polyphenols despite their poor bioavailability may be explained by the production of many polyphenol metabolites with biological activity. One of the most abundant metabolites secreted by the microbiota is trimethylamine N-oxide (TMAD). High levels of these metabolites have been suggested to be linked to an increased risk of developing an ischemic stroke (Roberts et al., 2018; Zhu et al., 2016). Caffeic acid, catechin, and epicatechin, among other compounds, reduced TMAD formation in vitro (Iglesias-Carres et al., 2022) and resveratrol, have been demonstrated to decrease the secretion of this metabolite in vivo, and therefore, potentially mitigate the risk of clot formation leading to ischemic stroke (Annunziata et al., 2019; Chen, Yi, et al., 2016).

Therefore the gut—brain axis constitutes a highly significant factor in the development of neurological pathologies such as ischemic stroke. This interaction encompasses all previously mentioned factors, along with pathways implicated in hormonal regulation and the functioning of the autonomic nervous system (Peh et al., 2022).

Preclinical studies

Two model types of experimental stroke are used for the study of new therapies for brain ischemia, mainly in mice and rats. Among them, the most common method used is the focal ischemia model by the transient occlusion of the middle cerebral artery, aimed at mimicking the early reperfusion after stroke. Many polyphenols have been studied for their effect in brain ischemia-reperfusion *in vivo*. The majority of them have been shown to decrease the infarcted area within the brain, as well as improve cognitive functions, when administered both prior to ischemia, as is the case with caffeic acid (Zhou et al., 2006), vanillic acid (Wang et al., 2018), or Kaempferol (Wang et al., 2020) and after ischemic event, like chlorogenic acid (Shah et al., 2021), epicatechin (Shah et al., 2010), or ferulic acid (Ren et al., 2017). Moreover, several polyphenol-enriched plant extracts, which contain different compounds that can act synergistically, were also reported to exhibit promising disease-modifying activities (García-Rodríguez et al., 2022; Ko, n.d.; Rabiei et al., 2012).

One of the defining characteristics of polyphenols is their potent antioxidant capacity (Rana et al., 2022). Accordingly, numerous studies have been conducted wherein the administration of polyphenols such as rutin (Khan et al., 2009), myricetin (Sun et al., 2018), or epicatechin (Leonardo et al., 2013) prior to the ischemic event considerably mitigate the oxidative stress induced by experimental stroke such as the production of ROS or the protein carbonylation. In addition, supplementation with polyphenols also exerts beneficial effects against the inflammation triggered by oxidative stress in experimental stroke. Caffeic acid (Liang et al., 2015), myricetin (Sun et al., 2018), and chrysin (Yao et al., 2014) were shown to directly downregulate NF-κB expression in the brain and when administered prior to ischemia, resulting in a reduction of proinflammatory cytokines release. Moreover, polyphenols can also inhibit the activation of proinflammatory cells in the brain after stroke in animal models (Liao et al., 2023; Yao et al., 2014).

In addition to their antioxidant and antiinflammatory effects, polyphenols can behave as anti-apoptotic agents. Most cells within the penumbra area die by apoptosis gradually and are potentially salvageable (Uzdensky, 2019). Some polyphenols including gallic acid (Sun et al., 2014), ellagic acid (Liu et al., 2017), naringin (Yang et al., 2020), or rutin (Khan et al., 2009) significantly decreased cell apoptosis in the infarcted area through various signaling pathways such as p53 inhibitor or the regulation of Bcl-2/Bax expression ratio.

Given that the maintenance of the functionality, and hence the structure, of the BBB is crucial to containing the damage inflicted by ischemia (Jiang et al., 2018), studies have been conducted to evaluate the effect of polyphenols on the preservation of BBB integrity. Both gallic acid

(Mirshekari Jahangiri et al., 2020) and quercetin (Jin et al., 2019) have been shown to enhance BBB function when administered prior to ischemia. Similarly, apigenin (Pang et al., 2018) and isorhamnetin (Zhao et al., 2016) exert the same effect when administered after the ischemic injury.

The effects of polyphenols on different disorders considered as risk factors for the development of vascular disease have also been analyzsed (Otto, 2016). Obesity constitutes one of the primary modifiable risk factors for the development of CV (Koenen et al., 2021). Certain polyphenols, such as caffeic acid, have been demonstrated to decrease body weight in mice. Furthermore, the pathology of obesity is frequently associated with an unhealthy lipid profile, and this polyphenol has also been shown to improve lipid profile parameters (Xu et al., 2020). The management of diabetes, which is considered a risk factor for experiencing an ischemic stroke (Bruno, 2022), has been also addressed through the treatment with polyphenols. Resveratrol and capsaicin alleviated the effects of diabetes in streptozotocin-nicotinamide-induced diabetic rats and in genetically obese mice, respectively, by reducing both blood glucose levels and glycated hemoglobin (Kang et al., 2011; Palsamy & Subramanian, 2008). Other effects of dietary polyphenols on neurovascular health include the regulation of blood pressure (Grosso et al., 2022) and atherosclerotic plaque formation (Loke et al., 2010).

In summary, polyphenols have proven to be an effective strategy for both the prevention and treatment of ischemic stroke and stroke-associated risk factors. However, the majority of in vivo studies are conducted using a prophylactic therapeutic approach that complicates the translation to human, in which most studies are restricted to the context of secondary prevention at the moment.

Evidences from human clinical and epidemiological studies

After stroke, many survivors will have long-lasting cognitive and functional deficits (Bellone et al., 2019). The use of polyphenols is a promising neuroprotective therapeutic approach because they modulate key processes in the secondary injury cascade, such as inflammation and oxidative stress, influencing a wide range of molecular events that contribute to recovery (Waters et al., 2021).

Epidemiological studies using nutritional questionnaires link higher polyphenol intake to the incidence of cardiovascular events, such as stroke. Cassidy et al. conducted a retrospective study of 69,622 women during which they performed questionnaires regarding lifestyle and medical history at baseline and follow-up visits. Over 14 years of follow-up, high flavanone intake was associated with a 19% lower risk of ischemic stroke, a finding that fits with existing data on the protective effect of citrus fruit consumption (Cassidy et al., 2012). On the other hand, Tresserra-Rimbau et al. followed for an average of 4.3 years 7,172 participants and revealed that higher intakes of polyphenols, especially lignans, flavonols and hydroxybenzoic acids, reduced the risk of CV by 46% (Tresserra-Rimbau et al., 2014). Extensive preclinical research supports a beneficial role of polyphenols in animal models of stroke, but the number of human trials with polyphenols remains limited (Bayes et al., 2023).

Clinical trials have been conducted to study the effect of specific polyphenols or polyphenol-rich plant extracts on cerebrovascular health. All the studies reviewed are placebo or active treatment controlled, which adds to the scientific quality. In general, these studies aim to improve the prognosis of patients in the acute (<2 weeks after stroke) and sub-acute (3–11 weeks after stroke) phase (Wu et al., 2015) of stroke either on their own or as adjunctive therapy to other classical therapies. Currently, reperfusion with thrombolytic recombinant tissue plasminogen activator (rt-PA) is the most effective therapeutic option of treatment against acute brain ischemic stroke, although it needs to be applied within 3 h after the symptoms onset to be safe and effective (thus making their clinical use extremely difficult and resulting in the majority of patients remaining untreated) (Qin et al., 2020). Excess or delayed rt-PA treatments beyond the initial 3 h were more likely to cause adverse responses as symptomatic intracerebral hemorrhage (SIH) or hyperperfusion. These adverse events lead to extracellular matrix (ECM) degradation and via upregulation of matrix metalloproteinases (MMPs), such as MMP-2 and MMP-9, as well as BBB disruption (Chen, Bai, et al., 2016). Hence, suppression of MMP upregulation might help to promote ideal therapeutic outcomes.

Several double-blind, randomized, and placebo-controlled trials revealed the benefits of simultaneous intravenous administration of rt-PA and polyphenols. Combined treatment of rt-PA with hesperidin (4 mg/kg body weight; 400 mg maximum; intravenously (i.v.)) (Qin et al., 2020), resveratrol (2.5 mg/kg body weight; maximum 250 mg; i.v.) (Chen, Bai, et al., 2016), fisetin (100 mg; i.v.) (Wang et al., 2019) and epigallocatechin gallate (Wang & You, 2017) (EGCG; 500 mg; i.v.) down-regulated serum levels of MMP-2 and MMP-9 as well as improved treatment outcomes of stroke patients with respect to neurological deficits according to the National Institutes of Health Stroke Scale (NIHSS). A positive correlation was detected between reductions in both MMPs and patient NIHSS scores (Wang & You, 2017). Also, hesperidin co-treatment enhanced middle cerebral artery blood flow according to transcranial Doppler ultrasonography (TCD) scores and up-regulated serum concentrations of transforming growth factor (TGF)-β1, which reportedly promotes ECM reconstruction and production, and inhibits MMP expression (MMP-9 in particular). On the other hand, fisetin co-treatment reduced levels of CRP in the serum, there highlighting a strong linear correlation between changes in the NIHSS scores and the CRP serum levels too. Recapping polyphenol co-treatment extends the clinical therapeutic window of r-tPA (Chen, Bai, et al., 2016; Wang & You, 2017), improves the treatment outcomes and reduces the occurrence of adverse events, especially in patients receiving late stroke treatment. Moreover, after initial administration, continued short-term (seven days) (Qin et al., 2020; Wang et al., 2019) and long-term (two months) (Qin et al., 2020) treatment with polyphenols improves the results of initial rt-PA therapy as indicated by TCD, Glasgow Outcome Scale, the Modified Rankin Scale (MRS) (Qin et al., 2020) and NIHSS (Qin et al., 2020; Wang et al., 2019).

Bellone et al. reported that acute/sub-acute stroke patients, after taking 1 g of polyphenol-rich extracts derived from whole pomegranate for one week, significantly improved their independence (improved caregiving and self-motivation according to the Functional Independence Measure scale) and were discharged earlier compared to those receiving placebo, and the pomegranate group improved cognitively over time as measured by the Repeatable Battery for the Assessment of Neuropsychological Status (RBANS) scale (total change), while the placebo group did not. These effects were driven by trends in the visuospatial/constructional and language domains (Bellone et al., 2019).

Treatment with purified isoflavone extracted from soya beans (80 mg/day) for 12 weeks in patients with clinically manifest atherosclerosis and impaired endothelial function who have suffered primary or recurrent ischemic stroke (>6 months) resulted in a vasoprotective effect. This effect was pronounced in patients with more severe endothelial dysfunction, as reflected in improved of brachial flow-mediated dilatation and decreased serum CRP levels. Based on results suggesting a vascular protective effect, the isoflavone could be used for secondary prevention in patients with CV (Chan et al., 2008).

Honuga et al. designed a pilot trial to evaluate the effect of Citrus depressa Hayata fruit extract (CFEx) on muscle mass and composition in subacute stroke patients with rehabilitation. After stroke the muscle mass and strength of the lower limbs decreases (especially in the paretic extremity) and at the same time intramuscular fat increases due to hemiparesis leading to gait difficulties and mobility problems. They found that at baseline the total and normal muscle area of the paretic muscle were inferior to the non-paretic. After 12 weeks of supplementation, there were no significant changes on the paretic side between the supplementation group and the placebo group, nor on the non-paretic side with respect to total muscle area, but there were significant changes in normal muscle area. They concluded that 12-week supplementation with CFEx appeared to improve non-paretic thigh muscle composition compared to placebo. CRP levels in serum were also reduced after treatment with CFEx. However, findings of this study need to be verified by a large-scale randomized trial (Honaga et al., 2022).

Human clinical trials suggest that polyphenol therapy in patients with acute or subacute stroke can prevent or reduce the debilitating effects of stroke on motor function, cognitive impairment, neurological disability, or blood flow to the brain. In addition, the combination of traditional treatments with polyphenols improves their outcomes and/or clinical therapeutic windows. Clinical trials have also been performed to study the role of polyphenols in primary and secondary prevention by controlling stroke risk factors such as diabetes mellitus II (Wong et al., 2016), atheroma plaque (Amato et al., 2015), hypertension, or vascular function (Keane et al., 2016). Many of the clinical studies have administered polyphenols orally, which can be a limitation of the use of the treatment in stroke patients. Patients with acute ischemic stroke may benefit from intravenous administration. Moreover, it is also plausible that polyphenols could be administered as intraarterial adjunctive treatment for acute ischemic stroke patients during mechanical thrombectomy.

Although some studies demonstrate the effect of specific subgroups of polyphenols, when it comes to polyphenol-rich plant extracts it is not clear which specific subgroup might be responsible for the effects exerted, it is likely that there is an additive and synergistic effect between the different types and amounts of polyphenols that is responsible for most of their health benefits (Bellone et al., 2019).

References

Abulizi, A., Simayi, J., Nuermaimaiti, M., Han, M., Hailati, S., Talihati, Z., Maihemuti, N., Nuer, M., Khan, N., Abudurousuli, K., Dilimulati, D., Nueraihemaiti, N., Moore, N., Zhou, W., & Wumaier, A. (2023). Quince extract resists atherosclerosis in rats by down-regulating the EGFR/PI3K/Akt/GSK-3β pathway. *Biomedicine & Pharmacotherapy*, *160*, 114330. Available from https://doi.org/10.1016/j.biopha.2023.114330.

Adams, H. P., Bendixen, B. H., Kappelle, L. J., Biller, J., Love, B. B., Gordon, D. L., & Marsh, E. E. (1993). Classification of subtype of acute ischemic stroke. Definitions for use in a multicenter clinical trial. TOAST. Trial of Org 10172 in acute stroke treatment. *Stroke; A Journal of Cerebral Circulation*, *24*(1), 35–41. Available from https://doi.org/10.1161/01.str.24.1.35.

Aguilar, G. F. G., Alquisiras-Burgos, I., Franco-Pérez, J., Pineda-Ramírez, N., Ortiz-Plata, A., Torres, I., Pedraza-Chaverri, J., & Aguilera, P. (2020). Resveratrol prevents GLUT3 up-regulation induced by middle cerebral artery occlusion. *Brain Sciences*, *10* (9), 1–17. Available from https://doi.org/10.3390/brainsci10090651, https://www.mdpi.com/2076-3425/10/9/651/pdf.

Ali, S. H., Madhana, R. M., K.V., A., Kasala, E. R., Bodduluru, L. N., Pitta, S., Mahareddy, J. R., & Lahkar, M. (2015). *Resveratrol ameliorates depressive-like behavior in repeated corticosterone-induced depression in mice. Steroids* (101, pp. 37–42). India: Elsevier Inc. Available from http://www.elsevier.com/locate/steroids, 10.1016/j.steroids.2015.05.010.

Amato, B., Compagna, R., Amato, M., Gallelli, L., de Franciscis, S., & Serra, R. (2015). *Aterofisiol® in carotid plaque evolution, Drug Design, Development and Therapy* (9, pp. 3877–3884). Italy: Dove Medical Press Ltd. Available from http://www.dovepress.com/getfile.php?fileID = 26105, 10.2147/DDDT.S87609.

Annunziata, G., Maisto, M., Schisano, C., Ciampaglia, R., Narciso, V., Tenore, G., & Novellino, E. (2019). Effects of grape pomace polyphenolic extract (Taurisolo®) in reducing TMAO serum levels in humans: preliminary results from a randomized, placebo-controlled, cross-over study. *Nutrients*, *11*(1), 139. Available from https://doi.org/10.3390/nu11010139.

Arfaoui, L. (2021). Dietary plant polyphenols: effects of food processing on their content and bioavailability. *Molecules*, *26*(10), 2959. Available from https://doi.org/10.3390/molecules26102959.

Arulselvan, P., Fard, M. T., Tan, W. S., Gothai, S., Fakurazi, S., Norhaizan, M. E., & Kumar, S. S. (2016). Role of antioxidants and natural products in inflammation. *Oxidative Medicine and Cellular Longevity*, *2016*, 1–15. Available from https://doi.org/10.1155/2016/5276130.

Atlas, S. A. (2007). The Renin-Angiotensin aldosterone system: Pathophysiological role and pharmacologic inhibition. *Journal of Managed Care Pharmacy*, *13*(8 Supp B), 9–20. Available from https://doi.org/10.18553/jmcp.2007.13.s8-b.9.

Bamji, S. X., Rico, B., Kimes, N., & Reichardt, L. F. (2006). BDNF mobilizes synaptic vesicles and enhances synapse formation by disrupting cadherin-β-catenin interactions. *Journal of Cell Biology*, *174*(2), 289–299. Available from https://doi.org/10.1083/jcb.200601087Canada, http://www.jcb.org/cgi/reprint/174/2/289.

Bayes, J., Bedaso, A., Peng, W., Adams, J., & Sibbritt, D. (2023). The effect of polyphenols in post stroke adults: A systematic review of randomised controlled trials. *Clinical Nutrition ESPEN*, *54*, 113–121. Available from https://doi.org/10.1016/j.clnesp.2023.01.013.

Bejot, Y. (2011). Circulating and brain BDNF levels in stroke rats. Relevance to clinical studies. *PLoS One*, *6*(12), 2011.

Bellone, J. A., Murray, J. R., Jorge, P., Fogel, T. G., Kim, M., Wallace, D. R., & Hartman, R. E. (2019). Pomegranate supplementation improves cognitive and functional recovery following ischemic stroke: A randomized trial. *Nutritional Neuroscience*, *22*(10), 738–743. Available from https://doi.org/10.1080/1028415X.2018.1436413, http://www.tandfonline.com/loi/ynns20#.VvukQLdf1Hh.

Bernatoniene, J., & Kopustinskiene, D. M. (2018). The role of catechins in cellular responses to oxidative stress. *Molecules*, *23*(4). Available from https://doi.org/10.3390/molecules23040965, http://www.mdpi.com/1420-3049/23/4/965/pdf.

Bhullar, K. S., Lassalle-Claux, G., Touaibia, M., & Rupasinghe, H. P. V. (2014). Antihypertensive effect of caffeic acid and its analogs through dual renin-angiotensin-aldosterone system inhibition. *European Journal of Pharmacology*, *730*(1), 125–132. Available from https://doi.org/10.1016/j.ejphar.2014.02.038, http://www.elsevier.com/locate/ejphar.

Braniste, V., Al-Asmakh, M., Kowal, C., Anuar, F., Abbaspour, A., Tóth, M., Korecka, A., Bakocevic, N., Ng, L. G., Kundu, P., Gulyás, B., Halldin, C., Hultenby, K., Nilsson, H., Hebert, H., Volpe, B. T., Diamond, B., & Pettersson, S. (2014). The gut microbiota influences blood-brain barrier permeability in mice. *Science Translational Medicine*, *6*(263). Available from https://doi.org/10.1126/scitranslmed.3009759.

Bruno, A. (2022). Pre-diabetes, diabetes, hyperglycemia, and stroke: Bittersweet therapeutic opportunities. *Current Neurology and Neuroscience Reports*, *22*(11), 781–787. Available from https://doi.org/10.1007/s11910-022-01236-0.

Béjot, Y., Prigent-Tessier, A., Cachia, C., Giroud, M., Mossiat, C., Bertrand, N., Garnier, P., & Marie, C. (2011). Time-dependent contribution of non neuronal cells to BDNF production after ischemic stroke in rats. *Neurochemistry International*, *58*(1), 102–111. Available from https://doi.org/10.1016/j.neuint.2010.10.019.

Campbell, B. C. V. (2019). Ischaemic stroke. *Nat Rev Dis Primers*, *5*, 2019.

Cassidy, A., Rimm, E. B., O'Reilly, É. J., Logroscino, G., Kay, C., Chiuve, S. E., & Rexrode, K. M. (2012). Dietary flavonoids and risk of stroke in women. *Stroke; A Journal of Cerebral Circulation*, *43*(4), 946–951. Available from https://doi.org/10.1161/strokeaha.111.637835.

Chan, Y.-H., Lau, K.-K., Yiu, K.-H., Li, S.-W., Chan, H.-T., Fong, D. Y.-T., Tam, S., Lau, C.-P., & Tse, H.-F. (2008). Reduction of C-reactive protein with isoflavone supplement reverses endothelial dysfunction in patients with ischaemic stroke. *European Heart Journal*, *29*(22), 2800–2807. Available from https://doi.org/10.1093/eurheartj/ehn409.

Chang, S. S., Lee, V. S. Y., Tseng, Y. L., Chang, K. C., Chen, K. B., Chen, Y. L., & Li, C. Y. (2012). Gallic acid attenuates platelet activation and platelet-leukocyte aggregation: Involving pathways of Akt and GSK3β. *Evidence-based Complementary and Alternative Medicine*, *2012*. Available from https://doi.org/10.1155/2012/683872.

Chen, J., Bai, Q., Zhao, Z., Sui, H., & Xie, X. (2016). Resveratrol improves delayed r-tPA treatment outcome by reducing MMPs. *Acta Neurologica Scandinavica*, *134*(1), 54–60. Available from https://doi.org/10.1111/ane.12511.

Chen, M. L., Yi, L., Zhang, Y., Zhou, X., Ran, L., Yang, J., Zhu, J. D., Zhang, Q. Y., & Mi, M. T. (2016). Resveratrol attenuates trimethylamine-N-oxide (TMAO)-induced atherosclerosis by regulating TMAO synthesis and bile acid metabolism via remodeling of the gut microbiota. *mBio*, 7(2). Available from https://doi.org/10.1128/mBio.02210-15, http://mbio.asm.org/content/7/2/e02210-15.full.pdf.

Collaborators, G. B. D. S. (2019). Global, regional, and national burden of stroke and its risk factors, 1990-2019: A systematic analysis for the Global Burden of Disease Study. *Lancet Neurology*, *20*(10), 795–820, 2019.

Comalada, M., Ballester, I., Bailón, E., Sierra, S., Xaus, J., Gálvez, J., Medina, F. S. de, & Zarzuelo, A. (2006). Inhibition of pro-inflammatory markers in primary bone marrow-derived mouse macrophages by naturally occurring flavonoids: Analysis of the structure—activity relationship. *Biochemical Pharmacology*, *72*(8), 1010—1021. Available from https://doi.org/10.1016/j.bcp.2006.07.016.

Cremonini, E., Daveri, E., Mastaloudis, A., Adamo, A. M., Mills, D., Kalanetra, K., Hester, S. N., Wood, S. M., Fraga, C. G., & Oteiza, P. I. (2019). Anthocyanins protect the gastrointestinal tract from high fat diet-induced alterations in redox signaling, barrier integrity and dysbiosis. *Redox Biology*, *26*, 101269. Available from https://doi.org/10.1016/j.redox.2019.101269.

Datta, A., Sarmah, D., Mounica, L., Kaur, H., Kesharwani, R., Verma, G., Veeresh, P., Kotian, V., Kalia, K., Borah, A., Wang, X., Dave, K. R., Yavagal, D. R., & Bhattacharya, P. (2020). Cell Death Pathways in Ischemic Stroke and Targeted Pharmacotherapy. *Translational Stroke Research*, *11*(6), 1185—1202. Available from https://doi.org/10.1007/s12975-020-00806-z.

De Silva, T. M., & Faraci, F. M. (2020). Contributions of aging to cerebral small vessel disease. Annual Reviews Inc., Australia. *Annual Review of Physiology*, *82*, 275—295. Available from https://doi.org/10.1146/annurev-physiol-021119-034338, http://www.annualreviews.org/journal/physiol.

Di Lorenzo, C, Colombo, f, Biella, S, Stockley, C, & Restani, P (2021). Polyphenols and human health: The role of bioavailability. *Nutrients*, *13*(1), 273. Available from https://doi.org/10.3390/nu13010273.

Ding, M. L., Ma, H., Man, Y. G., & Lv, H. Y. (2017). Protective effects of a green tea polyphenol, epigallocatechin-3-gallate, against sevoflurane-induced neuronal apoptosis involve regulation of CREB/BDNF/TrkB and PI3K/Akt/mTOR signalling pathways in neonatal mice. *Canadian Journal of Physiology and Pharmacology*, *95*(12), 1396—1405. Available from https://doi.org/10.1139/cjpp-2016-0333, http://www.nrc.ca/cgi-bin/cisti/journals/rp/rp_desy_e?cjpp.

Easton, J. D., Saver, J. L., Albers, G. W., Alberts, M. J., Chaturvedi, S., Feldmann, E., Hatsukami, T. S., Higashida, R. T., Johnston, S. C., Kidwell, C. S., Lutsep, H. L., Miller, E., & Sacco, R. L. (2009). Definition and evaluation of transient ischemic attack: A scientific statement for healthcare professionals from the American heart association/American stroke association stroke council; council on cardiovascular surgery and anesthesia; council on cardiovascular radiology and intervention; council on cardiovascular nursing; and the interdisciplinary council on peripheral vascular disease. *Stroke; a Journal of Cerebral Circulation*, *40*(6), 2276—2293. Available from https://doi.org/10.1161/STROKEAHA.108.192218.

Fan, Y., Li, Y., Yang, Y., Lin, K., Lin, Q., Luo, S., Zhou, X., Lin, Q., Zhang, F., & Shi, Z. (2022). Chlorogenic acid prevents microglia-induced neuronal apoptosis and oxidative stress under hypoxia-ischemia environment by regulating the MIR497HG/miR-29b-3p/SIRT1 axis. *Disease Markers*, *2022*, 1—18. Available from https://doi.org/10.1155/2022/1194742.

Favaron, M., Manev, R. M., Rimland, J. M., Candeo, P., Beccaro, M., & Manev, H. (1993). NMDA-stimulated expression of BDNF mRNA in cultured cerebellar granule neurones. *Neuroreport*, *4*(10), 1171—1174.

Feng, J., Chen, X., Lu, S., Li, W., Yang, D., Su, W., Wang, X., & Shen, J. (2018). Naringin attenuates cerebral ischemia-reperfusion injury through inhibiting peroxynitrite-mediated mitophagy activation. *Molecular Neurobiology*, *55*(12), 9029—9042. Available from https://doi.org/10.1007/s12035-018-1027-7.

Feng, Z., Wei, R.-B., Hong, Q., Cui, S.-Y., & Chen, X.-M. (2010). Grape seed extract enhances eNOS expression and NO production through regulating calcium-mediated AKT phosphorylation in H 2 O 2 -treated endothelium. *Cell Biology International*, *34* (10), 1055—1061. Available from https://doi.org/10.1042/cbi20100054.

Figueira, G., Garcia, R. C., Pimpão, A. P., Terrasso, I., Costa, A. F., Almeida, L., Tavares, T. F., Pais, P., Pinto, M. R., Ventura, A., Filipe, G. J., McDougall, D., Stewart, K. S., Kim, I., Palmela, D., Brites, M. A., Brito, C., Brito, C. N., & Santos. (2017). Polyphenols journey through blood-brain barrier towards neuronal protection. *Scientific Reports*, 7(1). Available from https://doi.org/10.1038/s41598-017-11512-6.

Flammer, A. J., Hermann, F., Sudano, I., Spieker, L., Hermann, M., Cooper, K. A., Serafini, M., Lüscher, T. F., Ruschitzka, F., Noll, G., & Corti, R. (2007). Dark chocolate improves coronary vasomotion and reduces platelet reactivity. *Circulation*, *116*(21), 2376–2382. Available from https://doi.org/10.1161/CIRCULATIONAHA.107.713867.

Frerichs, K. U., Lindsberg, P. J., Hallenbeck, J. M., & Feuerstein, G. Z. (1990). Increased cerebral lactate output to cerebral venous blood after forebrain ischemia in rats. *Stroke; a Journal of Cerebral Circulation*, *21*(4), 614–617. Available from https://doi.org/10.1161/01.str.21.4.614.

García-Rodríguez, P., Ma, F., Río, C. del, Romero-Bernal, M., Najar, A. M., Cádiz-Gurrea, M. de la L., Leyva-Jimenez, F. J., Ramiro, L., Menéndez-Valladares, P., Pérez-Sánchez, S., Segura-Carretero, A., & Montaner, J. (2022). Diet supplementation with polyphenol-rich salicornia ramosissima extracts protects against tissue damage in experimental models of cerebral ischemia. *Nutrients*, *14*(23), 5077. Available from https://doi.org/10.3390/nu14235077.

Grosso, G., Godos, J., Currenti, W., Micek, A., Falzone, L., Libra, M., Giampieri, F., Forbes-Hernández, T. Y., Quiles, J. L., Battino, M., La Vignera, S., & Galvano, F. (2022). The effect of dietary polyphenols on vascular health and hypertension: current evidence and mechanisms of action. *Nutrients*, *14*(3), 545. Available from https://doi.org/10.3390/nu14030545.

Guerrero, L., Castillo, J., Quiñones, M., Garcia-Vallvé, S., Arola, L., Pujadas, G., Muguerza, B., & Lu, J. R. (2012). Inhibition of angiotensin-converting enzyme activity by flavonoids: Structure-activity relationship studies. *PLoS One*, 7(11), e49493. Available from https://doi.org/10.1371/journal.pone.0049493.

Guven, M., Bozkurt Aras, A., Akman, T., Murat Sen, H., Ozkan, A., Salis, O., Sehitoglu, I., Kalkan, Y., Silan, C., Deniz, M., & Cosar, M. (2015). Neuroprotective effect of p-coumaric acid in rat model of embolic cerebral ischemia. *Iranian Journal of Basic Medical Sciences*, *18*(4), 356–363. Available from http://ijbms.mums.ac.ir/pdf_4284_8cc68e4e116f5850243d461555c154a3.html.

Han, B. H., D'Costa, A., Back, S. A., Parsadanian, M., Patel, S., Shah, A. R., Gidday, J. M., Srinivasan, A., Deshmukh, M., & Holtzman, D. M. (2000). BDNF blocks Caspase-3 activation in neonatal hypoxia–ischemia. *Neurobiology of Disease*, 7(1), 38–53. Available from https://doi.org/10.1006/nbdi.1999.0275.

Hernandez, V. G., Lechtenberg, K. J., Peterson, T. C., Zhu, L., Lucas, T. A., Owah, J. O., Dorsey, A. I., Gentles, A. J., & Buckwalter, M. S. (2023). Translatome analysis reveals microglia and astrocytes to be distinct regulators of inflammation in the hyperacute and acute phases after stroke. *bioRxiv*. Available from https://doi.org/10.1101/2023.02.14.520351, https://www.biorxiv.org.

Honaga, K., Otaka, Y., Kaji, K., Sakata, Y., Hamano, H., Nakamura, H., & Kondo, K. (2022). Effects of citrus depressa hayata fruit extract on thigh muscles mass and composition in subacute stroke patients: a double-blind, randomized, controlled pilot trial. *Journal of Integrative and Complementary Medicine*, *28*(5), 391–398. Available from https://doi.org/10.1089/jicm.2021.0299.

Hong, S. (2020). A comparative study on phenolic content. Antioxidant activity and anti-inflammatory capacity of aqueous and ethanolic extracts of sorghum in lipopolysaccharide-induced RAW 264.7 macrophages. *Antioxidants (Basel)*, *9*, 12. Available from https://doi.org/10.3390/antiox9121297.

Horch, H. W., Krüttgen, A., Portbury, S. D., & Katz, L. C. (1999). Destabilization of cortical dendrites and spines by BDNF. Cell Press, United States. *Neuron, 23*(2), 353–364. Available from https://doi.org/10.1016/S0896-6273(00)80785-0, http://www.neuron.org.

Huang, Y., Xu, M., Li, J., Chen, K., Xia, L., Wang, W., Ren, P., & Huang, X. (2021). Ex vivo to in vivo extrapolation of syringic acid and ferulic acid as grape juice proxies for endothelium-dependent vasodilation: Redefining vasoprotective resveratrol of the French paradox. *Food Chemistry, 363*, 130323. Available from https://doi.org/10.1016/j.foodchem.2021.130323.

Huang, Y., Tian, L., Liu, Y., Liu, J., Huang, J., & Ye, L. (2022). Ferulic acid protects endothelial cells from hypoxia-induced injury by regulating microRNA-92a. *Applied Bionics and Biomechanics, 2022*, 1–6. Available from https://doi.org/10.1155/2022/6148361.

Hwang, S. L., Lin, J. A., Shih, P. H., Yeh, C. T., & Yen, G. C. (2012). Pro-cellular survival and neuroprotection of citrus flavonoid: The actions of hesperetin in PC12 cells. *Food and Function, 3*(10), 1082–1090. Available from https://doi.org/10.1039/c2fo30100h, http://www.rsc.org/Publishing/Journals/FO/about.asp.

Iglesias-Carres, L., Krueger, E. S., Herring, J. A., Tessem, J. S., & Neilson, A. P. (2022). Potential of phenolic compounds and their gut microbiota-derived metabolites to reduce TMA formation: Application of an in vitro fermentation high-throughput screening model. *Journal of Agricultural and Food Chemistry, 70*(10), 3207–3218. Available from https://doi.org/10.1021/acs.jafc.2c00247.

Jang, I. A., Kim, E. N., Lim, J. H., Kim, M. Y., Ban, T. H., Yoon, H. E., Park, C. W., Chang, Y. S., & Choi, B. S. (2018). Effects of resveratrol on the renin-angiotensin system in the aging kidney. *Nutrients, 10*(11). Available from https://doi.org/10.3390/nu10111741, https://www.mdpi.com/2072-6643/10/11/1741/pdf.

Jantas, D., Szymanska, M., Budziszewska, B., & Lason, W. (2009). An involvement of BDNF and PI3-K/Akt in the anti-apoptotic effect of memantine on staurosporine-evoked cell death in primary cortical neurons. *Apoptosis: An International Journal on Programmed Cell Death, 14*(7), 900–912. Available from https://doi.org/10.1007/s10495-009-0370-6.

Jayaraj, R. L., Azimullah, S., Beiram, R., Jalal, F. Y., & Rosenberg, G. A. (2019). Neuroinflammation: Friend and foe for ischemic stroke. *Journal of Neuroinflammation, 16*(1). Available from https://doi.org/10.1186/s12974-019-1516-2, http://www.jneuroinflammation.com/home/.

Jemimah, S., Chabib, C. M. M., Hadjileontiadis, L., AlShehhi, A., & Al-Marzooq, F. (2023). Gut microbiome dysbiosis in Alzheimer's disease and mild cognitive impairment: A systematic review and meta-analysis. *PLoS One, 18*(5), e0285346. Available from https://doi.org/10.1371/journal.pone.0285346.

Jian-Bin, Z., Hong-Fang, Z., & Hong, G. (2005). Investigation on electrochemical behavior and scavenging superoxide anion ability of chrysin at mercury electrode. *Chinese Journal of Chemistry., 23*(8), 1042–1046. Available from https://doi.org/10.1002/cjoc.200591042.

Jiang, X., Andjelkovic, A. V., Zhu, L., Yang, T., Bennett, M. V. L., Chen, J., Keep, R. F., & Shi, Y. (2018). Blood-brain barrier dysfunction and recovery after ischemic stroke. *Progress in Neurobiology, 163-164*, 144–171. Available from https://doi.org/10.1016/j.pneurobio.2017.10.001.

Jin, T., Zhang, Y., Botchway, B. O. A., Zhang, J., Fan, R., Zhang, Y., & Liu, X. (2022). Curcumin can improve Parkinson's disease via activating BDNF/PI3k/Akt signaling pathways. *Food and Chemical Toxicology, 164*, 113091. Available from https://doi.org/10.1016/j.fct.2022.113091.

Jin, Y. R., Han, X. H., Zhang, Y. H., Lee, J. J., Lim, Y., Chung, J. H., & Yun, Y. P. (2007). Antiplatelet activity of hesperetin, a bioflavonoid, is mainly mediated by inhibition of PLC-γ2 phosphorylation and cyclooxygenase-1 activity. *Atherosclerosis, 194* (1), 144–152. Available from https://doi.org/10.1016/j.atherosclerosis.2006.10.011.

Jin, Z., Ke, J., Guo, P., Wang, Y., & Wu, H. (2019). Quercetin improves blood-brain barrier dysfunction in rats with cerebral ischemia reperfusion via Wnt signaling pathway. *American Journal of Translational Research, 11*(8), 4683–4695. Available from http://www.ajtr.org/files/ajtr0093449.pdf.

Jäckel, S., Kiouptsi, K., Lillich, M., Hendrikx, T., Khandagale, A., Kollar, B., Hörmann, N., Reiss, C., Subramaniam, S., Wilms, E., Ebner, K., Brühl, M.-L. von, Rausch, P., Baines, J. F., Haberichter, S., Lämmle, B., Binder, C. J., Jurk, K., Ruggeri, Z. M., … Reinhardt, C. (2017). Gut microbiota regulate hepatic von Willebrand factor synthesis and arterial thrombus formation via Toll-like receptor-2. *Blood, 130*(4), 542–553. Available from https://doi.org/10.1182/blood-2016-11-754416.

Kang, J. H., Tsuyoshi, G., Le Ngoc, H., Kim, H. M., Tu, T. H., Noh, H. J., Kim, C. S., Choe, S. Y., Kawada, T., Yoo, H., & Yu, R. (2011). Dietary capsaicin attenuates metabolic dysregulation in genetically obese diabetic mice. *Journal of Medicinal Food, 14* (3), 310–315. Available from https://doi.org/10.1089/jmf.2010.1367.

Kaur, G., Rao, L. V. M., Agrawal, A., & Pendurthi, U. R. (2007). Effect of wine phenolics on cytokine-induced C-reactive protein expression. *Journal of Thrombosis and Haemostasis, 5*(6), 1309–1317. Available from https://doi.org/10.1111/j.1538-7836.2007.02527.x.

Keane, K. M., Haskell-Ramsay, C. F., Veasey, R. C., & Howatson, G. (2016). Montmorency Tart cherries (*Prunus cerasus* L.) modulate vascular function acutely, in the absence of improvement in cognitive performance. *British Journal of Nutrition, 116* (11), 1935–1944. Available from https://doi.org/10.1017/s0007114516004177.

Kernan, W. N., Ovbiagele, B., Black, H. R., Bravata, D. M., Chimowitz, M. I., Ezekowitz, M. D., Fang, M. C., Fisher, M., Furie, K. L., Heck, D. V., Johnston, S. C., Kasner, S. E., Kittner, S. J., Mitchell, P. H., Rich, M. W., Richardson, D., Schwamm, L. H., & Wilson, J. A. (2014). Guidelines for the prevention of stroke in patients with stroke and transient ischemic attack: A guideline for healthcare professionals from the American Heart Association/American Stroke Association. Lippincott Williams and Wilkins, undefined. *Stroke; A Journal of Cerebral Circulation, 45*(7), 2160–2236. Available from https://doi.org/10.1161/STR.0000000000000024, http://stroke.ahajournals.org/.

Khan, M. M., Ahmad, A., Ishrat, T., Khuwaja, G., Srivastawa, P., Khan, M. B., Raza, S. S., Javed, H., Vaibhav, K., Khan, A., & Islam, F. (2009). Rutin protects the neural damage induced by transient focal ischemia in rats. *Brain Research, 1292*, 123–135. Available from https://doi.org/10.1016/j.brainres.2009.07.026.

Kleindorfer, D. O. (2021). Guideline for the prevention of stroke in patients with stroke and transient ischemic attack: A guideline from the American Heart Association/ American Stroke Association. *Stroke; A Journal of Cerebral Circulation, 52*, 2021.

Khaku, AS, & Tadi, P (2023). *Cerebrovascular Disease*. StatPearls.

KoG. (n.d.). Salvia miltiorrhiza alleviates memory deficit induced by ischemic brain injury in a transient MCAO mouse model by inhibiting ferroptosis. *Antioxidants*.

Koenen, M., Hill, M. A., Cohen, P., & Sowers, J. R. (2021). Obesity, adipose tissue and vascular dysfunction. *Circulation Research, 128*(7), 951–968. Available from https://doi.org/10.1161/circresaha.121.318093.

Kubo, T., Nonomura, T., Enokido, Y., & Hatanaka, H. (1995). Brain-derived neurotrophic factor (BDNF) can prevent apoptosis of rat cerebellar granule neurons in culture. *Developmental Brain Research, 85*(2), 249–258. Available from https://doi.org/10.1016/0165-3806(94)00220-t.

Lackland, D. T., Roccella, E. J., Deutsch, A. F., Fornage, M., George, M. G., Howard, G., Kissela, B. M., Kittner, S. J., Lichtman, J. H., Lisabeth, L. D., Schwamm, L. H., Smith, E. E., & Towfighi, A. (2014). Factors influencing the decline in stroke mortality a statement from the American Heart Association/American Stroke Association. *Stroke; A Journal of Cerebral Circulation*, *45*(1), 315–353. Available from https://doi.org/10.1161/01.str.0000437068.30550.cf.

Larussa, T., Oliverio, M., Suraci, E., Greco, M., Placida, R., Gervasi, S., Marasco, R., Imeneo, M., Paolino, D., Tucci, L., Gulletta, E., Fresta, M., Procopio, A., & Luzza, F. (2017). Oleuropein decreases cyclooxygenase-2 and interleukin-17 expression and attenuates inflammatory damage in colonic samples from ulcerative colitis patients. *Nutrients*, *9*(4), 391. Available from https://doi.org/10.3390/nu9040391.

Leikert, J. F., Räthel, T. R., Wohlfart, P., Cheynier, V., Vollmar, A. M., & Dirsch, V. M. (2002). Red wine polyphenols enhance endothelial nitric oxide synthase expression and subsequent nitric oxide release from endothelial cells. *Circulation*, *106*(13), 1614–1617. Available from https://doi.org/10.1161/01.CIR.0000034445.31543.43.

Leonardo, C. C., Agrawal, M., Singh, N., Moore, J. R., Biswal, S., & Doré, S. (2013). Oral administration of the flavanol (-)-epicatechin bolsters endogenous protection against focal ischemia through the Nrf2 cytoprotective pathway. *European Journal of Neuroscience*, *38*(11), 3659–3668. Available from https://doi.org/10.1111/ejn.12362, http://www.blackwell-synergy.com/loi/EJN.

Leong, X. F., Ng, C. Y., & Jaarin, K. (2015). *Animal models in cardiovascular research: Hypertension and atherosclerosis, BioMed Research International* (2015). Malaysia: Hindawi Publishing Corporation. Available from http://www.hindawi.com/journals/biomed/, https://doi.org/10.1155/2015/528757.

Lessmann, V., Gottmann, K., & Malcangio, M. (2003). Neurotrophin secretion: Current facts and future prospects. *Progress in Neurobiology*, *69*(5), 341–374. Available from https://doi.org/10.1016/S0301-0082(03)00019-4, http://www.elsevier.com/locate/pneurobio.

Li, M. T., Ke, J., Guo, S. F., Wu, Y., Bian, Y. F., Shan, L. L., Liu, Q. Y., Huo, Y. J., Guo, C., Liu, M. Y., Liu, Y. J., & Han, Y. (2021). *The protective effect of quercetin on endothelial cells injured by hypoxia and reoxygenation, Frontiers in pPharmacology* (12). China: Frontiers Media S.A. Available from http://www.frontiersin.org/Pharmacology, 10.3389/fphar.2021.732874.

Liang, G., Shi, B., Luo, W., & Yang, J. (2015). The protective effect of caffeic acid on global cerebral ischemia-reperfusion injury in rats. *Behavioral and Brain Functions*, *11*(1). Available from https://doi.org/10.1186/s12993-015-0064-x.

Liao, H., Huang, J., Liu, J., Chen, Y., Zhu, H., Li, X., Wen, J., Xiang, Q., & Yang, Q. (2023). Resveratrol inhibits activation of microglia after stroke through triggering translocation of smo to primary cilia. *Journal of Personalized Medicine*, *13*(2), 268. Available from https://doi.org/10.3390/jpm13020268.

Lindsberg, P. J., & Roine, R. O. (2004). Hyperglycemia in acute stroke. *Stroke; a Journal of Cerebral Circulation*, *35*(2), 363–364. Available from https://doi.org/10.1161/01.STR.0000115297.92132.84.

Liu, D., Wang, H., Zhang, Y., & Zhang, Z. (2020). *Protective effects of chlorogenic acid on cerebral ischemia/reperfusion injury rats by regulating oxidative stress-related nrf2 pathway. Drug design, development and therapy* (14, pp. 51–60). China: Dove Medical Press Ltd. Available from https://www.dovepress.com/getfile.php?fileID = 55186, https://doi.org/10.2147/DDDT.S228751.

Liu, Q. S., Deng, R., Li, S., Li, X., Li, K., Kebaituli, G., Li, X., & Liu, R. (2017). Ellagic acid protects against neuron damage in ischemic stroke through regulating the ratio of Bcl-2/Bax expression. *Applied Physiology, Nutrition and Metabolism*, *42*(8), 855–860.

Available from https://doi.org/10.1139/apnm-2016-0651, http://pubs.nrc-cnrc.gc.ca/rp-ps/volumes.jsp?jcode = apnm&lang = eng&exp34.

Liu, W., Wang, X., O'Connor, M., Wang, G., & Han, F. (2020). *Brain-derived neurotrophic factor and its potential therapeutic role in stroke comorbidities, Neural Plasticity* (2020). China: Hindawi Limited. Available from http://www.hindawi.com/journals/np/, https://doi.org/10.1155/2020/1969482.

Loke, W. M., Proudfoot, J. M., Hodgson, J. M., McKinley, A. J., Hime, N., Magat, M., Stocker, R., & Croft, K. D. (2010). Specific dietary polyphenols attenuate atherosclerosis in apolipoprotein e-knockout mice by alleviating inflammation and endothelial dysfunction. *Arteriosclerosis, Thrombosis, and Vascular Biology*, *30*(4), 749–757. Available from https://doi.org/10.1161/ATVBAHA.109.199687.

López Andrés, N, Tesse, A, Regault, V, Louis, H, Cattan, V, Thornton, S N, Labat, C, Kakau, A, Tual-Chalot, S, Faure, S, Challande, P, Osborne-Pellegrin, M, Martínez, M C, Lacolley, P, & Andriantsitohaina, R (2012). Increased microparticle production and impaired microvascular endothelial function in aldosterone-salt-treated rats: Protective effects of polyphenols. *PLoS One*, 7(e39235). Available from https://doi.org/10.1371/journal.pone.0039235.

Luengo-Fernandez, R., Violato, M., Candio, P., & Leal, J. (2020). Economic burden of stroke across Europe: A population-based cost analysis. *European Stroke Journal*, *5*(1), 17–25. Available from https://doi.org/10.1177/2396987319883160.

Léger, C. L., Carbonneau, M. A., Michel, F., Mas, E., Monnier, L., Cristol, J. P., & Descomps, B. (2005). A thromboxane effect of a hydroxytyrosol-rich olive oil wastewater extract in patients with uncomplicated type I diabetes. *European Journal of Clinical Nutrition*, *59*(5), 727–730. Available from https://doi.org/10.1038/sj.ejcn.1602133.

Manach, C., Scalbert, A., Morand, C., Rémésy, C., & Jiménez, L. (2004). Polyphenols: food sources and bioavailability. *The American Journal of Clinical Nutrition*, *79*(5), 727–747. Available from https://doi.org/10.1093/ajcn/79.5.727.

Martínez-González, M. A., Gea, A., & Ruiz-Canela, M. (2019). The mediterranean diet and cardiovascular health. *Circulation Research*, *124*(5), 779–798. Available from https://doi.org/10.1161/circresaha.118.313348.

Marunaka, Y., Marunaka, R., Sun, H., Yamamoto, T., Kanamura, N., Inui, T., & Taruno, A. (2017). Actions of quercetin, a polyphenol, on blood pressure. *Molecules*, *22*(2), 209. Available from https://doi.org/10.3390/molecules22020209.

Mirshekari Jahangiri, H., Sarkaki, A., Farbood, Y., Dianat, M., & Goudarzi, G. (2020). Gallic acid affects blood-brain barrier permeability, behaviors, hippocampus local EEG, and brain oxidative stress in ischemic rats exposed to dusty particulate matter. *Environmental Science and Pollution Research*, *27*(5), 5281–5292. Available from https://doi.org/10.1007/s11356-019-07076-9.

Mnafgui, K., Derbali, A., Sayadi, S., Gharsallah, N., Elfeki, A., & Allouche, N. (2015). Anti-obesity and cardioprotective effects of cinnamic acid in high fat diet-induced obese rats. *Journal of Food Science and Technology*, *52*(7), 4369–4377. Available from https://doi.org/10.1007/s13197-014-1488-2.

Nani, A., Murtaza, B., Sayed Khan, A., Khan, N. A., & Hichami, A. (2021). Antioxidant and anti-inflammatory potential of polyphenols contained in mediterranean diet in obesity: Molecular mechanisms. *Molecules*, *26*(4), 985. Available from https://doi.org/10.3390/molecules26040985.

Nassiri-Asl, M., Ghorbani, A., Salehisar, S., Asadpour, E., & Sadeghnia, H. R. (2020). Effect of rutin on oxidative DNA damage in PC12 neurons cultured in nutrients deprivation condition. *Iranian Journal of Basic Medical Sciences*, *23*(3), 390–395. Available from https://doi.org/10.22038/IJBMS.2020.31832.7657, http://ijbms.mums.ac.ir/article_14632_4b992aab73432b2579c70e17d2cf4401.pdf.

Nemzer, B. V., Rodriguez, L. C., Hammond, L., Disilvestro, R., Hunter, J. M., & Pietrzkowski, Z. (2011). Acute reduction of serum 8-iso-PGF2-alpha and advanced oxidation protein products in vivo by a polyphenol-rich beverage; A pilot clinical study with phytochemical and in vitro antioxidant characterization. *Nutrition Journal, 10*(1). Available from https://doi.org/10.1186/1475-2891-10-67.

Nignpense, B. E., Chinkwo, K. A., Blanchard, C. L., & Santhakumar, A. B. (2020). Black sorghum phenolic extract modulates platelet activation and platelet microparticle release. *Nutrients, 12*(6), 1−11. Available from https://doi.org/10.3390/nu12061760, https://www.mdpi.com/2072-6643/12/6/1760/pdf.

Nignpense, E. (2019). Polyphenols: Modulators of platelet function and platelet microparticle generation? *International Journal of Molecular Sciences, 21*(1).

Olas, B., Wachowicz, B., Stochmal, A., & Oleszek, W. (2012). The polyphenol-rich extract from grape seeds inhibits platelet signaling pathways triggered by both proteolytic and non-proteolytic agonists. *Platelets, 23*(4), 282−289. Available from https://doi.org/10.3109/09537104.2011.618562.

Orellana-Urzúa, S., Rojas, I., Líbano, L., & Rodrigo, R. (2020). Pathophysiology of ischemic stroke: Role of oxidative stress. *Current Pharmaceutical Design, 26*(34), 4246−4260. Available from https://doi.org/10.2174/1381612826666200708133912.

Otto, C. M. (2016). Heartbeat: Dietary polyphenols and vascular function. *Heart, 102*(17), 1337−1338. Available from https://doi.org/10.1136/heartjnl-2016-310315, http://heart.bmj.com/.

Pacifici, F., Rovella, V., Pastore, D., Bellia, A., Abete, P., Donadel, G., Santini, S., Beck, H., Ricordi, C., Daniele, N. D., Lauro, D., & Della-Morte, D. (2021). Polyphenols and ischemic stroke: insight into one of the best strategies for prevention and treatment. *Nutrients, 13*(6), 1967. Available from https://doi.org/10.3390/nu13061967.

Palsamy, P., & Subramanian, S. (2008). Resveratrol, a natural phytoalexin, normalizes hyperglycemia in streptozotocin-nicotinamide induced experimental diabetic rats. *Biomedicine & Pharmacotherapy, 62*(9), 598−605. Available from https://doi.org/10.1016/j.biopha.2008.06.037.

Pang, Q., Zhao, Y., Chen, X., Zhao, K., Zhai, Q., & Tu, F. (2018). Apigenin protects the brain against ischemia/reperfusion injury via caveolin-1/VEGF in vitro and in vivo. *Oxidative Medicine and Cellular Longevity, 2018*, 1−12. Available from https://doi.org/10.1155/2018/7017204.

Peh, A., O'Donnell, J. A., Broughton, B. R. S., & Marques, F. Z. (2022). Gut microbiota and their metabolites in stroke: A double-edged sword. *Stroke; A Journal of Cerebral Circulation, 53*(5), 1788−1801. Available from https://doi.org/10.1161/strokeaha.121.036800.

Qin, C., Yang, S., Chu, Y.-H., Zhang, H., Pang, X.-W., Chen, L., Zhou, L.-Q., Chen, M., Tian, D.-S., & Wang, W. (2022). Signaling pathways involved in ischemic stroke: Molecular mechanisms and therapeutic interventions. *Signal Transduction and Targeted Therapy, 7*(1). Available from https://doi.org/10.1038/s41392-022-01064-1.

Perrone, L., Sampaolo, S., & Melone, M. A. B. (2020). Bioactive phenolic compounds in the modulation of central and peripheral nervous system cancers: Facts and misdeeds. *Cancers (Basel), 12*(2). Available from https://doi.org/10.3390/cancers12020454.

Platzer, M. (2022). Radical scavenging mechanisms of phenolic compounds: A quantitative structure-property relationship (QSPR) study. *Frontiers in Nutrition*, 882458. Available from https://doi.org/10.3389/fnut.2022.882458.

Qin, Z., Chen, L., Liu, M., Tan, H., & Zheng, L. (2020). Hesperidin reduces adverse symptomatic intracerebral hemorrhage by promoting TGF-β1 for treating ischemic stroke using tissue plasminogen activator. *Neurological Sciences, 41*(1), 139−147. Available from https://doi.org/10.1007/s10072-019-04054-4.

Rabiei, Z., Bigdeli, M. R., Rasoulian, B., Ghassempour, A., & Mirzajani, F. (2012). The neuroprotection effect of pretreatment with olive leaf extract on brain lipidomics in rat stroke model. *Phytomedicine: International Journal of Phytotherapy and Phytopharmacology*, *19*(10), 940–946. Available from https://doi.org/10.1016/j.phymed.2012.06.003.

Rana, A, Samtiya, M, Dhewa, T, Mishra, V, & Aluko, RE (2022). Health benefits of polyphenols: A concise review. *Journal of Food Biochemistry*, *46*(10). Available from https://doi.org/10.1111/jfbc.14264.

Ren, Z., Zhang, R., Li, Y., Li, Y., Yang, Z., & Yang, H. (2017). Ferulic acid exerts neuroprotective effects against cerebral ischemia/reperfusion-induced injury via antioxidant and anti-apoptotic mechanisms in vitro and in vivo. *International Journal of Molecular Medicine*, *40*(5), 1444–1456. Available from https://doi.org/10.3892/ijmm.2017.3127.

Roberts, A. B., Gu, X., Buffa, J. A., Hurd, A. G., Wang, Z., Zhu, W., Gupta, N., Skye, S. M., Cody, D. B., Levison, B. S., Barrington, W. T., Russell, M. W., Reed, J. M., Duzan, A., Lang, J. M., Fu, X., Li, L., Myers, A. J., Rachakonda, S., ... Hazen, S. L. (2018). Development of a gut microbe–targeted nonlethal therapeutic to inhibit thrombosis potential. *Nature Medicine*, *24*(9), 1407–1417. Available from https://doi.org/10.1038/s41591-018-0128-1, http://www.nature.com/nm/index.html.

Rosignoli, P., Fuccelli, R., Fabiani, R., Servili, M., & Morozzi, G. (2013). Effect of olive oil phenols on the production of inflammatory mediators in freshly isolated human monocytes. *The Journal of Nutritional Biochemistry*, *24*(8), 1513–1519. Available from https://doi.org/10.1016/j.jnutbio.2012.12.011.

Sacco, R. L., Kasner, S. E., Broderick, J. P., Caplan, L. R., Connors, J. J., Culebras, A., Elkind, M. S. V., George, M. G., Hamdan, A. D., Higashida, R. T., Hoh, B. L., Janis, L. S., Kase, C. S., Kleindorfer, D. O., Lee, J. M., Moseley, M. E., Peterson, E. D., Turan, T. N., Valderrama, A. L., & Vinters, H. V. (2013). An updated definition of stroke for the 21st century: A statement for healthcare professionals from the American heart association/American stroke association. *Stroke; a Journal of Cerebral Circulation*, *44*(7), 2064–2089. Available from https://doi.org/10.1161/STR.0b013e318296aeca, http://stroke.ahajournals.org/.

Sarkhosh-Khorasani, S., & Hosseinzadeh, M. (2021). The effect of grape products containing polyphenols on C-reactive protein levels: A systematic review and meta-analysis of randomised controlled trials. *British Journal of Nutrition*, *125*(11), 1230–1245. Available from https://doi.org/10.1017/S0007114520003591, http://journals.cambridge.org/BJN.

Shah, M.-A., Kang, J.-B., Park, D.-J., Kim, M.-O., & Koh, P.-O. (2021). Chlorogenic acid alleviates neurobehavioral disorders and brain damage in focal ischemia animal models. *Neuroscience Letters*, *760*, 136085. Available from https://doi.org/10.1016/j.neulet.2021.136085.

Shah, Z. A., Li, R. C., Ahmad, A. S., Kensler, T. W., Yamamoto, M., Biswal, S., & Doré, S. (2010). The flavanol (-)-epicatechin prevents stroke damage through the Nrf2/HO1 pathway. *Journal of Cerebral Blood Flow and Metabolism*, *30*(12), 1951–1961. Available from https://doi.org/10.1038/jcbfm.2010.53.

Sharma, V., Mishra, M., Ghosh, S., Tewari, R., Basu, A., Seth, P., & Sen, E. (2007). Modulation of interleukin-1β mediated inflammatory response in human astrocytes by flavonoids: Implications in neuroprotection. *Brain Research Bulletin*, *73*(1-3), 55–63. Available from https://doi.org/10.1016/j.brainresbull.2007.01.016.

Shen, Y., Croft, K. D., Hodgson, J. M., Kyle, R., Lee, I.-L. E., Wang, Y., Stocker, R., & Ward, N. C. (2012). Quercetin and its metabolites improve vessel function by inducing eNOS activity via phosphorylation of AMPK. *Biochemical Pharmacology*, *84*(8), 1036–1044. Available from https://doi.org/10.1016/j.bcp.2012.07.016.

Singh, V., Roth, S., Llovera, G., Sadler, R., Garzetti, D., Stecher, B., Dichgans, M., & Liesz, A. (2016). Microbiota dysbiosis controls the neuroinflammatory response after stroke. *Journal of Neuroscience*, *36*(28), 7428−7440. Available from https://doi.org/10.1523/jneurosci.1114-16.2016.

Sobrino García, P., García Pastor, A., García Arratibel, A., Vicente Peracho, G., Rodriguez Cruz, P. M., Pérez Sánchez, J. R., Díaz Otero, F., Vázquez Alén, P., Villanueva Osorio, J. A., & Gil Núñez, A. (2013). Aetiological classification of ischaemic strokes: Comparison of the new A-S-C-O classification and the classification by the Spanish Society of Neurology's Cerebrovascular Disease Study Group. *Neurología (English Edition)*, *28*(7), 417−424. Available from https://doi.org/10.1016/j.nrleng.2012.07.018.

Stanne, T. M., Aberg, N. D., Nilsson, S., Jood, K., Blomstrand, C., Andreasson, U., Blennow, K., Zetterberg, H., Isgaard, J., Svensson, J., & Jern, C. (2016). Low circulating acute brain-derived neurotrophic factor levels are associated with poor long-term functional outcome after ischemic stroke. *Stroke; a Journal of Cerebral Circulation*, *47*(7), 1943−1945. Available from https://doi.org/10.1161/STROKEAHA.115.012383, http://stroke.ahajournals.org/.

Sun, J., Li, Y.-zi, Ding, Y.-hui, Wang, J., Geng, J., Yang, H., Ren, J., Tang, J.-yan, & Gao, J. (2014). Neuroprotective effects of gallic acid against hypoxia/reoxygenation-induced mitochondrial dysfunctions in vitro and cerebral ischemia/reperfusion injury in vivo. *Brain Research*, *1589*, 126−139. Available from https://doi.org/10.1016/j.brainres.2014.09.039.

Sun, L., Xu, P., Fu, T., Huang, X., Song, J., Chen, M., Tian, X., Yin, H., & Han, J. (2018). Myricetin against ischemic cerebral injury in rat middle cerebral artery occlusion model. *Molecular Medicine Reports*, *17*(2), 3274−3280. Available from https://doi.org/10.3892/mmr.2017.8212, https://www.spandidos-publications.com/mmr/17/2/3274/download.

Tai, T., Sawano, H., & Ito. (2012). Antioxidative properties of vanillic acid esters in multiple antioxidant assays. *Bioscience, Biotechnology, and Biochemistry*, *76*(2), 314−318. Available from https://doi.org/10.1271/bbb.110700Japan, http://www.jstage.jst.go.jp/article/bbb/76/2/314/_pdf.

Terra, X., Montagut, G., Bustos, M., Llopiz, N., Ardèvol, A., Bladé, C., Fernández-Larrea, J., Pujadas, G., Salvadó, J., Arola, L., & Blay, M. (2009). Grape-seed procyanidins prevent low-grade inflammation by modulating cytokine expression in rats fed a high-fat diet. *The Journal of Nutritional Biochemistry*, *20*(3), 210−218. Available from https://doi.org/10.1016/j.jnutbio.2008.02.005.

Thiriet, M. (2019). *Cardiovascular disease: An introduction* (pp. 1−90). Springer Science and Business Media LLC. Available from https://doi.org/10.1007/978-3-319-89315-0_1.

Timmis, A., Vardas, P., Townsend, N., Torbica, A., Katus, H., Smedt, D. D., Gale, C. P., Maggioni, A. P., Petersen, S. E., Huculeci, R., Kazakiewicz, D., Rubio, V. B., Ignatiuk, B., Raisi-Estabragh, Z., Pawlak, A., Karagiannidis, E., Treskes, R., Gaita, D., Beltrame, J. F., . . . Papasavvas, I. (2022). European Society of Cardiology: cardiovascular disease statistics 2021. *European Heart Journal*, *43*(8), 716−799. Available from https://doi.org/10.1093/eurheartj/ehab892.

Tresserra-Rimbau, E. B., Rimm, A., Medina-Remón, M. A., Martínez-González, R., de la Torre, D., Corella, J., Salas-Salvadó, E., Gómez-Gracia, J., Lapetra, F., Arós, M., Fiol, E., Ros, L., Serra-Majem, X., Pintó, G. T., Saez, J., Basora, J. V., Sorlí, J. A., Martínez, E., Vinyoles, V., . . . Lamuela-Raventós. (2014). Inverse association between habitual polyphenol intake and incidence of cardiovascular events in the PREDIMED study. *Nutrition, Metabolism and Cardiovascular Diseases*, *24*(6), 639−647. Available from https://doi.org/10.1016/j.numecd.2013.12.014.

Tsai, T., Klausmeyer, A., Conrad, R., Gottschling, C., Leo, M., Faissner, A., & Wiese, S. (2013). 7,8-Dihydroxyflavone leads to survival of cultured embryonic motoneurons by activating intracellular signaling pathways. *Molecular and Cellular Neuroscience*, *56*, 18–28. Available from https://doi.org/10.1016/j.mcn.2013.02.007.

Tu, F., Pang, Q., Chen, X., Huang, T., Liu, M., & Zhai, Q. (2017). Angiogenic effects of apigenin on endothelial cells after hypoxia-reoxygenation via the caveolin-1 pathway. *International Journal of Molecular Medicine*, *40*(6), 1639–1648. Available from https://doi.org/10.3892/ijmm.2017.3159, https://www.spandidos-publications.com/ijmm/40/6/1639/download.

Tu, P., Tang, Q., Wang, M., Chen, W., Ye, X., & Zheng, X. (2022). Protective role of bayberry extract: associations with gut microbiota modulation and key metabolites. *Food & Function*, *13*(10), 5547–5558. Available from https://doi.org/10.1039/d1fo04253j.

Turovskaya, M. V., Gaidin, S. G., Mal'tseva, V. N., Zinchenko, V. P., & Turovsky, E. A. (2019). Taxifolin protects neurons against ischemic injury in vitro via the activation of antioxidant systems and signal transduction pathways of GABAergic neurons. *Molecular and Cellular Neuroscience*, *96*, 10–24. Available from https://doi.org/10.1016/j.mcn.2019.01.005.

Ungvari, Z., Tarantini, S., Donato, A. J., Galvan, V., & Csiszar, A. (2018). Mechanisms of vascular aging. *Circulation Research*, *123*(7), 849–867. Available from https://doi.org/10.1161/circresaha.118.311378.

Unno, T., Hisada, T., & Takahashi, S. (2015). Hesperetin modifies the composition of fecal microbiota and increases cecal levels of short-chain fatty acids in rats. *Journal of Agricultural and Food Chemistry*, *63*(36), 7952–7957. Available from https://doi.org/10.1021/acs.jafc.5b02649, http://pubs.acs.org/journal/jafcau.

Uzdensky, A. B. (2019). Apoptosis regulation in the penumbra after ischemic stroke: expression of pro- and antiapoptotic proteins. *Apoptosis: an International Journal on Programmed Cell Death*, *24*(9-10), 687–702. Available from https://doi.org/10.1007/s10495-019-01556-6, http://link.springer.com/journal/10495.

Wang, J., Pae, M., Meydani, S. N., & Wu, D. (2013). Green tea epigallocatechin-3-gallate modulates differentiation of naïve CD4 + T cells into specific lineage effector cells. *Journal of Molecular Medicine*, *91*(4), 485–495. Available from https://doi.org/10.1007/s00109-012-0964-2.

Wang, J., Guo, Y., & Zhang, S. Y. (2018). Vanillic acid improve neural function after focal cerebral ischemia-reperfusion rats. *International Journal of Pharmacology*, *14*(4), 488–494. Available from https://doi.org/10.3923/ijp.2018.488.494.

Wang, J., Mao, J., Wang, R., Li, S., Wu, B., & Yuan, Y. (2020). Kaempferol protects against cerebral ischemia reperfusion injury through intervening oxidative and inflammatory stress induced apoptosis. *Frontiers in Pharmacology*, *11*. Available from https://doi.org/10.3389/fphar.2020.00424.

Wang, L., Cao, D., Wu, H., Jia, H., Yang, C., & Zhang, L. (2019). Fisetin prolongs therapy window of brain ischemic stroke using tissue plasminogen activator: A double-blind randomized placebo-controlled clinical trial. *Clinical and Applied Thrombosis/Hemostasis*, *25*. Available from https://doi.org/10.1177/1076029619871359, 107602961987135.

Wang, N., Zhang, Y., Wu, L., Wang, Y., Cao, Y., He, L., Li, X., & Zhao, J. (2014). Puerarin protected the brain from cerebral ischemia injury via astrocyte apoptosis inhibition. *Neuropharmacology*, *79*, 282–289. Available from https://doi.org/10.1016/j.neuropharm.2013.12.004.

Unnithan AKA, Das JM, Mehta P. (2023) Hemorrhagic Stroke. *StatPearls*PMID: 32644599

Wang, X., Qi, Y., & Zheng, H. (2022). Dietary polyphenol, gut microbiota, and health benefits. *Antioxidants (Basel), 11*(6). Available from https://doi.org/10.3390/antiox11061212.

Wang, X. H., & You, Y. P. (2017). Epigallocatechin gallate extends therapeutic window of recombinant tissue plasminogen activator treatment for brain ischemic stroke: A randomized double-blind and placebo-controlled trial. *Clinical Neuropharmacology, 40* (1), 24–28. Available from https://doi.org/10.1097/WNF.0000000000000197, http://journals.lww.com/clinicalneuropharm/pages/default.aspx.

Waters, E. S., Kaiser, E. E., Yang, X., Fagan, M. M., Scheulin, K. M., Jeon, J. H., Shin, S. K., Kinder, H. A., Kumar, A., Platt, S. R., Duberstein, K. J., Park, H. J., Xie, J., & West, F. D. (2021). *Intracisternal administration of tanshinone IIA-loaded nanoparticles leads to reduced tissue injury and functional deficits in a porcine model of ischemic stroke. IBRO Neuroscience Reports* (10, pp. 18–30). United States: Elsevier B.V. Available from https://www.journals.elsevier.com/ibro-neuroscience-reports, 10.1016/j.ibneur.2020.11.003.

Williamson, G. (2017). The role of polyphenols in modern nutrition. *Nutrition Bulletin, 42* (3), 226–235. Available from https://doi.org/10.1111/nbu.12278.

Wong, R. H. X., Nealon, R. S., Scholey, A., & Howe, P. R. C. (2016). Low dose resveratrol improves cerebrovascular function in type 2 diabetes mellitus. *Nutrition, Metabolism and Cardiovascular Diseases, 26*(5), 393–399. Available from https://doi.org/10.1016/j.numecd.2016.03.003.

Wu, F., Wu, J., Nicholson, A. D., Echeverry, R., Haile, W. B., Catano, M., An, J., Lee, A. K., Duong, D., Dammer, E. B., Seyfried, N. T., Tong, F. C., Votaw, J. R., Medcalf, R. L., & Yepes, M. (2012). Tissue-type plasminogen activator regulates the neuronal uptake of glucose in the ischemic brain. *Journal of Neuroscience, 32*(29), 9848–9858. Available from https://doi.org/10.1523/jneurosci.1241-12.2012.

Wu, P., Zeng, F., Li, Y.-xin, Yu, B.-li, Qiu, L.-hua, Qin, W., Li, J., Zhou, Y.-mei, & Fan-rong, L. (2015). Changes of resting cerebral activities in subacute ischemic stroke patients. *Neural Regeneration Research, 10*(5), 760. Available from https://doi.org/10.4103/1673-5374.156977.

Wu, Y., Fan, L., Wang, Y., Ding, J., & Wang, R. (2021). Isorhamnetin alleviates high glucose-aggravated inflammatory response and apoptosis in oxygen-glucose deprivation and reoxygenation-induced HT22 hippocampal neurons through Akt/SIRT1/Nrf2/HO-1 signaling pathway. *Inflammation, 44*(5), 1993–2005. Available from https://doi.org/10.1007/s10753-021-01476-1.

Wu, Z., Ying, F., & Gomez-Pinilla. (2006). Dietary curcumin counteracts the outcome of traumatic brain injury on oxidative stress, synaptic plasticity, and cognition. *Experimental Neurology, 197*(2), 309–317. Available from https://doi.org/10.1016/j.expneurol.2005.09.004.

Xia, G. H., You, C., Gao, X. X., Zeng, X. L., Zhu, J. J., Xu, K. Y., Tan, C. H., Xu, R. T., Wu, Q. H., Zhou, H. W., He, Y., & Yin, J. (2019). *Stroke dysbiosis index (SDI) in gut microbiome are associated with brain injury and prognosis of stroke, Frontiers in Neurology* (10). China: Frontiers Media S.A. Available from http://www.frontiersin.org/Neurology, https://doi.org/10.3389/fneur.2019.00397.

Yahfoufi, N., Alsadi, N., Jambi, M., & Matar, C. (2018). The immunomodulatory and anti-inflammatory role of polyphenols. *Nutrients, 10*(11), 1618. Available from https://doi.org/10.3390/nu10111618.

Yang, J., Yuan, L., Wen, Y., Zhou, H., Jiang, W., Xu, D., & Wang, M. (2020). Protective effects of naringin in cerebral infarction and its molecular mechanism. *Medical Science Monitor, 26*. Available from https://doi.org/10.12659/msm.918772.

Yang, Q., Han, L., Li, J., Xu, H., Liu, X., Wang, X., Pan, C., Lei, C., Chen, H., & Lan, X. (2019). Activation of Nrf2 by phloretin attenuates palmitic acid-induced endothelial cell oxidative stress via ampk-dependent signaling. *Journal of Agricultural and Food Chemistry*, *67*(1), 120–131. Available from https://doi.org/10.1021/acs.jafc.8b05025.

Yang, X., Xu, S., Qian, Y., & Xiao, Q. (2017). Resveratrol regulates microglia M1/M2 polarization via PGC-1α in conditions of neuroinflammatory injury. *Brain, Behavior, and Immunity*, *64*, 162–172. Available from https://doi.org/10.1016/j.bbi.2017.03.003.

Yao, Y., Chen, L., Xiao, J., Wang, C., Jiang, W., Zhang, R., & Hao, J. (2014). Chrysin protects against focal cerebral ischemia/reperfusion injury in mice through attenuation of oxidative stress and inflammation. *International Journal of Molecular Sciences*, *15*(11), 20913–20926. Available from https://doi.org/10.3390/ijms151120913.

Yao, Y., Chen, Y., Adili, R., McKeown, T., Chen, P., Zhu, G., Li, D., Ling, W., Ni, H., & Yang, Y. (2017). Plant-based food cyanidin-3-glucoside modulates human platelet glycoprotein VI signaling and inhibits platelet activation and thrombus formation. *The Journal of Nutrition*, *147*(10), 1917–1925. Available from https://doi.org/10.3945/jn.116.245944.

Youdim, K. A., Dobbie, M. S., Kuhnle, G., Proteggente, A. R., Abbott, N. J., & Rice-Evans, C. (2003). Interaction between flavonoids and the blood-brain barrier: In vitro studies. *Journal of Neurochemistry*, *85*(1), 180–192. Available from https://doi.org/10.1046/j.1471-4159.2003.01652.x, http://www.blackwellpublishing.com/journals/JNC.

Yuan, Y., Zhai, Y., Chen, J., Xu, X., & Wang, H. (2021). Kaempferol ameliorates oxygen-glucose deprivation/reoxygenation-induced neuronal ferroptosis by activating nrf2/slc7a11/gpx4 axis. *Biomolecules*, *11*(7). Available from https://doi.org/10.3390/biom11070923, https://www.mdpi.com/2218-273X/11/7/923/pdf.

Zha, X., Hu, Z., Ji, S., Jin, F., Jiang, K., Li, C., Zhao, P., Tu, Z., Chen, X., Di, L., Zhou, H., & Zhang, H. (2015). NFκB up-regulation of glucose transporter 3 is essential for hyperactive mammalian target of rapamycin-induced aerobic glycolysis and tumor growth. *Cancer Letters*, *359*(1), 97–106. Available from https://doi.org/10.1016/j.canlet.2015.01.001.

Zhang, J., Yu, Z., Yu, Z., Yang, Z., Zhao, H., Liu, L., & Zhao, J. (2011). rAAV-mediated delivery of brain-derived neurotrophic factor promotes neurite outgrowth and protects neurodegeneration in focal ischemic model. *International Journal of Clinical and Experimental Pathology*, *4*(5), 496–504. Available from http://www.ijcep.com/files/IJCEP1105003.pdf, China.

Zhang, S., Hu, X., Guo, S., Shi, L., He, Q., Zhang, P., Yu, S. Q., & Zhao, R. (2019). Myricetin ameliorated ischemia/reperfusion-induced brain endothelial permeability by improvement of eNOS uncoupling and activation eNOS/NO. *Journal of Pharmacological Sciences*, *140*(1), 62–72. Available from https://doi.org/10.1016/j.jphs.2019.04.009.

Zhao, J. J., Song, J. Q., Pan, S. Y., & Wang, K. (2016). Treatment with Isorhamnetin protects the brain against ischemic injury in mice. *Neurochemical Research*, *41*(8), 1939–1948. Available from http://www.wkap.nl/journalhome.htm/, https://doi.org/10.1007/s11064-016-1904-2.

Zhou, Y., Fang, S.-hua, Ye, Y.-lu, Chu, L.-sheng, Zhang, W.-ping, Wang, M.-ling, & Er-qing, W. E. I. (2006). Caffeic acid ameliorates early and delayed brain injuries after focal cerebral ischemia in rats. *Acta Pharmacologica Sinica*, *27*(9), 1103–1110. Available from https://doi.org/10.1111/j.1745-7254.2006.00406.x.

Zhu, W., Gregory, J. C., Org, E., Buffa, J. A., Gupta, N., Wang, Z., Li, L., Fu, X., Wu, Y., Mehrabian, M., Sartor, R. B., McIntyre, T. M., Silverstein, R. L., Tang, W. H. W., Di Donato, J. A., Brown, J. M., Lusis, A. J., & Hazen, S. L. (2016). Gut microbial metabolite TMAO enhances platelet hyperreactivity and thrombosis risk. *Cell, 165*(1), 111–124. Available from https://doi.org/10.1016/j.cell.2016.02.011.

Zhu, W., Lin, K., Li, K., Deng, X., & Li, C. (2018). Reshaped fecal gut microbiota composition by the intake of high molecular weight persimmon tannin in normal and high-cholesterol diet-fed rats. *Food & Function, 9*(1), 541–551. Available from https://doi.org/10.1039/c7fo00995j.

Xu, J., G, J., He, X., Sheng, Y. Zheng, S. Zhang, C., Xu, W., HuangK. (2020). Caffeic acid reduces body weight by regulating gut microbiota in diet-induced-obese mice. *Journal of Functional Foods*, 74 https://doi.org/10.1016/j.jff.2020.104061.

CHAPTER SEVEN

Polyphenols as neuromodulators of learning and memory processes

David Garcia-Burgos
Department of Psychobiology, Institute of Neurosciences, University of Granada, Granada, Spain

Introduction

There is increasing interest in the impact of dietary influences on brain functions throughout the lifespan (Flanagan et al., 2020); in particular from improving cognitive development in children to attenuating aging-related cognitive decline and reducing the risk of neurodegenerative diseases (Lamport et al., 2021). When it comes to functions like learning and memory, difficulties range from mild deficits, which are not clinically detectable, to dementia (Knopman et al., 2003). Importantly, deficits in these mental functions are crucial indicators of brain injury or disease, spanning conditions such as traumatic brain injury, stroke, multiple sclerosis, or Huntington's disease. Notably, memory loss and learning difficulties often emerge as primary concerns, underscoring the pivotal role of memory in evaluating neurological health, even among children and adolescents (Reynolds & Bigler, 1997). Furthermore, with population aging contributing to neurodegeneration and subsequent decline in neurological function, the prevalence of learning and memory deficits is on the rise (Pais et al., 2020). Consequently, there is considerable interest in enhancing these brain functions. However, it is essential to clarify what we mean by learning and memory.

Psychological and biological basis of learning and memory

In a changing environment, organisms are continually engaged in the process of behavioral adaptation. The acquisition of knowledge

Bioactive Polyphenols for Health and Pathology Treatment.
DOI: https://doi.org/10.1016/B978-0-443-13821-8.00007-X

through experiences is a valuable asset that enhances their prospects for survival. Within the context of this adaptation, learning and memory facilitate our ability to meaningfully interact with the world. Whereas learning refers to the ability for acquiring new information (i.e., encoding of experience-dependent information), memory is defined as the ability for consolidation, maintenance, off-line transformation, and retrieval information over time. Neither the former nor the latter are unitary processes (D'Souza and Avati, 2021). Instead, both mental functions can be broadly categorized into different types. Regarding learning, for example, non-associative and associative learning (also called conditioning) are two of the most basic forms (Graham, 1981). Non-associative learning is that in which organisms alter their response upon exposure to a single event or stimulus. Thus behavioral responses become attenuated or augmented after repeated or prolonged stimulation. For instance, if you reside near an airport, you might become accustomed to the sounds of planes arriving and departing, while visitors may question how you manage to tolerate living there. By contrast, associative learning is when learning occurs through the link between two or more events (Pearce & Bouton, 2001). Such association sometimes results in a reward or punishment, which in turn modifies our behavior. For example, during instrumental conditioning, when you are training a dog to sit on command. Every time the dog sits in response to the command, it receives a food reward. Over time, the dog learns to associate sitting with receiving the reward and is more likely to sit when given the command in anticipation of such reward.

In biological terms, learning relies on dynamic processes that entail modifications in the biochemistry and physiology of the nervous system. At the molecular level, it is governed by synaptic plasticity (Middei et al., 2014). Thus short-term synaptic plasticity, characterized by activity-dependent modifications in synaptic transmission strength at existing synapses, plays a crucial role in assimilating transient experiences during learning, such as those encountered in classical or instrumental conditioning tasks, which gradually intensify synaptic connections (Fig. 7.1).

Similarly, memory can be split into several partially independent types and mediated by disparate brain regions (Squire, 2004). For instance, it can temporally be classified into short-term (lasting minutes) and long-term (persisting for at least 24 hours to a lifetime). Thus the term working memory denotes processes involved in the temporary storage of knowledge, like retaining phone numbers while dialing. Long-term memory includes declarative and non-declarative forms according to the nature of

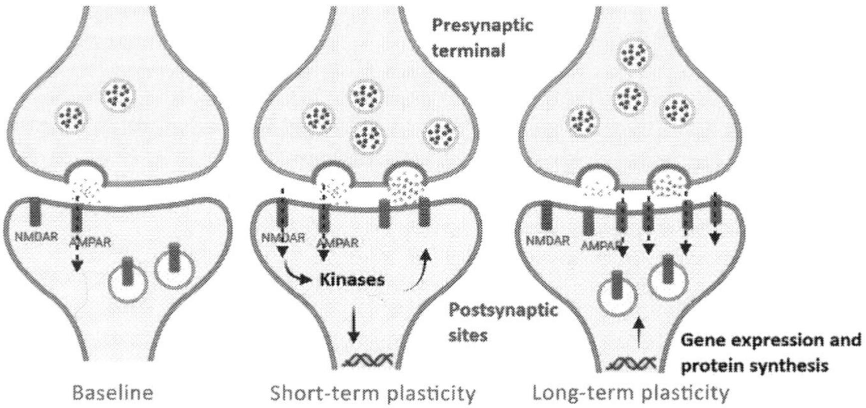

Figure 7.1 Mechanisms of synaptic plasticity during learning: short-term and long-term changes. During the learning process, synaptic strength undergoes dynamic changes influenced by alterations in neurotransmitter release and/or the availability of postsynaptic receptors. These changes modulate the flow of electrical current through ion channels, thereby regulating synaptic transmission. Short-term synaptic plasticity plays a pivotal role in this process by facilitating activity-dependent adjustments in synaptic transmission strength at existing synapses. This mechanism enables the integration of transient experiences during learning. Activation of NMDA-type glutamate receptors, typically blocked until strong synaptic input occurs, is crucial for short-term plasticity. In contrast, long-term plasticity involves the synthesis of new mRNAs and proteins. This process requires activated kinases to modify local protein synthesis and gene expression, ultimately increasing the overall level of synaptic proteins. This augmentation allows for greater synaptic response and underpins the consolidation of learning experiences and memory over time. *NMDAR,* N-methyl-D-aspartate receptor (also known as NMDA receptor); *AMPAR,* α-amino-3-hydroxy-5-methyl-4-isoxazolepropionic acid receptor (also known as AMPA receptor).

information. Declarative memory (also known as explicit memory) is well adapted for the rapid learning of events and encompasses the conscious recollection of meanings (semantic), facts, and events (episodic). Here, the key structures are the hippocampus and the adjacent entorhinal, perirhinal, and parahippocampal cortices. Conversely, non-declarative memory (also known as implicit memory) operates unconsciously and encompasses a range of motor skills or habits. Those types of long-term memory typically require a large number of learning trials and engage the basal ganglia, cerebellum, or amygdala (Squire & Dede, 2015).

In cellular terms, short-term synaptic plasticity is also evident in brief memory forms triggered by short bursts of presynaptic activity. However, for memory lasting hours, days, or even years, long-term potentiation (LTP) emerges as an important mechanism (Malenka & Nicoll, 1999).

Unlike short-term memory, which involves modifications of existing proteins, long-term memory necessitates the synthesis of new mRNAs and proteins (Asok et al., 2019; Goelet et al., 1986). Memory consolidation entails transcriptional processes within the nucleus, leading to gene expression-dependent conversion of information into a durable form that persists for months to years within neural circuits (Fig. 7.2).

LTP enables presynaptic neurons to elicit robust postsynaptic responses upon stimulation and is a process crucial for synaptic strengthening and consolidation in brain regions like the hippocampus (Lamprecht & LeDoux, 2004). For example, prototypic forms of synaptic plasticity like the LTP observed in the CA1 region of the hippocampus are triggered by activation of N-methyl-D-aspartate (NMDA) receptors (NMDARs) and mediated by an increase in postsynaptic alfa-amino-3-hydroxy-5-methyl-4-isoxazolepropionic acid (AMPA)

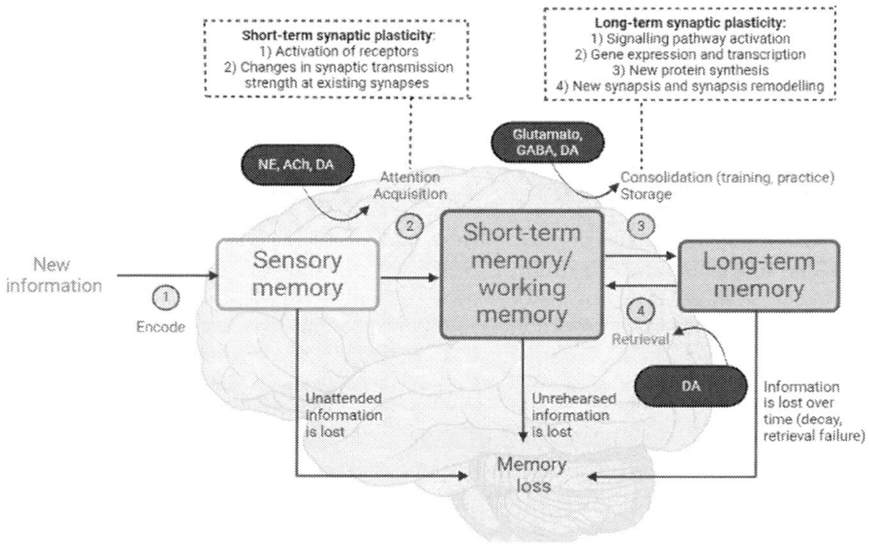

Figure 7.2 Brain information processing model: stages of memory and synaptic plasticity. The brain's information processing model based on a modification of the Atkinson—Shiffrin memory model, depicted here in color-coded segments representing four key stages: (1) encoding, (2) attention and acquisition, (3) consolidation and storage, and (4) retrieval. Memory loss may result from inattention, lack of rehearsal or retrieval failure. Dotted boxes highlight examples of short and long-term synaptic plasticity processes at the molecular level, while filled rounded bubbles indicate some of the major neurotransmitters and their role during the acquisition, consolidation, and retrieval of information. *DA*, dopamine; *NE*, norepinephrine; *GABA*, gamma-aminobutyric acid. *ACh,* Acetylcholine.

receptors (AMPARs) (Citri & Malenka, 2008; Ramirez-Amaya, 2007). Briefly, when a presynaptic neuron fires, it releases glutamate, activating AMPAR on the postsynaptic neuron, causing depolarization. Simultaneously, glutamate binds to NMDA receptors, usually blocked by magnesium ions. If the postsynaptic neuron is sufficiently depolarized by the AMPA receptor activation, the magnesium block is removed, allowing calcium ions to enter the neuron through NMDA receptors. Calcium influx triggers intracellular signaling pathways, leading to AMPA receptor phosphorylation and insertion into the postsynaptic membrane. Thus the NMDA receptor acts as a detector of coincident pre- and postsynaptic activity. This strengthens synaptic connections, facilitating LTP and enhancing synaptic transmission (Cooke & Bliss, 2006; Kennedy, 2016).

As a result, when glutamate is released from the presynaptic membrane again, it exerts a more potent excitatory effect on the postsynaptic cell. Activated NMDA receptors produce a strong calcium influx into the postsynaptic cell, which can activate a wide range of signaling pathways including calcium-calmodulin protein kinase (CaMK), cAMP-dependent protein kinase A (also known as protein kinase A, PKA), protein kinase C (PKC), protein kinase B (PKB, also known as Akt) or extracellular signal-regulated kinase (ERK) (Fig. 7.3). For instance, during the early phase of LTP, the influx of calcium activates calcium-calmodulin protein kinase (CaMKII), which phosphorylates AMPARs. Phosphorylation of these receptors increases their permeability to sodium and potassium ions, increasing the excitability of the postsynaptic neuron. Furthermore, additional AMPARs are inserted into the postsynaptic membrane from a cytoplasmic pool (Cooke & Bliss, 2006; Dunning & During, 2003; Lombroso & Ogren, 2009).

During the late phase, the increased activity of CaMKII stimulates the activation of other kinases responsible for the regulation of gene expression, such as ERK. The subsequent changes in gene transcription and translation lead to the increased synthesis of new AMPARs, which can be then inserted into the postsynaptic membrane. Additional AMPA receptors will facilitate the entry of more positive ions into the cell and strengthen the synapse, making the postsynaptic neuron more likely to fire in response to neurotransmitter release from the presynaptic neuron. These genomic changes continue for hours to days. Both phases ultimately ensure that the presynaptic neuron can evoke a greater postsynaptic response when stimulated. For example, phosphorylation of the kinases

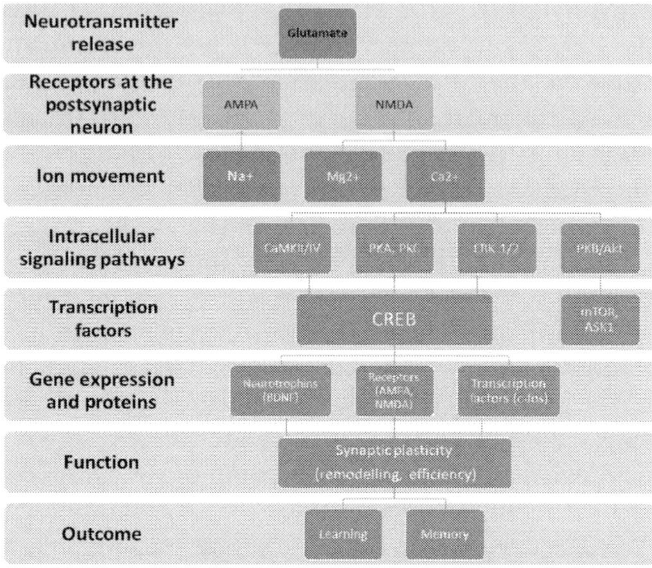

Figure 7.3 Levels of integration and plasticity in learning and memory. Integration of neuronal synapse processes involves a cascade of events including neurotransmitter release, receptor activation at the postsynaptic neuron, ion movement, intracellular signaling pathways, and gene expression regulation. These mechanisms orchestrate synaptic functions critical for learning and memory processes, particularly in brain regions like the hippocampus. Activation of various signaling pathways, including protein kinase A (PKA), protein kinase C (PKC), extracellular signal-regulated kinase 1/2 (ERK1/2), and calcium-calmodulin kinase (CaMK), converges to modulate the activity of the transcription factor cAMP response element-binding protein (CREB). CREB, in turn, plays a pivotal role in regulating the transcription of numerous genes that promote synaptic plasticity, including synaptic remodeling and changes in neuronal morphology. Ultimately, these processes underlie the acquisition and consolidation of new information, essential for learning and memory formation and memory (see text for full explanation).

leads to the modulation of synaptic re-wiring and its efficacy. Specifically, the initiation of cAMP response element-binding protein (CREB)-dependent gene expression triggered by learning seems crucial for the stabilization of memories (Alberini & Kandel, 2015), extending beyond the initial one-hour timeframe.

The CREB-mediated response to external stimuli can be activated by several kinases (e.g., PKA, CaMK, mitogen-activated protein kinase [MAPK], PKC) (Kandel, 2012; Puzzo et al., 2016). Upon activation, CREB-dependent transcription initiates the regulation of the expression of multiple genes that encode key products such as neurotrophins (e.g., brain-derived neurotrophic factor [BDNF]; proteins affecting cell signaling, structure, and metabolism) (Tao et al., 1998) as well as other transcription factors like c-fos, whose induction may catalyze a secondary wave of changes in gene expression (Bekinschtein et al., 2007; Kandel et al., 2015). For instance, BDNF appears to regulate neuronal function and memory, leading to an increase in translational efficacy and capacity. The importance of CREB activation in the induction of long-lasting changes in plasticity and memory is highlighted by studies that show that disruption of CREB activity specifically blocks the formation of long-term memory, whereas agents that increase the amount or activity of CREB accelerate the process (Bourtchuladze et al., 1994; Tully et al., 2003).

Why might polyphenols be relevant to learning and memory?

A promising avenue for modulating and enhancing brain functions, particularly learning and memory, involves harnessing the power of phytochemicals (Kennedy & Wightman, 2011; Kennedy, 2014a). Phytochemicals are compounds produced by plants, encompassing a diverse array of active components that contribute to biochemical and physiological reactions. As secondary metabolites, they perform vital functions in the plant's interactions with its environment. Found naturally in fruits, vegetables, and other dietary sources, phytochemicals also offer a rich source of compounds with potential cognitive benefits in humans. Structurally, phytochemicals fall into categories such as phenolics, terpenes, and alkaloids; each with unique properties and mechanisms of action. While terpenes and alkaloids tend more toward toxic deterrence, including via interactions with the nervous and hormonal systems of animals, phenolics tend toward exerting a global, beneficial effect on human physiological function (Kennedy, 2014b).

Focusing on the phenolics group is categorized based on their phenol ring quantity and linkage type, influencing their bioavailability. Thus they can be divided into simple phenolics (which incorporate one phenyl aromatic hydrocarbon ring within their structure) and polyphenols (which incorporate two aromatic rings) (Bowsher & Tobin, 2008; Singla et al., 2019; Subhan et al., 2024). In brief, the primary classes of polyphenols are flavonoids (the largest group of polyphenols including flavones, flavonols, isoflavones, flavanones, and anthocyanins), phenolic acids (hydroxybenzoic and hydroxycinnamic acids), lignans (sesamol, pinoresinol, sinol, and enterodiol) and stilbenes (resveratrol, piceatannol).

Their increasing popularity is underpinned by a substantial body of research elucidating their advantageous effects on human health and their potential utility in averting, ameliorating, and treating maladies. Notably, the majority of polyphenols' effects on mammalian cells occur within components of cellular signal-transduction cascades encoded by approximately 3000 conserved genes shared between plants and humans (Kennedy, 2014b). Their potential benefits are evident across a spectrum of presentations, including conventional food sources as well as functional, fortified, enriched/enhanced food products, dietary supplements, or nutraceuticals. These compounds appear to offer antioxidant, anti-inflammatory, and other biological roles, potentially guarding against chronic ailments like cancer, cardiovascular disease, or diabetes (Rana et al., 2022; Rasouli et al., 2017); but also improve cognitive function in healthy middle-aged and reduce cognitive impairment and cognitive decline in elderly populations (Kennedy, 2019; Pitt & Leung, 2016; Ramezani et al., 2023).

Particularly relevant to brain function is that, once ingested, these compounds enter the bloodstream and traverse the blood–brain barrier to reach the central nervous system (Matias et al., 2016), where they play various roles, including safeguarding vulnerable neurons, enhancing existing neuronal function or promoting neuronal regeneration. Moreover, the molecular effects encompass direct interactions with neurotransmitter receptors, modulation of signaling pathways, transcription factors and gene and/or protein expression such as the synthesis of neurotrophins crucial for synaptic plasticity—the biological foundation of learning and memory (Rendeiro et al., 2012; Spencer, 2009; Vauzour, 2012).

Interestingly, compared to other dietary components or therapeutic options, polyphenols have several advantages such as accessibility, targeted response, and low toxicity (Rasouli et al., 2017). Although they may demonstrate limited potential for certain activities in vivo (such as antioxidant effects) due to their poor bioavailability and extensive metabolism in the gut and liver, they can still exert pharmacological activity within neurons at lower levels typically obtained through dietary sources (Vauzour, 2017; Williams et al., 2004). However, the rapid metabolism, complex composition, and low bioavailability of polyphenols pose challenges in precisely determining their specific biological actions in the brain and the cognitive outcomes in humans.

Consequently, this chapter aims to explore the effects of orally administered polyphenols on learning and memory performance while considering biochemical parameters (such as levels of neurotransmitters, proteins, or other molecules involved in cellular function), electrophysiological or neurophysiological processes (i.e., activity within the nervous system, including the firing patterns of neurons, synaptic transmission or responses of neurons to stimuli) and behavioral processes (i.e., observable actions and responses in experimental or natural settings). To achieve this, we will focus on molecular and cellular changes characterized by neurotransmitter release, receptor activation regulation, modulation of signaling pathways (e.g., ERK and Akt pathways), activation of transcription factors like CREB, and regulation of CREB gene and protein expression in neurons. By influencing these pathways, polyphenols are expected to have the potential to stimulate new protein synthesis in neurons, facilitating morphological changes that directly impact the acquisition, consolidation, and storage of information.

Polyphenols, neurotransmitters, and receptor activation during learning and memory

Neurotransmitters are pivotal signaling molecules released by neurons to influence other target cells (e.g., another nerve cell, a muscle cell or a gland). They are stored in vesicles in the axon terminals of neurons and released into the synaptic cleft. Once in the synaptic cleft, neurotransmitters interact with receptors on target cells. Essential for the nervous system's proper functioning, key neurotransmitters include acetylcholine, glutamate, gamma-aminobutyric acid (GABA), serotonin, and dopamine (Fig. 7.2). Although acetylcholine and glutamate have received extensive attention in the study of learning and memory, evidence also suggests the involvement of GABA, dopamine, serotonin and norepinephrine (Bairy & Kumar, 2019).

In this context, polyphenols have been shown to regulate neurotransmission systems, particularly those related to acetylcholine, dopamine, and GABA. Indeed, different recent reviews have discussed the role of polyphenols and changes in the probability of neurotransmitter release during learning (Bairy & Kumar, 2019; Ramezani et al., 2023; Rebas et al., 2020). In the subsequent sections, we will explore this in detail.

Polyphenols and the cholinergic system

Illustrative instances are evident within the cholinergic system, denoting receptors sensitive to the neurotransmitter acetylcholine, known for its role in enhancing memory encoding following sensory input (Hasselmo, 2006). Animal studies have largely demonstrated that cholinergic agonists like physostigmine bolster memory formation, whereas cholinergic antagonists such as scopolamine significantly hinder the acquisition and retention of memories (Aigner & Mishkin, 1987; Kim et al., 2010). The cholinergic system is classified into two types according to whether they are stimulated by the drug nicotine or muscarine. Both muscarinic and nicotinic receptors play crucial roles in the encoding of new memories. Here, positive effects on learning and memory have been observed in studies involving rodents with experimentally induced memory decline. For instance, oral administration of *Pinus densiflora* Bark Extract at doses of 15–30 mg/kg/day body weight (rich in procyanidins and catechins) for 3 weeks demonstrated spatial learning and memory enhancement in scopolamine-induced amnesic rats, as evidenced by Y-maze[1], passive avoidance and Morris water maze tests.[2] Additionally, the rats treated with the extract showed a significant reduction in acetylcholinesterase activity in hippocampal lysate compared to the control group (Kim et al., 2022).

In another study involving Alzheimer's disease induction, the authors injected a specific fragment of beta-amyloid protein into the CA1 area of the bilateral hippocampus. However, the rats treated with doses of

[1] Used to assess spatial working memory. It is measured thorough the willingness of the experimental animals to explore new places by recording the number of arm entries and calculating the percentage of alternation among arms. The test apparatus is shaped like the letter "Y" and consists of three equal arms. First, animals undergo a training phase in which they are allowed to explore the entire maze except for a blocked "novel arm." Then, the test begins with the subject being placed in one arm and allowed to freely explore the entire maze for a specific duration (experimenters can adapt the protocol). The sequence and number of arm entries are recorded. Normal rodents tend to exhibit an innate preference for novelty and will explore the least recently visited arm more frequently. Impaired spatial working memory is indicated by repetitive entries into the same arm or random arm choices.

[2] Designed to test spatial memory, long-term memory and learning by evaluating scape latency, thigmotaxis duration (tendency to remain close the walls), distance moved and swimming velocity. It involves placing animals in a pool of water that is made opaque by adding powdered non-fat milk or nontoxic paint. The animals have to swim to a concealed escape platform, which they cannot see due to the opaque water. They must depend on external cues surrounding the maze. As the animals become more familiar with the task, they gradually improve their ability to locate the platform more quickly.

100–625 mg/kg/day of epigallocatechin-3-gallate (≥ 90%; from *Teavigo*) via gavage exhibited increased acetylcholine content by reducing acetylcholinesterase activity in a dose-dependent manner (Nan et al., 2021). Furthermore, in Streptozotocin-induced diabetic mice, chronic treatment with genistein for 30 days at doses of 2.5, 5.0, and 10.0 mg/kg, administered intraperitoneally once daily, significantly improved cognitive performance by modulating acetylcholinesterase activity (Rajput & Sarkar, 2017). Additionally, genistein seems to modulate cholinergic neurotransmission via the G protein-coupled receptor, further enhancing acetylcholine regulation (Liu et al., 2018). Reducing acetylcholine esterase activity was also observed in middle-aged rats orally administered with grape seed proanthocyanidin extract (400 mg/kg) for 14 weeks. Results revealed that supplementation improved acquisition and working memory while reducing acetylcholine esterase activity in the medial prefrontal cortex and hippocampus. Additional analyses showed increased levels of muscarinic M1 acetylcholine receptor protein and mRNA in the CA1 region of the hippocampus (Abhijit et al., 2017). These results illustrate the polyphenol modulation of cholinergic neurotransmission by regulating the balance between acetylcholinesterase and choline acetyltransferase activities, increasing acetylcholine levels in the hippocampus (Liu et al., 2018; Lu et al., 2018; Rajput & Sarkar, 2017).

Polyphenols and glutamatergic system

Another pivotal excitatory neurotransmitter in the central nervous system is glutamate. At the core of this system are glutamate receptors, which are divided into two main classes: ionotropic glutamate receptors (voltage sensitive) and metabotropic glutamate receptors (ligand sensitive). Ionotropic receptors are fast-acting signaling and can produce large changes in current flow once opened, even if the voltage difference across the membrane is small. After glutamate binds to an ionotropic receptor, the receptor's channel undergoes a conformational change to allow an immediate influx of extracellular sodium among other ions and an efflux of potassium ions. This triggers membrane depolarization in the postsynaptic cell sufficient to induce signal transmission. Two of the main glutamate ionotropic receptors are NMDARs and AMPARs.

The glutamatergic pathway assumes a central role in learning and memory, particularly through the mechanism of LTP (Martinez-Lozada & Ortega, 2023). NMDARs are essential for eliciting LTP. Notably, the role of NMDA in LTP is further emphasized by studies demonstrating ethanol-induced complete blockade of LTP through NMDA inhibition and GABA activation in the hippocampal CA1 area (Villarreal et al., 2002). On the other hand, both excessively high and abnormally low levels of glutamate have been associated with disruptions in learning and memory functions, particularly affecting LTP in the hippocampus. Excess release of glutamate followed by overactivated glutamate receptors (NMDAR and AMPAR) has been observed in many neurodegenerative diseases (e.g., Alzheimer's disease or Parkinson's disease due to more intensive calcium influx) (Lewerenz & Maher, 2015). On the other hand, low concentrations of glutamate or diminished activities of glutamate receptors are reported to disturb LTP in the hippocampus, impairing learning and memory functions (Katagiri et al., 2001).

Interestingly, certain phytochemicals possess the ability to modulate glutamatergic signaling. For instance, epigallocatechin gallate, along with others found in green tea, has been shown to restore spatial learning and memory impairments induced by ethanol. This effect is attributed to their enhancement of the expression of NMDAR1 and CREB, critical for synaptic plasticity and memory formation (Zhang et al., 2018). On the other hand, fisetin and baicalein have also been shown to facilitate LTP induction in the CA1 region of the hippocampus, which is dependent on NMDA glutamate receptors (He et al., 2018; Wang et al., 2011).

Human data, albeit harder to find, also support the memory-enhancing effects of polyphenols. For instance, a recent review suggests that green tea or green tea extracts, l-theanine and epigallocatechin gallate ameliorate working memory (Mancini et al., 2017). Likewise, it has been proposed that green tea extract might enhance fronto-parietal connectivity during working memory processing by modulating the synaptic plasticity of NMDAR, which would be measurable using functional magnetic resonance imaging methods (Schmidt et al., 2014).

Polyphenols and GABAergic system

GABA is the major inhibitory neurotransmitter in the central nervous system that provides its effects by acting on two distinct classes of receptors based on their physiological and pharmacological properties (Olsen & DeLorey, 1999): GABA type A receptor ($GABA_AR$) is a ligand-gated chloride channel which mediates fast inhibitory signals through rapid postsynaptic membrane hyperpolarization, whereas the metabotropic

GABA type B receptor ($GABA_BR$) produces slow and prolonged inhibitory signals via G proteins and second messengers. For example, $GABA_AR$ activation is one of the mechanisms by which alcohol induces LTP block, inhibiting the transfer of information from short-term to long-term memory. On the other hand, both ionotropic $GABA_AR$ and metabotropic $GABA_BR$ receptors have been involved in the stable development of learning processes such as conditioned learning inhibition (Heaney & Kinney, 2016; Schummers & Browning, 2001; Zyablitseva et al., 2009).

Numerous plant-derived compounds have been identified as regulators of GABAergic signaling, employing distinct mechanisms to modulate $GABA_AR$ activity. However, the functional diversity of $GABA_AR$s, contingent upon factors such as brain region, cell type, developmental stage and physiological or pathophysiological circumstances, underscores the complexity of GABAergic modulation. Interestingly, the distribution of $GABA_AR$ subtypes dictates their inhibitory effects. In this context, the $\alpha5\beta\gamma2$ receptors appear to be those more related to learning and memory processes (Collinson et al., 2002; Rudolph & Möhler, 2006). Despite the exploration of several polyphenols, such as epigallocatechin gallate from green tea, baicalein from *Scutellaria baicalensis*, apigenin, or amentoflavone, for their neuroprotective, anxiolytic, sedative, or hypnotic properties through $GABA_AR$ engagement (Ríos et al., 2022), their impact on learning and memory remain uncertain.

Polyphenols and dopaminergic system

Brain dopamine receptors regulate movement and locomotion, motivation, and memory. Five subtypes of mammalian dopamine receptors have been identified and are divided into D1-like (D1, D5) or D2-like (D2, D3, D4) groups (Neve, 2013). Dopaminergic neurotransmission is essential in memory, which goes from reward signaling or working memory to memory consolidation (Kourosh-Arami et al., 2023).

Various flavonoids have been observed to enhance learning and memory processes by increasing the levels of two key dopaminergic neurotransmitter actors: dopamine transporter and tyrosine hydroxylase. This augmentation leads to elevated levels of dopamine metabolites. Regarding the elevation of dopamine transporter levels, baicalein from the root of *S. baicalensis* Georgi has been identified. By promoting the recycling of dopamine back to the presynaptic terminal, baicalein contributes to the regulation of dopaminergic signaling, essential for various memory processes (Wu et al., 2006). Another example is nobiletin from *Citrus depressa*, which has demonstrated potential benefits in combating memory dysfunction by increasing tyrosine hydroxylase levels. Specifically, experimental studies using animal models have revealed that nobiletin administration can enhance tyrosine hydroxylase phosphorylation in key brain regions implicated in memory, such as the striatum and hippocampal CA1 region; which has been associated with improvements in cognitive impairment induced by neurotoxic substances like 1-methyl-4-phenyl-1,2,3,6-tetrahydropyridine. Behavioral tasks, such as step-through passive avoidance and novel object recognition test,[3] have demonstrated the efficacy of nobiletin in ameliorating memory deficits induced by this neurotoxin, underscoring its potential as a therapeutic agent for cognitive disorders (Yabuki et al., 2014).

Polyphenols and modulation of signaling pathways during learning and memory

Among the signaling pathways during learning and memory, five hold particular significance: (a) PKA, (b) PKB or PI3 kinase/Akt, (c) PKC, (d) CaMK, and (e) extracellular signal-regulated kinase 1/2 (ERK1/2). Activation of these pathways converges to signal to the CREB family of transcription factors; which, in turn, regulate genes essential for synaptic remodeling, increase neuronal spine density and facilitate synaptic plasticity (Lynch, 2004).

Through modulation of signaling pathways, polyphenols have exhibited potential in mitigating age-related memory decline in animal models. For example, research on senescence-accelerated mice demonstrated that oral administration of green tea catechins (0.05% and 0.1% w/v) in drinking water for 6 months prevented spatial memory decline as observed in the Morris water maze test; accompanied by a notable increase in the activity of the PKA/CREB pathway. Furthermore, chronic consumption

[3] Test for evaluating learning and memory, which consists of three phases. After one habituation day, the training phase is carried out: the animal is allowed to explore two equal objects. Then, on the test day, one of the objects is replaced by a novel one. Since the animals have natural preference by novelty, if they recognize the familiar object, they will spend more time exploring the new one. The measures recorded are the time spent exploring the objects and the discrimination index.

of green tea catechins prevented the reduction of three key proteins associated with synaptic function and structure such as brain-derived neurotrophic factor (BDNF) and CaMKII (Li et al., 2009). Similarly, the flavonoid genistein has revealed positive effects on learning and memory. In rat models of isoflurane-induced neuronal apoptosis, oral genistein administration at doses of 20, 40, or 80 mg/kg from postnatal day 3 to 15 led to elevated cAMP levels and phosphorylation of CREB, activating the cAMP/CREB-BDNF-TrkB signaling pathway. Additionally, the PI3K/Akt signaling pathway was significantly upregulated, resulting in enhanced learning and memory, as evidenced by behavioral assessments by fear conditioning tests and the spatial memory in the Morris water maze test (Jiang et al., 2017).

Furthermore, epigallocatechin-3-gallate (*Camellia sinensis*) at doses of 25, 50, or 75 mg/kg b.w. from postnatal day 3 to 21 seems to effectively mitigate sevoflurane-induced neurotoxicity in C57BL/6mice. This also enhanced learning and memory retention in the Morris water maze test. This neuroprotective effect was attributed to the activation of the CREB/BDNF/TrkB−PI3K/Akt signaling pathway (Ding et al., 2017). Furthermore, small molecules, such as a pyrazole derivative of curcumin administered at a dose of 10 mg/kg b.w., were found to activate CaMKII via a cAMP-independent mechanism (which plays a role in synaptic facilitation), thereby enhancing hippocampal LTP and memory in a rat object recognition test (Maher et al., 2010). Considering the potential of polyphenols to modulate Akt activity, it becomes crucial to facilitate increased translation of specific mRNA subpopulations, including the activity-regulated cytoskeletal-associated protein. This modulation promotes changes in synaptic strength and induces morphological alterations in dendritic spine density and outgrowth (Vauzour, 2017).

Regarding the ERK1/2 pathway, key members of the mitogen-activated protein kinase (MAPK) superfamily, are activated by various extracellular stimuli, including neurotransmitters and growth factors. This activation initiates a cascade of cellular responses, ranging from gene transcription to cell survival, through translocation to the nucleus and interaction with specific downstream substrates. Multiple studies utilizing animal models have suggested an enhancement in learning and memory with the potential involvement of the MEK/ERK1/2 pathway in mice (Moosavi et al., 2015). For instance, adult Wistar rats subjected to aluminum treatment exhibited impaired performance in a passive avoidance test, concurrent with decreased levels of monoamines and glutamate in the hippocampus. Additionally, levels of ERK1/2 were elevated, along with increased acetylcholinesterase activity. However, concurrent administration of a high dose of leaf extract from *Harrisonia abyssinica* (200 mg/kg) via oral gavage for 3 weeks restored nearly normal levels of catecholamines, ERK, acetylcholinesterase and glutamate content in the hippocampus of these aluminum-treated rats. In the passive avoidance behavioral test, the high dose of the extract reinstated normal entrance latency at 24 and 48 hours (Anwar et al., 2021).

Another study investigated the potential effects of grape powder treatment, which encompasses resveratrol, flavans (including catechin), flavonols (including quercetin), anthocyanins, and simple phenolics derived from fresh red, green, and blue-black California grapes. The treatment involved dissolving the grape powder in tap water at a concentration of 15 g/L for 3 weeks, targeting memory impairment induced by oxidative stress in Sprague-Dawley rats. Oxidative stress was induced using L-buthionine-(S, R)-sulfoximine (BSO), a known generator of oxidative stress. Results indicated that rats treated with BSO exhibited pronounced memory deficits compared to control rats or those solely treated with grape powder. However, rats concurrently treated with grape powder and BSO showed a significant reduction in memory deficits compared to those treated with BSO alone. Rats treated solely with grape powder demonstrated similar learning abilities and short- and long-term memory performance to control rats in a radial arm water maze, indicating that grape powder treatment alone did not affect memory performance compared to control rats. However, it effectively mitigated BSO-induced cognitive impairment. Further analysis suggested that grape powder may mitigate oxidative stress-induced anxiety, memory impairment, and hypertension in rats by modulating the levels of specific proteins involved in the oxidative stress response and neuronal function, including ERK1/2, CAMKIV, CREB, and BDNF (Allam et al., 2013).

Similarly, the flavonoid fisetin has been found to activate ERK signaling pathways and promote CREB phosphorylation in C57BL6J mice through a combination of biochemical, electrophysiological and behavioral assays. This activation facilitates LTP in hippocampal slices and enhances object recognition. Fisetin solutions were orally administered 60 minutes before training at a volume of 10 mL/kg b.w. (Maher et al., 2006).

Additionally, supplementation with a blueberry-enriched diet (2% w/w; 500 mg/day of lyophilized blueberry per day) for 12 weeks appears to enhance the spatial working memory performance of aged Lister-hooded rats. This improvement was noticeable within just 3 weeks of supplementation and persisted throughout the testing period. Notably, memory performance correlated positively with the activation of CREB and the elevation of both pro- and mature levels of BDNF in the hippocampus. These changes in CREB and BDNF levels were accompanied by an increase in the phosphorylation state of ERK1/2 rather than calcium-calmodulin kinases (CaMKII and CaMKIV) or PKA (Williams et al., 2008).

While establishing direct causal relationships among polyphenols, biochemical parameters, and behavior remains challenging, emerging evidence suggests a potential link between polyphenols (and flavonoids in particular) effects on the signaling pathway and improvements in memory among impaired animals. This underscores the promising role of polyphenol-rich interventions in addressing age-related cognitive decline.

Polyphenols and activation of transcription factors like cAMP response element-binding protein during learning and memory

Polyphenols have been implicated in enhancing learning and memory not only by directly influencing kinase activity but also by modulating downstream signaling cascades, such as transcription factor activation and binding to promoter sequences (Jeremy and Spencer, 2008; Spencer, 2007). Among these factors, CREB emerges as particularly pivotal. Its activation and phosphorylation play essential roles in forming long-term memories and memory consolidation (Tully et al., 2003). In this sense, several studies have suggested the potential of polyphenols to enhance learning and memory using green tea polyphenols, fisetin, genistein, or curcumin, which have all been shown to influence CREB phosphorylation and subsequent memory processes in animal models (Jiang et al., 2017; Lee et al., 2023; Maher et al., 2006; Zhang et al., 2018).

For example, in 8-week-old male Sprague-Dawley rats, green tea polyphenols were found to enhance hippocampal CREB activity, thereby alleviating ethanol-induced spatial learning and memory impairments. Ethanol intake was observed to impair spatial memory and reduce hippocampal NMDAR expression, CREB phosphorylation and neuronal density. Indeed, rats treated with green tea polyphenols, receiving a polyphenol solution dissolved in 56% (v/v) ethanol at a dosage of 0.25 g/kg of body weight (administered at 15 mL/kg per rat), demonstrated mitigation of memory deficits. Additionally, this treatment increased NMDAR expression and CREB phosphorylation while preserving neuronal morphology in ethanol-exposed rats (Zhang et al., 2018).

On the other hand, the flavonoid fisetin has demonstrated the ability to trigger phosphorylation of CREB in the rat hippocampus. This activation contributes to LTP and proved essential for its memory-enhancing effects when orally administered to male C57BL/6J mice before training, at doses of 10 and 25 mg/kg. Notably, fisetin's presence in commonly consumed foods underscores its potential for delivering long-term cognitive benefits with minimal side effects (Maher et al., 2006). Encouragingly, studies have revealed that older rats fed a diet enriched in strawberry extract for 8 weeks exhibited enhanced cognitive performance in the Morris water maze, aligning with the presence of fisetin in strawberries and other commonly ingested foods (Joseph et al., 1999).

Likewise, it is related to the protective effects of genistein against isoflurane-induced neurotoxicity and cognitive impairments in rats. Exposure to isoflurane in neonatal rats led to neuroapoptosis. However, administration of genistein significantly mitigated apoptosis in the hippocampus and enhanced the expression of both proapoptotic and antiapoptotic factors. Moreover, genistein elevated cAMP levels and promoted the phosphorylation of CREB, thereby activating the cAMP/CREB-BDNF-TrkB signaling pathway. Notably, at a dosage of 80 mg, genistein boosted the expression levels of total CREB to 101.0%. Behavioral assessments further demonstrated that genistein enhanced overall behavior, fear responses and spatial learning and memory in rats exposed to isoflurane (Jiang et al., 2017). Recent clinical trials are in agreement with these findings, showing concurrent changes in serum BDNF levels and global cognition scores following high polyphenol consumption, suggesting a role for BDNF in polyphenol-induced cognitive improvements (Neshatdoust et al., 2016).

Research on the impact of a blueberry-rich diet on memory in aged animals found that supplementation (administered orally at a concentration of 2% w/w for 12 weeks) significantly enhanced spatial working memory within 3 weeks and throughout the 12-week testing period. This improvement was associated with increased activation of CREB and higher BDNF levels in the hippocampus. Specifically, CREB activation was tied to ERK1/2 phosphorylation. Flavonoids, abundant in blueberries, were proposed to modulate neuronal function via the ERK-CREB-

BDNF pathway, potentially enhancing memory. The study also observed increased phosphorylation of Akt and mTOR, indicating involvement in protein synthesis, and elevated levels of Arc/Arg3.1, a protein linked to synaptic plasticity and memory consolidation (Williams et al., 2008).

Regarding the impact of curcumin (a polyphenolic compound found in curry spice) on adult hippocampal neurogenesis and cognitive function, a recent work aimed to investigate whether oral administration of curcumin would promote hippocampal neurogenesis and improve neurocognitive function in young mice. The researchers administered different doses of curcumin orally to mice for 14 days and observed its effects on hippocampal neurogenesis and cognitive function. Surprisingly, low-dose curcumin (0.4 mg/kg) increased the proliferation and survival of newly generated cells in the hippocampus, while higher doses (2 or 10 mg/kg) did not have the same effect. Moreover, low-dose curcumin elevated levels of BDNF and phosphorylated CREB, both associated with enhanced hippocampal neurogenesis. Behavioral tests revealed that low-dose curcumin improved memory retention in mice. Overall, the findings suggest that low-dose curcumin has neurogenic effects and may mitigate age and neurodegenerative disease-related cognitive deficits (Lee et al., 2023).

Despite these preliminary findings, translating the benefits observed in animal models to human clinical studies poses challenges. While studies have shown the presence of polyphenols in the brain following supplementation and oral administration, further research is needed to establish causal relationships between polyphenol intake, molecular changes, neuronal morphology and behavior. Future investigations should aim to elucidate the temporal dynamics of kinase activation and CREB elevation in response to supplementation to provide deeper insights into the mechanisms underlying the observed cognitive benefits.

Polyphenols, regulation of gene and protein expression during learning and memory

The activation of CREB is especially relevant in regulating the expression of genes crucial for synaptic plasticity and memory consolidation, among which BDNF stands out as essential for neuronal survival and synaptic function (Alberini & Kandel, 2015; Ortega-Martínez, 2015). In particular, key CREB target genes related to memory consolidation and LTP include c-fos, activity-regulated cytoskeleton-associated protein (Arc), and BDNF.

In a recent investigation, researchers explored the behavioral validation of an optogenetics model to probe fear memory recall and its potential enhancement through dietary polyphenols. This study involved injecting c-fos-tTA mice with either AAV9-TRE-ChR2-mCherry or AAV9-TRE-mCherry (control) in the dentate gyrus of the hippocampus, followed by the implantation of optical fibers above the injection site for stimulation. After surgical recovery, mice underwent fear conditioning in Context B, and fear memory recall was subsequently tested in Context A under light-off and light-on periods. Mice injected with AAV9-TRE-ChR2-mCherry displayed significantly higher freezing levels during light-on periods, indicating light stimulation-induced recall of fear memory, while those injected with AAV9-TRE-mCherry showed no increase in freezing during light-on periods. Further exploration of dietary polyphenols' effect on fear memory recall revealed that mice treated with a blend of polyphenols exhibited a greater increase in freezing compared to vehicle-treated mice. Immunohistochemical analysis demonstrated an increased number of c-Fos- and ChR2-mCherry-expressing neurons in BDPP-treated mice, suggesting that BDPP treatment enhances c-Fos expression, potentially leading to increased neuronal recruitment to memory engrams (Smith et al., 2018).

In a separate investigation, resveratrol (a polyphenolic phytoalexin renowned for its neuroprotective properties) was assessed for its ability to mitigate oxidative stress and memory impairment in obese (ob/ob) mice and Aβ-treated H19-7 rat hippocampal cell line (Rege, 2016). Resveratrol was orally administered at a dose of 25 mg/kg daily for 3 weeks to lean and obese (ob/ob) mice. Additionally, cultured rat hippocampal H19-7 neuronal cell line was pretreated with 75 μM of resveratrol for 2 hrs followed by 25 μM of Aβ (1−40) for 24 hours. Despite no alterations in body weight or blood glucose levels in ob/ob mice, resveratrol treatment significantly decreased lipid peroxide levels in the brains of obese mice and Aβ-treated H19-7 cells. Moreover, resveratrol increased enzymic and nonenzymic antioxidant activities, upregulated the expression of memory-associated proteins, and attenuated the expression of Tau and phosphorylated forms of tau and glial fibrillary acidic protein. These findings underscore the neuroprotective potential of resveratrol in mitigating oxidative damage and memory impairment both in vivo and in vitro (Rege, 2016).

In the realm of BDNF modulation, ovarian hypofunction leads to decreased BDNF, impacting cognitive and memory function. While hormone replacement therapy is common, it poses risks, necessitating safer alternatives. Phytoestrogens like *Hibiscus sabdariffa* L. are promising candidates due to their natural properties. In an ovariectomized rat model, the phytoestrogenic effect of *H. sabdariffa* was evaluated, focusing on spatial memory and BDNF expression. Female Wistar rats were divided into groups, with some receiving *H. sabdariffa* extract orally at doses of 50 or 100 mg/kg b.w. Memory and BDNF expression were assessed after 42 days. Ovariectomy reduced memory performance and BDNF expression, but *H. sabdariffa* extract reversed these effects, especially at lower doses. The extract acted as a phytoestrogen, enhancing spatial memory and BDNF expression (Lorenzana-Martínez et al., 2022).

Additionally, the efficacy and safety of a combination treatment using intranasal administration of rifampicin and resveratrol in mouse models of neurodegenerative dementia has been explored. Resveratrol, a natural polyphenol, was chosen for its hepatoprotective actions and potential to induce BDNF expression. Rifampicin and resveratrol were administered intranasally to mice at doses of 0.02 mg/day and either 0.02 mg/day or 0.1 mg/day, respectively. The combinatorial medicine consisted of 0.02 mg rifampicin + 0.02 mg resveratrol/day, administered 5 days a week for 1 month. BDNF plays a crucial role in synaptic plasticity and memory. Resveratrol alone and combinatorial medicine increased BDNF levels in the hippocampus, while rifampicin alone did not show this effect. The Morris water maze test evaluated cognitive function. Rifampicin alone and resveratrol alone showed some memory improvement, but not significantly. However, the combination significantly improved mouse memory compared to single-drug treatments, supported by probe tests. Synaptophysin levels were restored by the combination treatment, indicating improved synaptic integrity. In summary, the intranasal combination of rifampicin and resveratrol showed promising results in improving memory, reducing oligomer-related pathologies, and enhancing BDNF levels (Umeda et al., 2021).

In human studies, the landscape becomes more intricate. For instance, researchers have delved into the combined effects of eicosapentaenoic acid (EPA) and docosahexaenoic acid (DHA) with cocoa flavanols (OM3FLAV) on cognitive performance and brain structures among older adults with memory complaints. This entailed a randomized placebo-controlled trial involving 259 older adults presenting subjective cognitive impairment or mild cognitive impairment. Participants were administered a daily dose of DHA-rich fish oil providing 1.1 g/d DHA and 0.4 g/d EPA, alongside flavanol-rich dark chocolate offering 500 mg/d flavan-3-ols. Evaluations took place at baseline, 3 months, and 12 months, with the primary outcome focused on the number of false positives on a picture recognition task. Secondary outcomes included various cognitive and mood measures, plasma lipids, BDNF, and glucose levels. Additionally, a subset of 110 participants underwent structural neuroimaging at baseline and 12 months. Despite confirming compliance to the intervention, evidenced by changes in plasma EPA and DHA levels and urinary flavonoid metabolite concentrations, the combined intervention failed to significantly affect cognitive outcomes in the OM3FLAV group, nor did it impact BDNF levels. Thus co-supplementation with omega-3 polyunsaturated fatty acids and cocoa flavanols over 12 months did not yield improved cognitive outcomes in older adults with cognitive impairment (Vauzour et al., 2023).

Further exploration has honed in on the intricate interplay between supplementation with whole coffee cherry extract and cognitive function among older adults grappling with mild cognitive decline. The potential cognitive enhancements associated with coffee cherry extract supplementation may stem from its rich composition of polyphenols and other phytochemicals. Among these compounds, polyphenols stand out for their potent antioxidant and anti-inflammatory properties. Of particular interest is their interaction with BDNF, a pivotal neuroprotein crucial for neurogenesis and cognitive improvement. Employing a robust double-blind, randomized, placebo-controlled 28-day regimen, researchers meticulously scrutinized the impact of coffee cherry extract on cognitive performance, especially in tasks involving working memory processes. In terms of administration, participants were directed to consume encapsulated supplements containing either coffee cherry extract or a placebo twice daily—once in the morning and once in the evening. The study comprised four treatment groups: placebo at both ingestion time points, coffee cherry extract in the morning followed by placebo in the evening, placebo in the morning followed by coffee cherry extract in the evening, and coffee cherry extract in the morning followed by coffee cherry extract in the evening. Each capsule contained 100 mg of either micro-cellulose (placebo) or coffee cherry extract. Crucially, the administration protocol was meticulously designed to optimize absorption, with participants advised to take the capsules on an empty stomach, 15–30 minutes before a meal, and at bedtime, at least 1 hour after their last meal. This timing strategy aimed to mitigate chronotherapeutic effects and ensure consistency across participants. While the study did not uncover a significant main effect on accuracy, discernible trends hinted at enhanced accuracy among those receiving coffee cherry extract. Nonetheless, further investigation is warranted to unravel the intricate mechanisms underpinning these effects and to refine optimal dosage and administration protocols for maximizing cognitive benefits (Robinson et al., 2020).

Altogether, these findings provide valuable insights into the design of future interventions aimed at optimizing memory in older adults. However, further research is warranted to elucidate the specific mechanisms underlying the cognitive benefits of polyphenols, refine optimal dosage and administration protocols, and explore potential synergies with other therapeutic strategies. By continuing to explore the therapeutic potential of dietary polyphenols, we may uncover novel approaches for maintaining cognitive vitality and enhancing the quality of life in aging populations.

Additional considerations about polyphenols and mental functions

As highlighted in a recent study by Naomi et al. (2023), dietary intake of polyphenols at recommended dosages has been suggested as a potential remedy for reversing cognitive dysfunction, with considerations for gender differences, lifestyle factors, and underlying causes of cognitive decline. Unfortunately, despite promising findings in preclinical studies regarding their benefits for learning and memory, unresolved challenges persist, necessitating confirmation through randomized clinical trials.

When evaluating the therapeutic potential of dietary polyphenols, the ability to penetrate the central nervous system is of paramount importance. Factors such as molecular weight, lipophilicity, and the presence of polar groups can significantly influence a compound's capacity to cross the blood–brain barrier. For example, polyphenols like curcumin may exhibit poor central nervous system penetrability due to their low lipophilicity and high molecular weight, while others like epigallocatechin can readily traverse the blood–brain barrier even at low doses (Shimazu et al., 2021).

Another critical factor to consider is the bioavailability of dietary polyphenols, which refers to the proportion of a nutrient that undergoes digestion, absorption, and utilization at the site of action. Many dietary polyphenolic compounds demonstrate low oral bioavailability and are primarily excreted following gastrointestinal passage (Teng & Chen, 2019). Furthermore, high dietary polyphenol intake does not necessarily correlate with high bioavailability, as it depends on various factors such as polyphenol type, food matrix, processing methods, digestive enzymes, intestinal microenvironment, metabolism rate, and gut microbiota composition.

Dosage effects of polyphenols can also vary based on the type as well as the age and strain differences in animals. For example, in human experimental and epidemiological studies, some cognitive benefits from polyphenol consumption have been suggested, albeit with small effects observed. Declarative and spatial memory were identified as particularly sensitive to polyphenols, with effects potentially varying depending on the source. Berry fruit juice consumption improved immediate verbal memory, while isoflavone interventions showed benefits for delayed spatial memory and executive function. However, there is uncertainty regarding polyphenol dose-response relationships, intervention duration and specific populations studied (Lamport et al., 2012).

Excessive consumption of polyphenols can lead to adverse effects, as evidenced by toxicity evaluations in animal studies. Continuous intake of quercetin at high doses induced chronic nephropathy and carcinogenic symptoms in rats (Lu et al., 2006), while other compounds like caffeic acid and green tea catechins also showed carcinogenic potential (Lambert et al., 2010). Some polyphenols may interfere with iron absorption (Lopez-Garcia et al., 2006) or digestive enzyme levels (Gonzalez et al., 2015), while others exhibit pro-oxidant effects (Rietjens et al., 2005) or alter drug pharmacokinetics (Spencer et al., 2001). Overall, while certain polyphenols may offer health benefits like antioxidant and anti-inflammatory effects, others may pose risks like carcinogenic or genotoxic effects at high doses. Evaluating the potential toxicity of polyphenols can be challenging due to variability in types and amounts found in foods, as well as the transformation of polyphenols by gut bacteria into metabolites with different effects. Despite potential benefits, more research is needed to determine optimal doses and potential risks associated with long-term polyphenol use.

Specific polyphenolic compounds and human learning and memory

While comprehensive human studies combining biochemical, electrophysiological/neurophysiological, and behavioral mechanisms are currently lacking, further exploration of pure polyphenolic compounds is warranted to identify those with cognitive benefits. Despite most research focusing on the consumption of whole polyphenol-rich extracts, these studies do not provide insight into the specific compounds responsible for these benefits. Investigating pure compounds presents challenges, necessitating significant time and resources. In this sense, current findings point out the potential of specific polyphenolic compounds to support cognitive health, highlighting the need for further research to elucidate their mechanisms and optimize dosages for broader therapeutic applications (Anton et al., 2018; Evans et al., 2016, 2017; Huhn et al., 2018; Soleas et al., 2001; Thaung Zaw et al., 2021a; Viña et al., 2022; Yimam et al., 2016; Zortea et al., 2016).

For instance, resveratrol has garnered significant attention for its potential health advantages. Despite initial doubts regarding its efficacy due to low bioavailability, recent studies have reignited interest in its therapeutic potential, particularly the trans isomer. Human studies targeting postmenopausal women, the elderly, and schizophrenia patients have yielded mixed results. While some studies reported improved cerebrovascular responsiveness and cognitive performance, others found no significant cognitive improvement with resveratrol supplementation (Soleas et al., 2001). Moreover, studies by Evans et al. (2016, 2017) evaluated resveratrol's effects on cognitive function and cerebrovascular function, showing mixed results as well. While some studies reported improved cerebrovascular responsiveness, others demonstrated significant enhancements in cognitive performance, particularly in verbal memory and overall cognitive function. In addition, a study by Thaung Zaw et al. (2021a) found no significant improvement in cognitive performance compared to placebo, although certain subgroups showed potential benefits. In the elderly, conflicting findings were observed in studies by Anton et al. (2018) and Huhn et al. (2018). While Anton et al. found no significant cognitive improvement with resveratrol supplementation, Huhn et al. reported enhanced overall cognitive performance. Studies in schizophrenia patients, such as that by Zortea et al. (2016), yielded inconclusive results regarding resveratrol's effects on cognitive function.

In a double-blind, placebo-controlled clinical trial involving 24 prodromal Alzheimer's disease patients, daily oral supplementation of 120 mg of genistein, a chemically defined polyphenol, for 12 months was evaluated. The study aimed to delay the progression from minimal cognitive impairment to Alzheimer's dementia, utilizing a nutritional supplement approach. Results showed significant improvements in the Verbal Learning Test and Rey Complex Figure Test. Analysis of amyloid-beta deposition indicated that genistein-treated patients did not exhibit increased uptake in the anterior cingulate gyrus, while placebo-treated patients did. These findings may suggest that genistein supplementation may delay the onset of Alzheimer's dementia in prodromal Alzheimer's disease patients (Viña et al., 2022).

A randomized, double-blind, placebo-controlled study investigated the effects of Shentai Tea Polyphenols on memory in healthy middle-aged and elderly individuals. 240 subjects with normal cognitive function were randomly assigned to receive either Shentai Tea Polyphenols or placebo capsules for 90 days. Neuropsychological assessments and near-infrared brain imaging data were collected before and after intervention. Results showed significant improvements in delayed recall, recognition memory, cognitive assessment scores, and reductions in anxiety and sleep disturbances in the experimental group compared to baseline and the control group. Brain imaging data indicated significant changes in functional connectivity, particularly involving the forehead peak network and its connections with other brain networks. These findings suggest that Shentai Tea Polyphenols may enhance cognitive function and mood while modulating brain network connections in healthy elderly adults (Ni et al., 2020).

Another promising candidate is UP326, a proprietary blend of polyphenol extracts derived from *S. baicalensis* and *Acacia catechu*. This study investigates the cognitive effects of UP326, focusing on memory and learning, using both animal models and human trials with an emphasis on oral administration. In an 11-week experiment with aged F344 rats, UP326 administration at varying doses resulted in significant improvements in memory performance as assessed by radial arm water maze and contextual fear conditioning tests. Furthermore, a separate human clinical trial demonstrated notable enhancements in cognitive function, specifically in speed and accuracy of processing complex information, following oral supplementation of 300 mg of UP326 twice daily for 30 days. These findings underscore the potential of UP326 as a natural supplement to support cognitive health, highlighting the need for further research to elucidate its mechanisms and optimize dosages for broader therapeutic applications (Yimam et al., 2016).

Final comments

The exploration of polyphenols as potential modulators of learning and memory represents a fascinating area of research. While there is promising evidence suggesting that certain polyphenols, such as resveratrol or flavonoids, may exert beneficial effects on learning and memory functions, the current literature lacks consistency and robustness.

The complexity of the relationship between polyphenols and cognitive health underscores the need for further high-quality research to elucidate their true potential. Standardized methodologies, larger sample sizes, and controlled studies are necessary to overcome methodological limitations and generate reliable conclusions. Moreover, the multifaceted nature of the psychological, biochemical, and physiological underpinnings of each type of learning and memory process poses additional challenges. Understanding the precise mechanisms by which polyphenols exert their effects, optimizing dosage and administration protocols, and addressing potential risks are essential considerations for future research.

While the potential cognitive benefits of polyphenols are promising, it is crucial to approach their use with caution. Careful consideration of the implications, including the risk-benefit analysis and potential long-term effects, is necessary; especially in vulnerable populations such as older adults and individuals with cognitive impairments.

Overall, the investigation of polyphenols and their impact on learning and memory represents an exciting frontier in neuroscience and nutritional science. By advancing our understanding of the complex interplay between diet and cognitive health, we may uncover novel strategies for promoting brain health and enhancing mental functions across the lifespan.

Acknowledgment

This work has been funded by the Andalusian Regional Government (grant number P21_00001; Consejería de Universidad, Investigación e Innovación, Junta de Andalucía, Spain).

AI disclosure

During the preparation of this work, the author used Chat GTP and CoPilot in order to improve the readability and language of the work. After using this tool/service, the author reviewed and edited the content as needed and takes full responsibility for the content of the publication.

References

Abhijit, S., Subramanyam, M. V. V., & Devi, S. A. (2017). Grape seed proanthocyanidin and swimming exercise protects against cognitive decline: A study on M1 acetylcholine receptors in aging male rat brain. *Neurochemical Research*, *42*(12), 3573–3586. Available from https://doi.org/10.1007/s11064-017-2406-6, http://www.wkap.nl/journalhome.htm/0364-3190.

Aigner, T. G., & Mishkin, M. (1987). The effects of physostigmine and scopolamine on recognition memory in monkeys. *Alzheimer Disease & Associated Disorders*, *1*(1), 48. Available from https://doi.org/10.1097/00002093-198701000-00025.

Alberini, C. M., & Kandel, E. R. (2015). The regulation of transcription in memory consolidation. *Cold Spring Harbor Perspectives in Biology*, 7(1). Available from https://doi.org/10.1101/cshperspect.a021741, http://cshperspectives.cshlp.org/content/7/1/a021741.full.pdf.

Allam, F., Dao, A. T., Chugh, G., Bohat, R., Jafri, F., Patki, G., Mowrey, C., Asghar, M., Alkadhi, K. A., & Salim, S. (2013). Grape powder supplementation prevents oxidative stress-induced anxiety-like behavior, memory impairment, and high blood pressure in rats. *Journal of Nutrition*, *143*(6), 835–842. Available from https://doi.org/10.3945/jn.113.174649, http://jn.nutrition.org/content/143/6/835.full.pdf.

Anton, S. D., Ebner, N., Dzierzewski, J. M., Zlatar, Z. Z., Gurka, M. J., Dotson, V. M., Kirton, J., Mankowski, R. T., Marsiske, M., & Manini, T. M. (2018). Effects of 90 days of resveratrol supplementation on cognitive function in elders: A pilot study. *Journal of Alternative and Complementary Medicine*, *24*(7), 725–732. Available from https://doi.org/10.1089/acm.2017.0398, http://www.liebertonline.com/acm.

Anwar, H. M., Georgy, G. S., Hamad, S. R., Badr, W. K., El Raey, M. A., Abdelfattah, M. A. O., Wink, M., & Sobeh, M. (2021). A leaf extract of *Harrisonia abyssinica* ameliorates neurobehavioral, histological and biochemical changes in the hippocampus of rats with aluminum chloride-induced Alzheimer's disease. *Antioxidants*, *10*(6). Available from https://doi.org/10.3390/antiox10060947, https://www.mdpi.com/2076-3921/10/6/947/pdf.

Asok, A., Leroy, F., Rayman, J. B., & Kandel, E. R. (2019). Molecular mechanisms of the memory trace. *Trends in Neurosciences*, *42*(1), 14–22. Available from https://doi.org/10.1016/j.tins.2018.10.005, http://www.elsevier.com/locate/tins.

Bairy, L. K., & Kumar, S. (2019). Neurotransmitters and neuromodulators involved in learning and memory. *International Journal of Basic & Clinical Pharmacology*, *8*(12), 2777. Available from https://doi.org/10.18203/2319-2003.ijbcp20195296.

Bekinschtein, P., Katche, C., Slipczuk, L. N., Igaz, L. M., Cammarota, M., Izquierdo, I., & Medina, J. H. (2007). mTOR signaling in the hippocampus is necessary for memory formation. *Neurobiology of Learning and Memory*, *87*(2), 303–307. Available from https://doi.org/10.1016/j.nlm.2006.08.007.

Bourtchuladze, R., Frenguelli, B., Blendy, J., Cioffi, D., Schutz, G., & Silva, A. J. (1994). Deficient long-term memory in mice with a targeted mutation of the cAMP-responsive element-binding protein. *Cell*, *79*(1), 59–68. Available from https://doi.org/10.1016/0092-8674(94)90400-6.

Bowsher, C. S., & Tobin, M. (2008). *Plant biochemistry*. Garland Science.

Citri, A., & Malenka, R. C. (2008). Synaptic plasticity: Multiple forms, functions, and mechanisms. *Neuropsychopharmacology: Official Publication of the American College of Neuropsychopharmacology*, *33*(1), 18–41. Available from https://doi.org/10.1038/sj.npp.1301559.

Collinson, N., Kuenzi, F. M., Jarolimek, W., Maubach, K. A., Cothliff, R., Sur, C., Smith, A., Otu, F. M., Howell, O., Atack, J. R., McKernan, R. M., Seabrook, G. R., Dawson, G. R., Whiting, P. J., & Rosahl, T. W. (2002). Enhanced learning and memory and altered GABAergic synaptic transmission in mice lacking the α5 subunit of the GABAA receptor. *Journal of Neuroscience*, *22*(13), 5572–5580. Available from https://doi.org/10.1523/jneurosci.22-13-05572.2002, http://www.jneurosci.org.

Cooke, S. F., & Bliss, T. V. P. (2006). Plasticity in the human central nervous system. *Brain*, *129*(7), 1659–1673. Available from https://doi.org/10.1093/brain/awl082, http://brain.oxfordjournals.org/.

D'Souza, D., & Avati, A. (2021). *Memory and learning: Basic concepts. Neuro-systemic applications in learning* (pp. 227–240). India: Springer International Publishing. Available from https://link.springer.com/book/10.1007/978-3-030-72400-9, https://doi.org/10.1007/978-3-030-72400-9_11.

Ding, M. L., Ma, H., Man, Y. G., & Lv, H. Y. (2017). Protective effects of a green tea polyphenol, epigallocatechin- 3-gallate, against sevoflurane-induced neuronal apoptosis involve regulation of CREB/BDNF/TrkB and PI3K/Akt/mTOR signalling pathways in neonatal mice. *Canadian Journal of Physiology and Pharmacology*, *95*(12), 1396–1405. Available from https://doi.org/10.1139/cjpp-2016-0333, http://www.nrc.ca/cgi-bin/cisti/journals/rp/rp_desy_e?cjpp.

Dunning, J., & During, M. J. (2003). Molecular mechanisms of learning and memory. *Expert Reviews in Molecular Medicine*, *5*(25). Available from https://doi.org/10.1017/S1462399403006707, http://journals.cambridge.org.

Evans, H., Howe, P., & Wong, R. (2016). Clinical evaluation of effects of chronic resveratrol supplementation on cerebrovascular function, cognition, mood, physical function and general well-being in postmenopausal women—rationale and study design. *Nutrients*, *8*(3), 150. Available from https://doi.org/10.3390/nu8030150.

Evans, H., Howe, P., & Wong, R. (2017). Effects of resveratrol on cognitive performance, mood and cerebrovascular function in post-menopausal women: A 14-week randomised placebo-controlled intervention trial. *Nutrients*, *9*(1), 27. Available from https://doi.org/10.3390/nu9010027.

Flanagan, E., Lamport, D., Brennan, L., Burnet, P., Calabrese, V., Cunnane, S. C., de Wilde, M. C., Dye, L., Farrimond, J. A., Emerson Lombardo, N., Hartmann, T., Hartung, T., Kalliomäki, M., Kuhnle, G. G., La Fata, G., Sala-Vila, A., Samieri, C., Smith, A. D., Spencer, J. P. E., Thuret, S., Tuohy, K., Turroni, S., Vanden Berghe, W., Verkuijl, M., Verzijden, K., Yannakoulia, M., Geurts, L., & Vauzour, D. (2020). Nutrition and the ageing brain: Moving towards clinical applications. *Ageing Research Reviews*, *62*. Available from https://doi.org/10.1016/j.arr.2020.101079, http://www.elsevier.com/locate/arr.

Goelet, P., Castellucci, V. F., Schacher, S., & Kandel, E. R. (1986). The long and the short of long-term memory - A molecular framework. *Nature*, *322*(6078), 419–422. Available from https://doi.org/10.1038/322419a0.

Gonzalez, C. A., Salas-Salvadó, J., & Ros, E. (2015). The potential of nuts in the prevention of cancer. *British Journal of Nutrition*, *113*(S2).

Graham, D. (1981). *Animal learning and conditioning*. University Park Press.

Hasselmo, M. E. (2006). The role of acetylcholine in learning and memory. *Current Opinion in Neurobiology*, *16*(6), 710–715. Available from https://doi.org/10.1016/j.conb.2006.09.002.

He, W. B., Abe, K., & Akaishi, T. (2018). Oral administration of fisetin promotes the induction of hippocampal long-term potentiation in vivo. *Journal of Pharmacological Sciences, 136*(1), 42–45. Available from https://doi.org/10.1016/j.jphs.2017.12.008, http://www.jstage.jst.go.jp/browse/.

Heaney, C. F., & Kinney, J. W. (2016). Role of GABAB receptors in learning and memory and neurological disorders. *Neuroscience and Biobehavioral Reviews, 63*, 1–28. Available from https://doi.org/10.1016/j.neubiorev.2016.01.007, http://www.elsevier.com/locate/neubiorev.

Huhn, S., Beyer, F., Zhang, R., Lampe, L., Grothe, J., Kratzsch, J., Willenberg, A., Breitfeld, J., Kovacs, P., Stumvoll, M., Trampel, R., Bazin, P. L., Villringer, A., & Witte, A. V. (2018). Effects of resveratrol on memory performance, hippocampus connectivity and microstructure in older adults – A randomized controlled trial. *Neuroimage, 174*, 177–190. Available from https://doi.org/10.1016/j.neuroimage.2018.03.023, http://www.elsevier.com/inca/publications/store/6/2/2/9/2/5/index.htt.

Jeremy, P. E., & Spencer. (2008). Flavonoids: modulators of brain function? *British Journal of Nutrition, 99*(E-S1), ES60. Available from https://doi.org/10.1017/s0007114508965776.

Jiang, T., Wang, X.-qin, Ding, C., & Xue-lian, D. (2017). Genistein attenuates isoflurane-induced neurotoxicity and improves impaired spatial learning and memory by regulating cAMP/CREB and BDNF-TrkB-PI3K/Akt signaling. *The Korean Journal of Physiology & Pharmacology, 21*(6), 579. Available from https://doi.org/10.4196/kjpp.2017.21.6.579.

Joseph, J. A., Shukitt-Hale, B., Denisova, N. A., Bielinski, D., Martin, A., McEwen, J. J., & Bickford, P. C. (1999). Reversals of age-related declines in neuronal signal transduction, cognitive, and motor behavioral deficits with blueberry, spinach, or strawberry dietary supplementation. *Journal of Neuroscience, 19*(18), 8114–8121. Available from https://doi.org/10.1523/jneurosci.19-18-08114.1999, http://www.jneurosci.org.

Kandel, E. R. (2012). The molecular biology of memory: CAMP, PKA, CRE, CREB-1, CREB-2, and CPEB. *Molecular Brain, 5*(1). Available from https://doi.org/10.1186/1756-6606-5-14.

Kandel, E. R., Dudai, Y., & Mayford, M. R. (2015). *Learning and memory*. Cold Spring Harbor Laboratory Press.

Katagiri, H., Tanaka, K., & Manabe, T. (2001). Requirement of appropriate glutamate concentrations in the synaptic cleft for hippocampal LTP induction. *European Journal of Neuroscience, 14*(3), 547–553. Available from https://doi.org/10.1046/j.0953-816X.2001.01664.x.

Kennedy, D. O. (2014a). Polyphenols and the human brain: Plant\Secondary Metabolite \ecologic roles and endogenous signaling functions drive benefits. *Advances in Nutrition, 5*(5), 515–533. Available from https://doi.org/10.3945/an.114.006320, http://advances.nutrition.org/content/5/5/515.full.pdf.

Kennedy, D. O. (2014b). *Plants and the human brain*. Oxford University Press.

Kennedy, D. O. (2019). Phytochemicals for improving aspects of cognitive function and psychological state potentially relevant to sports performance. *Sports Medicine, 49*, 39–58. Available from https://doi.org/10.1007/s40279-018-1007-0, http://rd.springer.com/journal/40279.

Kennedy, D. O., & Wightman, E. L. (2011). Herbal extracts and phytochemicals: Plant secondary metabolites and the enhancement of human brain function. *Advances in Nutrition, 2*(1), 32–50. Available from https://doi.org/10.3945/an.110.000117, http://advances.nutrition.org/content/2/1/32.full.pdf.

Kennedy, M. B. (2016). Synaptic signaling in learning and memory. *Cold Spring Harbor Perspectives in Biology, 8*(2), 1–16. Available from https://doi.org/10.1101/cshperspect.a016824, http://cshperspectives.cshlp.org/content/8/2/a016824.full.pdf.

Kim, K. J., Hwang, E. S., Kim, M. J., Rha, C. S., Song, M. C., Maeng, S., Park, J. H., & Kim, D. O. (2022). Effects of phenolic-rich pinus densiflora extract on learning, memory, and hippocampal long-term potentiation in scopolamine-induced amnesic rats. *Antioxidants*, *11*(12). Available from https://doi.org/10.3390/antiox11122497, http://www.mdpi.com/journal/antioxidants.

Kim, Y. H., Lee, Y., Kim, D., Jung, M. W., & Lee, C. J. (2010). Scopolamine-induced learning impairment reversed by physostigmine in zebrafish. *Neuroscience Research*, *67* (2), 156–161. Available from https://doi.org/10.1016/j.neures.2010.03.003.

Knopman, D. S., Boeve, B. F., & Petersen, R. C. (2003). Essentials of the proper diagnoses of mild cognitive impairment, dementia, and major subtypes of dementia. *Mayo Clinic Proceedings*, *78*(10), 1290–1308. Available from https://doi.org/10.4065/78.10.1290, http://www.journals.elsevier.com/mayo-clinic-proceedings.

Kourosh-Arami, M., Komaki, A., & Zarrindast, M. R. (2023). Dopamine as a potential target for learning and memory: contributing to related neurological disorders. *CNS and Neurological Disorders - Drug Targets*, *22*(4), 558–576. Available from https://doi.org/10.2174/1871527321666220418115503, http://www.eurekaselect.com/646/journal/cns-amp-neurological-disorders-drug-targets.

Lambert, J. D., Sang, S., & Yang, C. S. (2010). Possible controversy over dietary polyphenols: Benefits vs risks. *Chemical Research in Toxicology*, *23*(3), 595–605.

Lamport, D. J., Dye, L., Wightman, J. L. D., & Lawton, C. L. (2012). The effects of flavonoid and other polyphenol consumption on cognitive performance: A systematic research review of human experimental and epidemiological studies. *Nutrition and Aging*, *1*(1), 5–25. Available from https://doi.org/10.3233/NUA-2012-0002.

Lamport, D. J., Williams, C. M., & Rendeiro, C. (2021). Polyphenols and cognition in humans: an overview of current evidence from recent systematic reviews and meta-analyses. *Brain Plasticity*, *6*(2), 139–153. Available from https://doi.org/10.3233/bpl-200111.

Lamprecht, R., & LeDoux, J. (2004). Structural plasticity and memory. *Nature Reviews. Neuroscience*, *5*(1), 45–54. Available from https://doi.org/10.1038/nrn1301, http://www.nature.com/nrn/.

Lee, Y., Park, H. R., Lee, J. Y., Kim, J., Yang, S., Lee, C., Kim, K., Kim, H. S., Chang, S. C., & Lee, J. (2023). Low-dose curcumin enhances hippocampal neurogenesis and memory retention in young mice. *Archives of Pharmacal Research*, *46*(5), 423–437. Available from https://doi.org/10.1007/s12272-023-01440-7, https://www.springer.com/journal/12272.

Lewerenz, J., & Maher, P. (2015). Chronic glutamate toxicity in neurodegenerative diseases-What is the evidence? *Frontiers in Neuroscience*, *9*. Available from https://doi.org/10.3389/fnins.2015.00469, http://journal.frontiersin.org/article/10.3389/fnins.2015.00469/full.

Li, Q., Zhao, H. F., Zhang, Z. F., Liu, Z. G., Pei, X. R., Wang, J. B., & Li, Y. (2009). Long-term green tea catechin administration prevents spatial learning and memory impairment in senescence-accelerated mouse prone-8 mice by decreasing Aβ1-42 oligomers and upregulating synaptic plasticity-related proteins in the hippocampus. *Neuroscience*, *163*(3), 741–749. Available from https://doi.org/10.1016/j.neuroscience.2009.07.014.

Liu, E. Y. L., Xu, M. L., Jin, Y., Wu, Q., Dong, T. T. X., & Tsim, K. W. K. (2018). Genistein, a phytoestrogen in soybean, induces the expression of acetylcholinesterase via G protein-coupled receptor 30 in PC12 cells. *Frontiers in Molecular Neuroscience*, *11*. Available from https://doi.org/10.3389/fnmol.2018.00059, http://journal.frontiersin.org/article/10.3389/fnmol.2018.00059/full.

Lombroso, P., & Ogren, M. (2009). Learning and memory, part II: Molecular mechanisms of synaptic plasticity. *Journal of the American Academy of Child and Adolescent Psychiatry*, *48*(1), 5–9. Available from https://doi.org/10.1097/CHI.0b013e318190c4b3, http://www.jaacap.com.

Lopez-Garcia, E., Van Dam, R. M., Willett, W. C., Rimm, E. B., Manson, J. A. E., Stampfer, M. J., Rexrode, K. M., & Hu, F. B. (2006). Coffee consumption and coronary heart disease in men and women: A prospective cohort study. *Circulation, 113*(17), 2045–2053. Available from https://doi.org/10.1161/CIRCULATIONAHA.105.598664.

Lorenzana-Martínez, G., Santerre, A., Andrade-González, I., & Bañuelos-Pineda, J. (2022). Effects of *Hibiscus sabdariffa* calyces on spatial memory and hippocampal expression of BDNF in ovariectomized rats. *Nutritional Neuroscience, 25*(4), 670–680. Available from https://doi.org/10.1080/1028415X.2020.1804095, http://www.tandfonline.com/loi/ynns20#.VvukQLdf1Hh.

Lu, C., Wang, Y., Xu, T., Li, Q., Wang, D., Zhang, L., Fan, B., Wang, F., & Liu, X. (2018). Genistein ameliorates scopolamine-induced amnesia in mice through the regulation of the cholinergic neurotransmission, antioxidant system and the ERK/CREB/BDNF signaling. *Frontiers in Pharmacology, 9*. Available from https://doi.org/10.3389/fphar.2018.01153, http://www.frontiersin.org/Pharmacology.

Lu, K., Gray, M. A., Oliver, C., Liley, H. G., Harrison, D. J., Kay, C. D., & Brattin, D. (2006). Chronic administration of quercetin in drinking water reduces atherosclerosis in apo E-deficient mice. *Journal of Nutrition, 136*(11), 2882–2887.

Lynch, M. A. (2004). Long-Term Potentiation and Memory. *Physiological Reviews, 84*(1), 87–136. Available from https://doi.org/10.1152/physrev.00014.2003.

Maher, P., Akaishi, T., & Abe, K. (2006). Flavonoid fisetin promotes ERK-dependent long-term potentiation and enhances memory. *Proceedings of the National Academy of Sciences, 103*(44), 16568–16573. Available from https://doi.org/10.1073/pnas.0607822103.

Maher, P., Akaishi, T., Schubert, D., & Abe, K. (2010). A pyrazole derivative of curcumin enhances memory. *Neurobiology of Aging, 31*(4), 706–709. Available from https://doi.org/10.1016/j.neurobiolaging.2008.05.020.

Malenka, R. C., & Nicoll, R. A. (1999). Long-term potentiation – A decade of progress? *Science (New York, N.Y.), 285*(5435), 1870–1874. Available from https://doi.org/10.1126/science.285.5435.1870.

Mancini, E., Beglinger, C., Drewe, J., Zanchi, D., Lang, U. E., & Borgwardt, S. (2017). Green tea effects on cognition, mood and human brain function: A systematic review. *Phytomedicine: International Journal of Phytotherapy and Phytopharmacology, 34*, 26–37. Available from https://doi.org/10.1016/j.phymed.2017.07.008, http://www.urbanfischer.de/journals/phytomed.

Martinez-Lozada, Z., & Ortega, A. (2023). Milestone review: Excitatory amino acid transporters – Beyond their expected function. *Journal of Neurochemistry, 165*(4), 457–466. Available from https://doi.org/10.1111/jnc.15809, http://onlinelibrary.wiley.com/journal/10.1111/(ISSN)1471-4159.

Matias, I., Buosi, A. S., & Gomes, F. C. A. (2016). Functions of flavonoids in the central nervous system: Astrocytes as targets for natural compounds. *Neurochemistry International, 95*, 85–91. Available from https://doi.org/10.1016/j.neuint.2016.01.009, http://www.elsevier.com/locate/neuint.

Middei, S., Ammassari-Teule, M., & Marie, H. (2014). Synaptic plasticity under learning challenge. Academic Press Inc., Italy. *Neurobiology of Learning and Memory, 115*, 108–115. Available from https://doi.org/10.1016/j.nlm.2014.08.001, http://www.elsevier.com/inca/publications/store/6/2/2/9/2/4/index.htt.

Moosavi, F., Hosseini, R., Saso, L., & Firuzi, O. (2015). Modulation of neurotrophic signaling pathways by polyphenols. *Drug Design, Development and Therapy, 10*, 23–42. Available from https://doi.org/10.2147/DDDT.S96936, https://www.dovepress.com/getfile.php?fileID = 28451.

Nan, S., Wang, P., Zhang, Y., & Fan, J. (2021). Epigallocatechin-3-gallate provides protection against alzheimer's disease-induced learning and memory impairments in rats. *Drug*

Design, Development and Therapy, *15*, 2013–2024. Available from https://doi.org/10.2147/DDDT.S289473, https://www.dovepress.com/getfile.php?fileID = 69362.

Naomi, R., Yazid, M. D., Teoh, S. H., Balan, S. S., Shariff, H., Kumar, J., Bahari, H., & Embong, H. (2023). Dietary polyphenols as a protection against cognitive decline: evidence from animal experiments; mechanisms and limitations. *Antioxidants*, *12*(5). Available from https://doi.org/10.3390/antiox12051054, http://www.mdpi.com/journal/antioxidants.

Neshatdoust, S., Saunders, C., Castle, S. M., Vauzour, D., Williams, C., Butler, L., Lovegrove, J. A., & Spencer, J. P. E. (2016). High-flavonoid intake induces cognitive improvements linked to changes in serum brain-derived neurotrophic factor: Two randomised, controlled trials. *Nutrition and Healthy Aging*, *4*(1), 81–93. Available from https://doi.org/10.3233/nha-1615.

Neve, K. A. (2013). *Dopamine receptors. Encyclopedia of biological chemistry: Second Edition* (pp. 169–173). United States: Elsevier Inc. Available from http://www.sciencedirect.com/science/book/9780123786319, https://doi.org/10.1016/B978-0-12-378630-2.00326-1.

Ni, L., Hu, Z., Yang, Q., Liu, Y., Lin, L., Niu, H., & Han, Y. (2020). Improvement of memory in healthy elderly subjects by Shentai Tea Polyphenols. *Chinese Medical Association, China Chinese Journal of Neuromedicine*, *19*(11), 1142–1148. Available from https://doi.org/10.3760/cma.j.cn115354-20201016-00816, http://zhsjyxzz.yiigle.com/?lang = En.

Olsen, R. W., & DeLorey, T. M. (1999). *GABA receptor physiology and pharmacology*. In Siegel G. J., Agranoff B. W., & Albers R. W. (Eds.) et al., *Basic Neurochemistry: Molecular, Cellular and Medical Aspects (6th edition)*. Lippincott-Raven.

Ortega-Martínez, S. (2015). A new perspective on the role of the CREB family of transcription factors in memory consolidation via adult hippocampal neurogenesis. *Frontiers in Molecular Neuroscience*, *8*(AUGUST). Available from https://doi.org/10.3389/fnmol.2015.00046, http://journal.frontiersin.org/article/10.3389/fnmol.2015.00046/pdf.

Pais, R., Ruano, L., P. Carvalho, O., & Barros, H. (2020). Global cognitive impairment prevalence and incidence in community dwelling older adults—A systematic review. *Geriatrics*, *5*(4), 84. Available from https://doi.org/10.3390/geriatrics5040084.

Pearce, J. M., & Bouton, M. E. (2001). Theories of associative learning in animals. *Annual Review of Psychology*, *52*, 111–139. Available from https://doi.org/10.1146/annurev.psych.52.1.111.

Pitt, J., & Leung, Y. (2016). *Cognitive effects of nutraceuticals. Nutraceuticals: Efficacy, safety and toxicity* (pp. 29–48). United States: Elsevier Inc. Available from http://www.sciencedirect.com/science/book/9780128021477, https://doi.org/10.1016/B978-0-12-802147-7.00003-6.

Puzzo, D., Fiorito, J., Purgatorio, R., Gulisano, W., Palmeri, A., Arancio, O., & Nicholls, R. (2016). *Molecular mechanisms of learning and memory genes. Environment and Alzheimer's Disease* (pp. 1–27). Italy: Elsevier Inc. Available from http://www.sciencedirect.com/science/book/9780128028513, https://doi.org/10.1016/B978-0-12-802851-3.00001-2.

Rajput, M. S., & Sarkar, P. D. (2017). Modulation of neuro-inflammatory condition, acetylcholinesterase and antioxidant levels by genistein attenuates diabetes associated cognitive decline in mice. *Chemico-Biological Interactions*, *268*, 93–102. Available from https://doi.org/10.1016/j.cbi.2017.02.021, http://www.elsevier.com/locate/chembioint.

Ramezani, M., Meymand, A. Z., Khodagholi, F., Kamsorkh, H. M., Asadi, E., Noori, M., Rahimian, K., Shahrbabaki, A. S., Talebi, A., Parsaiyan, H., Shiravand, S., & Darbandi, N. (2023). A role for flavonoids in the prevention and/or treatment of cognitive dysfunction, learning, and memory deficits: a review of preclinical and clinical studies. *Nutritional Neuroscience*, *26*(2), 156–172. Available from https://doi.

org/10.1080/1028415X.2022.2028058, http://www.tandfonline.com/loi/ynns20#.VvukQLdf1Hh.

Ramirez-Amaya, V. (2007). *Molecular mechanisms of synaptic plasticity underlying long-term memory formation. Neural Plasticity and Memory: From Genes to Brain Imaging* (pp. 47–66). Mexico: CRC Press. Available from https://www.taylorfrancis.com/books/e/9781420008418.

Rana, A., Samtiya, M., Dhewa, T., Mishra, V., & Aluko, R. E. (2022). Health benefits of polyphenols: A concise review. *Journal of Food Biochemistry*, *46*(10). Available from https://doi.org/10.1111/jfbc.14264, http://onlinelibrary.wiley.com/journal/10.1111/(ISSN)1745-4514.

Rasouli, H., Farzaei, M. H., & Khodarahmi, R. (2017). Polyphenols and their benefits: A review. *International Journal of Food Properties*, *20*, 1700–1741. Available from https://doi.org/10.1080/10942912.2017.1354017, http://www.tandf.co.uk/journals/titles/10942912.asp.

Rebas, E., Rzajew, J., Radzik, T., & Zylinska, L. (2020). Neuroprotective polyphenols: A modulatory action on neurotransmitter pathways. *Current Neuropharmacology*, *18*(5), 431–445. Available from https://doi.org/10.2174/1570159X18666200106155127, http://www.eurekaselect.com/178126/article.

Rege, S. D. (2016). *Neuroprotective effects of resveratrol against oxidative stress and memory impairment in vivo and in vitro* (Doctoral dissertation). Auburn University.

Rendeiro, C., Guerreiro, J. D. T., Williams, C. M., & Spencer, J. P. E. (2012). Flavonoids as modulators of memory and learning: molecular interactions resulting in behavioural effects. *Proceedings of the Nutrition Society*, *71*(2), 246–262. Available from https://doi.org/10.1017/s0029665112000146.

Reynolds, C. R., & Bigler, E. D. (1997). *Clinical Neuropsychological Assessment of Child and Adolescent Memory with the Test of Memory and Learning* (pp. 296–319). Springer Science and Business Media LLC. Available from https://doi.org/10.1007/978-1-4757-5351-6_15.

Rietjens, I. M. C. M., Boersma, M. G., Van Der Woude, H., Jeurissen, S. M. F., Schutte, M. E., & Alink, G. M. (2005). Flavonoids and alkenylbenzenes: Mechanisms of mutagenic action and carcinogenic risk. *Mutation Research - Fundamental and Molecular Mechanisms of Mutagenesis*, *574*(1-2), 124–138. Available from https://doi.org/10.1016/j.mrfmmm.2005.01.028.

Ríos, J. L., Schinella, G. R., & Moragrega, I. (2022). Phenolics as GABAA receptor ligands: An updated review. *Molecules*, *27*(6). Available from https://doi.org/10.3390/molecules27061770, https://www.mdpi.com/1420-3049/27/6/1770/pdf.

Robinson, J. L., Hunter, J. M., Reyes-Izquierdo, T., Argumedo, R., Brizuela-Bastien, J., Keller, R., & Pietrzkowski, Z. J. (2020). Cognitive short- and long-term effects of coffee cherry extract in older adults with mild cognitive decline. *Aging, Neuropsychology, and Cognition*, *27*(6), 918–934. Available from https://doi.org/10.1080/13825585.2019.1702622, http://www.tandf.co.uk/journals/titles/13825585.asp.

Rudolph, U., & Möhler, H. (2006). GABA-based therapeutic approaches: GABAA receptor subtype functions. *Current Opinion in Pharmacology*, *6*(1), 18–23. Available from https://doi.org/10.1016/j.coph.2005.10.003.

Schmidt, A., Hammann, F., Wölnerhanssen, B., Meyer-Gerspach, A. C., Drewe, J., Beglinger, C., & Borgwardt, S. (2014). Green tea extract enhances parieto-frontal connectivity during working memory processing. *Psychopharmacology*, *231*(19), 3879–3888. Available from https://doi.org/10.1007/s00213-014-3526-1, http://link.springer.de/link/service/journals/00213/index.htm.

Schummers, J., & Browning, M. D. (2001). Evidence for a role for GABAA and NMDA receptors in ethanol inhibition of long-term potentiation. *Molecular Brain Research*, *94*(1-2), 9–14. Available from https://doi.org/10.1016/S0169-328X(01)00161-9.

Shimazu, R., Anada, M., Miyaguchi, A., Nomi, Y., & Matsumoto, H. (2021). Evaluation of blood-brain barrier permeability of polyphenols, anthocyanins, and their metabolites. *Journal of Agricultural and Food Chemistry*, *69*(39), 11676–11686. Available from https://doi.org/10.1021/acs.jafc.1c02898, http://pubs.acs.org/journal/jafcau.

Singla, R. K., Dubey, A. K., Garg, A., Sharma, R. K., Fiorino, M., Ameen, S. M., Haddad, M. A., & Al-Hiary, M. (2019). Natural polyphenols: Chemical classification, definition of classes, subcategories, and structures. *Journal of AOAC International*, *102*(5), 1397–1400. Available from https://doi.org/10.5740/jaoacint.19-0133, https://www.ingentaconnect.com/content/aoac/jaoac/2019/00000102/00000005/art00021?crawler = true&mimetype = application/pdf.

Smith, C., Frolinger, T., Brathwaite, J., Sims, S., & Pasinetti, G. M. (2018). Dietary polyphenols enhance optogenetic recall of fear memory in hippocampal dentate gyrus granule neuron subpopulations. *Communications Biology*, *1*(1). Available from https://doi.org/10.1038/s42003-018-0043-5, http://nature.com/commsbio/.

Soleas, G. J., Diamandis, E. P., & Goldberg, D. M. (2001). The world of resveratrol. *Advances in Experimental Medicine and Biology*, *492*, 159–182. Available from https://doi.org/10.1007/978-1-4615-1283-7_13.

Spencer, J. P. E. (2007). The interactions of flavonoids within neuronal signalling pathways. *Genes and Nutrition*, *2*(3), 257–273. Available from https://doi.org/10.1007/s12263-007-0056-z.

Spencer, J. P. E. (2009). The impact of flavonoids on memory: Physiological and molecular considerations. *Chemical Society Reviews*, *38*(4), 1152–1161. Available from https://doi.org/10.1039/b800422f.

Spencer, J. P. E., Schroeter, H., Rechner, A. R., & Rice-Evans, C. (2001). Bioavailability of flavan-3-ols and procyanidins: Gastrointestinal tract influences and their relevance to bioactive forms in vivo. *Antioxidants and Redox Signaling*, *3*(6), 1023–1039. Available from https://doi.org/10.1089/152308601317203558, http://www.liebertonline.com/ars.

Squire, L. R. (2004). Memory systems of the brain: A brief history and current perspective. *Neurobiology of Learning and Memory*, *82*(3), 171–177. Available from https://doi.org/10.1016/j.nlm.2004.06.005.

Squire, L. R., & Dede, A. J. O. (2015). Conscious and unconscious memory systems. *Cold Spring Harbor Perspectives in Biology*, 7(3). Available from https://doi.org/10.1101/cshperspect.a021667, http://cshperspectives.cshlp.org/content/7/3/a021667.full.pdf.

Subhan, S., Zeeshan, M., Rahman, A. U., & Yaseen, M. (2024). Fundamentals of polyphenols: Nomenclature, classification and properties. *Science and Engineering of Polyphenols: Fundamentals and Industrial Scale Applications*.

Tao, X., Finkbeiner, S., Arnold, D. B., Shaywitz, A. J., & Greenberg, M. E. (1998). Ca2 + influx regulates BDNF transcription by a CREB family transcription factor-dependent mechanism. *Neuron*, *20*(4), 709–726. Available from https://doi.org/10.1016/S0896-6273(00)81010-7, http://www.neuron.org.

Teng, H., & Chen, L. (2019). Polyphenols and bioavailability: an update. *Critical Reviews in Food Science and Nutrition*, *59*(13), 2040–2051. Available from https://doi.org/10.1080/10408398.2018.1437023, https://www.tandfonline.com/loi/bfsn20.

Thaung Zaw, J. J., Howe, P. R. C., & Wong, R. H. X. (2021a). Long-term effects of resveratrol on cognition, cerebrovascular function and cardio-metabolic markers in postmenopausal women: A 24-month randomised, double-blind, placebo-controlled, crossover study. *Clinical Nutrition*, *40*(3), 820–829. Available from https://doi.org/10.1016/j.clnu.2020.08.025.

Thaung Zaw, J. J., Howe, P. R., & Wong, R. H. (2021b). Long-term effects of resveratrol on cognition, cerebrovascular function and cardio-metabolic markers in postmenopausal women: A 24-month randomised, double-blind, placebo-controlled, crossover study. *Clinical Nutrition*, *40*(3), 820–829. Available from https://doi.org/10.1016/j.clnu.2020.08.025, http://www.elsevier-international.com/journals/clnu/.

Tully, T., Bourtchouladze, R., Scott, R., & Tallman, J. (2003). Targeting the creb pathway for memory enhancers. *Nature Reviews. Drug Discovery*, *2*(4), 267–277. Available from https://doi.org/10.1038/nrd1061, http://www.nature.com/nrd/index.html.

Umeda, T., Sakai, A., Shigemori, K., Yokota, A., Kumagai, T., & Tomiyama, T. (2021). Oligomer- targeting prevention of neurodegenerative dementia by intranasal rifampicin and resveratrol combination – A preclinical study in model mice. *Frontiers in Neuroscience*, *15*. Available from https://doi.org/10.3389/fnins.2021.763476, https://www.frontiersin.org/journals/neuroscience#.

Vauzour, D. (2017). Polyphenols and brain health. *OCL – Oilseeds and fats, Crops and Lipids*, *24*(2). Available from https://doi.org/10.1051/ocl/2017008, http://www.ocl-journal.org/.

Vauzour, D., Scholey, A., White, D. J., Cohen, N. J., Cassidy, A., Gillings, R., Irvine, M. A., Kay, C. D., Kim, M., King, R., Legido-Quigley, C., Potter, J. F., Schwarb, H., & Minihane, A. M. (2023). A combined DHA-rich fish oil and cocoa flavanols intervention does not improve cognition or brain structure in older adults with memory complaints: results from the CANN randomized, controlled parallel-design study. *American Journal of Clinical Nutrition*, *118*(2), 369–381. Available from https://doi.org/10.1016/j.ajcnut.2023.06.008, https://www.sciencedirect.com/journal/the-american-journal-of-clinical-nutrition/about/aims-and-scope.

Vauzour, D. (2012). Dietary polyphenols as modulators of brain functions: biological actions and molecular mechanisms underpinning their beneficial effects. *Oxidative Medicine and Cellular Longevity*, *2012*, 1–16. Available from https://doi.org/10.1155/2012/914273.

Villarreal, D. M., Do, V., Haddad, E., & Derrick, B. E. (2002). NMDA receptor antagonists sustain ltp and spatial memory: Active processes mediate LTP decay. *Nature Neuroscience*, *5*(1), 48–52. Available from https://doi.org/10.1038/nn776.

Viña, J., Escudero, J., Baquero, M., Cebrián, M., Carbonell-Asíns, J. A., Muñoz, J. E., Satorres, E., Meléndez, J. C., Ferrer-Rebolleda, J., Cózar-Santiago, M. P., Santabárbara-Gómez, J. M., Jové, M., Pamplona, R., Tarazona-Santabalbina, F. J., & Borrás, C. (2022). Genistein effect on cognition in prodromal Alzheimer's disease patients. The GENIAL clinical trial. *Alzheimer's Research & Therapy*, *14*(1). Available from https://doi.org/10.1186/s13195-022-01097-2.

Wang, W., Wang, F., Yang, Y. J., Hu, Z. L., Long, L. H., Fu, H., Xie, N., & Chen, J. G. (2011). The flavonoid baicalein promotes NMDA receptor-dependent long-term potentiation and enhances memory. *British Journal of Pharmacology*, *162*(6), 1364–1379. Available from https://doi.org/10.1111/j.1476-5381.2010.01143.x.

Williams, C. M., El Mohsen, M. A., Vauzour, D., Rendeiro, C., Butler, L. T., Ellis, J. A., Whiteman, M., & Spencer, J. P. E. (2008). Blueberry-induced changes in spatial working memory correlate with changes in hippocampal CREB phosphorylation and brain-derived neurotrophic factor (BDNF) levels. *Free Radical Biology and Medicine*, *45*(3), 295–305. Available from https://doi.org/10.1016/j.freeradbiomed.2008.04.008.

Williams, R. J., Spencer, J. P. E., & Rice-Evans, C. (2004). Flavonoids: Antioxidants or signalling molecules? *Free Radical Biology and Medicine*, *36*(7), 838–849. Available from https://doi.org/10.1016/j.freeradbiomed.2004.01.001.

Wu, P. H., Shen, Y. C., Wang, Y. H., Chi, C. W., & Yen, J. C. (2006). Baicalein attenuates methamphetamine-induced loss of dopamine transporter in mouse striatum. *Toxicology*, *226*(2-3), 238–245. Available from https://doi.org/10.1016/j.tox.2006.06.015.

Yabuki, Y., Ohizumi, Y., Yokosuka, A., Mimaki, Y., & Fukunaga, K. (2014). Nobiletin treatment improves motor and cognitive deficits seen in MPTP-induced parkinson model mice. *Neuroscience*, *259*, 126–141. Available from https://doi.org/10.1016/j.neuroscience.2013.11.051, http://www.elsevier.com/locate/neuroscience.

Yimam, M., Burnett, B. P., Brownell, L., & Jia, Q. (2016). Clinical and preclinical cognitive function improvement after oral treatment of a botanical composition composed of extracts from *Scutellaria baicalensis* and *Acacia catechu*. *Behavioural Neurology*, *2016*. Available from https://doi.org/10.1155/2016/7240802, http://www.hindawi.com/journals/bn/.

Zhang, Y., He, F., Hua, T., & Sun, Q. (2018). Green tea polyphenols ameliorate ethanol-induced spatial learning and memory impairments by enhancing hippocampus NMDAR1 expression and CREB activity in rats. *Neuroreport*, *29*(18), 1564–1570. Available from https://doi.org/10.1097/WNR.0000000000001152, http://journals.lww.com/neuroreport.

Zortea, K., Franco, V. C., Guimarães, P., & Belmonte-de-Abreu, P. S. (2016). Resveratrol supplementation did not improve cognition in patients with schizophrenia: results from a randomized clinical trial. *Frontiers in Psychiatry*, 7. Available from https://doi.org/10.3389/fpsyt.2016.00159.

Zyablitseva, E. A., Kositsyn, N. S., & Shul'Gina, G. I. (2009). The effects of agonists of ionotropic GABAA and metabotropic GABAB receptors on learning. *Spanish Journal of Psychology*, *12*(1), 12–20. Available from https://doi.org/10.1017/S1138741600001438, http://www.ucm.es/BUCM/revistasBUC/portal/modulos.php?name = Revistas2&id = SJOP.

SECTION 4

Polyphenol effects in cancer

CHAPTER EIGHT

Antigenotoxic and antitumor properties of polyphenols

Vipendra Kumar Singh[1], Vikas Kumar Singh[2], Arjun Kumar Mehara[3], Vikky Kumar Mahto[3] and Ankit Kumar Singh[3]

[1]School of Biosciences and Bioengineering, Indian Institute of Technology Mandi, VPO Kamand, Mandi, Himachal Pradesh, India
[2]Department of Science & Humanities, Supaul College of Engineering, Supaul, Bihar, India
[3]University Department of Botany, Lalit Narayan Mithila University, Darbhanga, Bihar, India

Introduction

Polyphenol is a class of organic compounds containing numerous phenolic rings. Naturally occurring polyphenol is a secondary metabolite of plants and its rich sources are olive oil, cocoa, nuts, peanuts, blueberries, grapes, apples, and other fruits, vegetables, cereals, etc. (Roszkowski, 2023). However, the distribution of polyphenolic compounds in plants is not uniform at the cellular and subcellular levels, soluble phenolic compounds are found in vacuoles whereas insoluble phenols are mainly present in the cell walls (Dini & Grumetto, 2022). In plant cells, phenolic compounds are derived from pentose phosphate, shikimate, and phenylpropanoid pathways that maintain the physiology of plants such as defense, pigmentation, lignifications, etc. (Kumar et al., 2014). Plant phenolic compounds have strong antimicrobial properties therefore widely used as food preservatives in the industry. Moreover, polyphenols act as building blocks in cosmetics, functional food, supplements, and drug industries (Becker et al., 2017; Li & Chemat, 2019). Presently, polyphenolic compounds have gained a lot of attention in the scientific community due to their potential use as prophylactic, antioxidant as well as therapeutic agents in many diseases (Singh & Yadav, 2022). Polyphenolic compounds act as free radical scavengers and protect against several diseases such as cancer, diabetes, cardiovascular and neurodegenerative diseases, autoimmune disorders, and some inflammatory diseases (Dini et al., 2020; Vladimir-Knežević et al., 2012). Among these,

Bioactive Polyphenols for Health and Pathology Treatment.
DOI: https://doi.org/10.1016/B978-0-443-13821-8.00008-1

Cancer is one the most dangerous and life-threatening diseases affecting 10 million people globally. According to a WHO report the risk of cancer disease will increase from 14 million to 27 million by the year 2050 (Sharma et al., 2018). The risk of cancer diseases is global but people in low and middle-income countries are more affected because of the lack of proper healthcare accessibility and advanced medical treatment (Farghadani & Naidu, 2023). The higher rate of cancer-related death may be due to a lack of timely diagnosis, adequate screening services, and inappropriate treatment (Pramesh et al., 2022). Cancer disease consists of different stages such as initiation, promotion, progression, invasion, and metastasis (Asif et al., 2021). However, cancer-related mortality is mainly due to metastasis. In the present scenario, the capability of cancer treatment has been not fully developed although several combinations of medication and chemotherapy drugs such as taxane, fluorouracil, cyclophosphamide, anthracycline, and cisplatin have been used for the treatment of cancer (Yin et al., 2020). However, the use of several medications and chemotherapy can only help the patient for a month or year. Still, the side effects of these treatments adversely affect the normal cells and develop drug resistance in the patient (Sharma et al., 2022). Currently, plant-based medicine and remedies have gained more attention due to their excellent role in the prevention and treatment of various diseases including cancer. Anticancer drugs derived from phytochemical and natural resources have been gaining more interest in recent years due to fewer side effects and excellent efficacy (Sharifi-Rad et al., 2021). Among the plant-based anticancer drugs, polyphenolic compounds attracted the scientific community because they can be used either individually or combined in a group to design medicine for the treatment of cancer with excellent efficacy (Azmi et al., 2006). Moreover, plant-based polyphenolic compounds can also be used in chemotherapy and radiotherapy for the treatment of cancer because they have negligible side effects on normal cells and other organs (Berdeja et al., 2021). Polyphenolic compounds have the potential ability against cancer cells because they can produce the expression of p21, p53, and p27 genes which can induce cell cycle arrest at the G1, S, and G2 phases (Dai & Mumper, 2010).

This chapter mainly focused on the antigenotoxic and antitumor potential of different plant-based polyphenols such as curcumin, ellagic acid, luteolin, epigallocatechin gallate (EGCG), quercetin, genistein, apigenin, protocatechuic acid, silibinin, and tannic acid (Brimson et al., 2021; Majrashi et al., 2023) (as shown in Fig. 8.1). Apart from this, this chapter also paid considerable attention to the mechanism of action of different polyphenolic

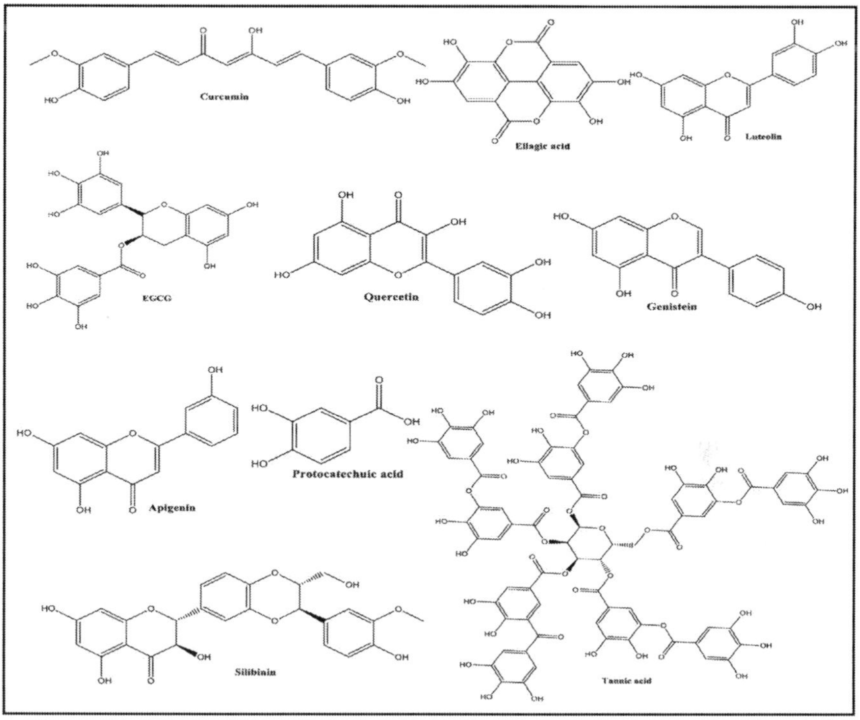

Figure 8.1 Chemical structure of some polyphenolic compounds.

compounds for antigenotoxic and antitumor effects and opened a new avenue for the scientific community the develop antigenotoxic and anticancer polyphenolic drugs from plant-based secondary metabolites.

The antigenotoxic potential of polyphenolic compounds

The genotoxic agents are chemicals or free radicals capable of causing damage to deoxyribonucleic acid (DNA), chromosomes, cell membrane, and cellular components such as lipids and proteins involved in the repair, condensation, and decondensation process of DNA in the chromosomes, mitotic spindle, and separation of chromosomes during cell division (Hartwig et al., 2020; Phillips & Arlt, 2009). During the biological processes, UV light radiation, environmental pollution, tobacco smoking, and consumption of contaminated food and water, cells and tissues are constantly exposed to oxidative stress such as reactive oxygen species (ROS) [such as superoxide ($.O_2$), hydrogen peroxide (H_2O_2) and hydroxyl radical (.OH)] and reactive nitrogen species (RNS) [such as nitric oxide (NO)] have to be kept under strict control by inner enzymatic antioxidant systems (such as catalases, superoxide dismutase [SOD], and glutathione peroxidases) otherwise the loss of equilibrium may lead to the development of serious diseases like cancers, cardiovascular, hepatic, neurodegenerative, diabetes, arthritis, chronic inflammation, and aging (Afzal et al., 2023; Juan et al., 2021; Singh & Sharma, 2020).

The plant contains polyphenolic compounds that show antioxidant activity and protect cells against the harmful effects of ROS. These compounds donate electrons to ROS, chelating metal ions stimulating antioxidant enzymes, and protecting cellular components such as lipids, proteins, and DNA from oxidative damage (Fig. 8.2) (Kasote et al., 2015; Rudrapal et al., 2022). Therefore in recent years, the exploration of natural resources such as antioxidants, antitumor, and antigenotoxic compounds has become a popular research area worldwide (Izquierdo-Vega et al., 2017; Marino et al., 2022). Traditionally many plants and herbs are used as medicine around the globe, plant products are considered healthy due to their natural origin and

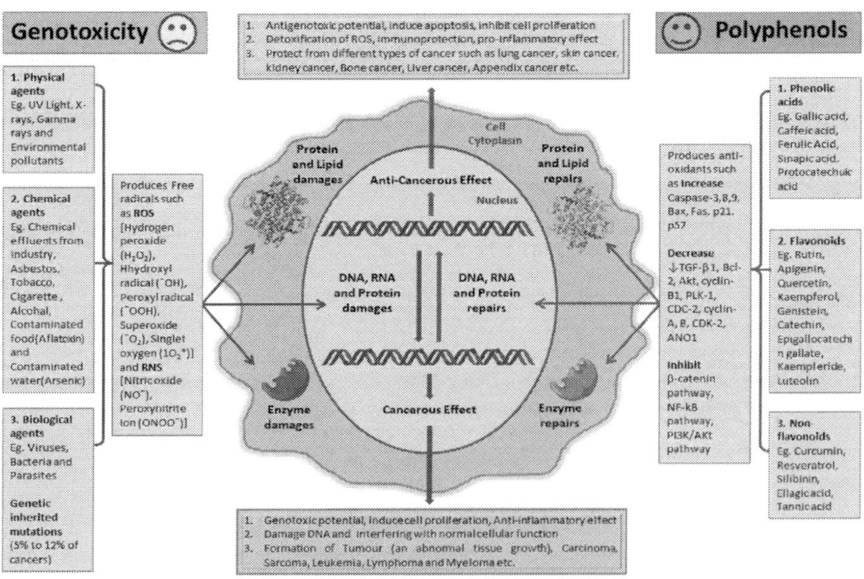

Figure 8.2 Mechanism of action of polyphenolic compounds to overcome genotoxicity.

their cultivation is eco-friendly (Chen et al., 2016; Ekor, 2014). In recent years, the use of plant-derived herbal products and medicines has constantly grown in the market, due to their beneficial effect on human health. It's time for researchers to focus on the polyphenolic compounds for their health-beneficial uses, and their role in controlling multiple serious diseases (Rahman et al., 2021; Rudrapal et al., 2022). Phenolic compounds (or polyphenols) are a large group of secondary metabolites widely distributed in various plant-based products such as flowers, vegetables, fruits, spices, grains, legumes, nuts, etc. Polyphenols comprise a large group of flavonoids and non-flavonoid compounds. In recent years studies, have shown that multiple numbers of phenolic compounds present in *Dendrobii herba*, *Taraxacum coreanum* which contain many natural antioxidant, antimicrobial, antimutagenic, anticarcinogenic, and anti-inflammatory properties. Here, we are going to discuss the antigenotoxic and antioxidant properties of some plant extracts polyphenols which have been demonstrated by reports previously.

Preclinical evidence of the antigenotoxic potential of polyphenolic compounds

Genotoxicity is the potential of distinct chemicals to induce damage to genetic materials. However, the damage caused in the genetic materials is mainly not only DNA but also the other cellular components associated with the usefulness and behavior of chromosomes within the cell. The classical example related to this is the proteins participating in the repair, decondensation of DNA in the chromosomes, condensation, or several other structures like the mitotic spindle, associated with the distribution of the chromosomes at the time of cell division. The agents efficient in causing genetic toxicity are defined as genotoxins or genotoxic. Genotoxic agents are directly associated with the initiation and promotion of almost every solid tumor, the importance of novel polyphenolic compounds plays a crucial role in counteracting the negative impacts of mutagenic and carcinogenic compounds is now seeking attention. Plant-based bioactive polyphenolic drugs that suppress the mutagenicity of chemical and physical mutagens are known as antimutagens. However, because of that, all mutagens are genotoxic, but in reverse, not all genotoxic agents are mutagenic, all the compounds that decrease the DNA damage caused by genotoxic substances are also known as antigenotoxic compounds. In recent years, various studies have been conducted to identify antigenotoxic compounds that might have a protective role against DNA damage and its consequences. There are huge efforts worldwide to explore antigenotoxic plant-derived phenolic compounds that are cost-effective and cause low cytotoxicity. García et al. showed that green tea polyphenol extracts induced antigenotoxic and apoptotic activity in chromium-induced DNA damage in the peripheral blood of CD-1 mice (García-Rodríguez et al., 2013). Resveratrol has shown antigenotoxic effects in HL60 cell lines and Swiss albino mice against the genotoxic effect induced by genotoxins nitroquinoline-1-oxide and mitomycin-C and diepoxybutane, methyl methanesulfonate (MMC) and procarbazine (Abraham et al., 2016). Quercetin has shown an antigenotoxic effect against VPA-induced genetic damage induced by valproic acid (VPA). This study demonstrated that polyphenolic compounds such as oral dietary supplements or quercetin as formulations may reduce VPA-induced toxicity (Canbolat et al., 2024). Galangin, a polyphenolic flavonoid, is present in several medicinal plants such as *Propolis* and *Alpinia officinarum* have antigenotoxic activity. In recent past, in vitro, and in vivo studies demonstrated that galangin reduced the genotoxic potential of chemicals by enhancing anti-oxidative and free radical scavenging potentials and altering enzyme activities (Heo et al., 2001). Novel lignin-derived polyphenolic composition (BP-C2) with ammonium molybdate showed antimutagenic and antigenotoxic potential in the bone marrow, blood, and liver cells of BALB/c mice against genotoxic chemicals such as dioxidine and cyclophosphamide (Zhanataev et al., 2022). In vitro-cultivated shoots of *Polyscias filicifolia* have high levels of phenolic compounds and triterpenoid saponins, contributing to their antioxidant properties. These extracts have shown significant antigenotoxic effects against 2-amino anthracene with metabolic activation, reduced genotoxicity induced by mitomycin-C, but increased genotoxicity with 4-nitroquinoline-N-oxide. The phenolic acids had lower activity levels compared to the extracts. Overall, these shoots have promising properties for antimutagenic, photo genotoxicity prevention, and antioxidant therapies (Figat et al., 2020). The ethanolic extract of the aerial part of the *Polygonum maritimum* L. (PME) showed strong antioxidant and antigenotoxic properties containing high levels of polyphenols like quercetin, catechin, myricetin glycosides, (epi)catechin dimer, and trimers. PME scavenged 2,2-Diphenyl-1-picrylhydrazyl (DPPH) radicals, improved yeast viability under oxidative stress, protect against DNA damage, and exhibited an antigenotoxic effect against H_2O_2-induced oxidative stress in yeast, indicating its potential to alleviate

DNA damage caused by oxidative stress (Oliveira et al., 2023). Abiola et al. examined the genotoxic effects of monosodium glutamate (MSG) and the potential antigenotoxic effects of *Vernonia amygdalina* leaf extract. The extract was analyzed for its compounds and tested in-silico against various proteins [NF-kB 50, Bcl-2, superoxide dismutase (SOD), DNA polymerase lambda, catalase, glutathione peroxidase (GPX), and DNA ligase]. In the experiment, rats were divided into three groups and administered MSG or MSG followed by *V. amygdalina* extract. The extract contained alkaloids, saponins, flavonoids, phenols, and tannins. In-silico studies showed strong binding interactions between vernoniosides-D and E and the proteins tested. The treated group showed a decrease in micro nucleated cells, eosinophils, and neutrophils, and an increase in lymphocyte count compared to the MSG group suggesting that *V. amygdalina* leaf extract may have antigenotoxic properties that could be beneficial in preventing diseases like cancer (Abiola et al., 2023). The nontoxic concentration of Hispolon a bioactive polyphenol and its derivative showed significant protection against 4 Gy radiation-induced damage of DNA as evaluated by classical genotoxic methods such as γ-H2AX and micronuclei in Hamster ovary cells (Chethna et al., 2018). Zor et al. investigated the antigenotoxic potential of *Paliurus* spina-christi Mill fruits and its bioactive components such as gallocatechin, rutin, and catechins in Chinese hamster lung fibroblast (V79) cell lines. They found that compounds significantly reduced H_2O_2-induced DNA damage (Zor et al., 2017). Carranza et al. reported that non-cytotoxic doses of polyphenolic compounds such as chlorogenic acid, rosmarinic acid, and caffeic acid protected kidney tissue against mercury-induced DNA damage in Vero cells (Carranza-Torres et al., 2019). Ben et al. reported that the flower extract of *Narcissus tazetta* is effective against the oxidative stress induced by CdCl2 in Swiss albino mice (Ben-Abdallah et al., 2021). The organic solvent extract of pricky pear peel from *Opuntia ficus-indica* (L.) Mill showed antigenotoxic activity against mitomycin-C and showed significant genoprotective potential on human lymphocytes (Dormousoglou et al., 2023). The flower extract of *S. nigra* showed antigenotoxic potential against RKO cells (Ferreira-Santos et al., 2021). Veras et al. investigated the genotoxic/antigenotoxic activity of chalcone (E)-1-(2-hydroxyphenyl)-3-(4-methylphenyl)-prop-2-en-1-one (2HMC) and the coumarin-chalcone hybrid - methoxy-3-(*E*)-3-(3,4,5-trimethoxyphenyl) acryloyl-2*H*-cromen-2-one (4-MET) in *Drosophila melanogaster* using the somatic mutation and recombination test (SMART), they found that 2HMC and 4-MET showed anti-recombinogenic and antimutagenic activity (Véras et al., 2022). Most recently Dereli et al. reported that resveratrol seems to reduce the H_2O_2-induced DNA damage in human limbal explant cultures (Dereli Can et al., 2023).

Preclinical evidence of the antitumor potential of polyphenolic compounds

Tumor progression has been described as phenotypic alterations in earlier-established neoplastic lesions. Thus progression is a natural habit of tumors to become malignant or aggressive with time. In the last few decades, a lot of synthetic drugs have been developed to treat various cancers or reduce tumor burden. Based on different mechanisms of action a variety of synthetic drugs have been commonly used to enhance antigenotoxic and antitumor burden and decrease cytotoxicity to normal cells. The major drawback associated with the available drugs is the recurrence of tumors due to metabolic reprogramming, multidrug resistance, increased efflux of drugs, etc. Drug resistance leads to a minimal therapeutic outcome and to improve the clinical outcomes, plant-derived polyphenolic compounds have been seeking significant attention in the last few years due to minimal side effects and easy availability and cost-effectiveness. Recently, Xie et al. discovered that lung cancer stem cells (LCSCs) in human lung cancer tissues are positively correlated with IL-6. In vitro experiments showed a significant increase in IL-6 levels in lung cancer sphere-forming cells, which enhanced LCSC stemness. This led to increased tumor sphere formation ability and the expression of specific markers (CD133, ALDH1A1, and Nanog) for lung cancer stem-like cells. Resveratrol (RSV) (trans-3, 5, 4'-trihydroxystilbene) effectively reduced the activation of Wnt/β-catenin signaling caused by lithium chloride (LiCl) and reversed the promotion of LCSC properties and Wnt/β-catenin signaling induced by IL-6. These findings collectively demonstrated that RSV exerts its inhibitory effects on lung cancer by targeting LCSCs and IL-6 within the tumor microenvironment (TME) (Xie et al., 2023). Rahman et al. screened the efficacy of resveratrol-loaded cationic solid lipid nanoparticles (RSV-c-SLNs), in the treatment of hepatocellular carcinoma. RSV-c-SLNs showed higher cytotoxicity on HepG2 cell lines compared to RSV solution and RSV-SLN. The optimized formulation showed uniform size, positive zeta potential, and stability over 12 weeks of storage. In vivo, preclinical investigation on a rat model of HCC showed RSV-c-SLNs significantly reduced tumor volume and elevated tumor tissue accumulation. They also down-regulated pro-inflammatory cytokines and balanced antioxidant enzymes. Histopathological investigation revealed reduced hepatic nodules, necrosis, inflammatory cell infiltration, blood vessel inflammation, and cell swelling (Rahman et al., 2020). Tong and Wu, observed that curcumin treatment might inhibit colon cancer cell proliferation and induce apoptosis by regulating *miR-206* (microRNA) expression. This, in turn, affects colon cancer cell function, enhancing apoptosis and inhibiting PD-L1 expression. Curcumin also increases T cell killing efficiency by suppressing PD-L1 through the JAK/STAT3 pathway (Fig. 8.3). Patients with high *miR-206* expression had better survival rates (Tong & Wu, 2024). Lu et al. studied the potential of curcumin in overcoming resistance to 5-FU in HCT-116 cells. A resistant cell line was established by exposing HCT-116 cells to 5-FU, revealing that curcumin inhibited cell proliferation, induced apoptosis, and arrested the cell cycle at the G0/G1 phase in the resistant cells. It also inhibited the expression of *TET1* and *NKD2* genes at higher doses. Curcumin suppressed the WNT signaling pathway and inhibited

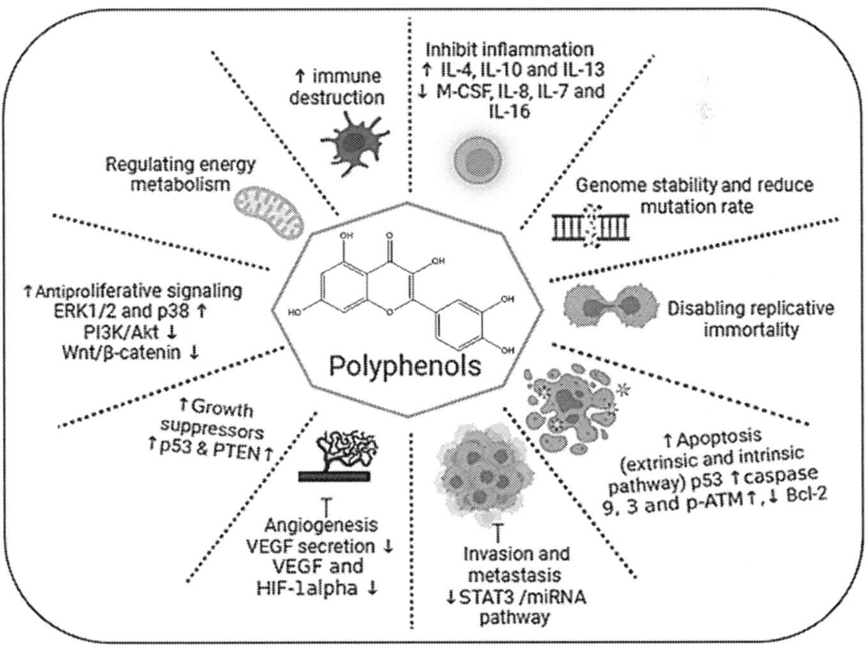

Figure 8.3 Mechanism of action of Polyphenolic compounds against tumor cells.

epithelial-mesenchymal transition (EMT) in the resistant cells. Curcumin effects were verified and consistent with *Pax-6*, *TET1*, and *NKD2* expression patterns. Curcumin may regulate the TET1-NKD2-WNT signaling pathway to inhibit EMT progress and overcome resistance to 5-FU (Lu et al., 2020). Howells et al. conducted a phase IIa trial demonstrating the safety and tolerability of combining FOLFOX chemotherapy with curcumin for metastatic colorectal cancer patients. Curcumin, when combined with FOLFOX, effectively inhibits the growth of FOLFOX-resistant colon cancer cells, thereby improving overall survival in metastatic colorectal cancer patients and demonstrating its safety and tolerability as an adjunct to FOLFOX chemotherapy. The molecular mechanisms of curcumin in colorectal cancer include its effects on EGFR and IGF-1R signaling pathways and its ability to decrease their expression and activation. The potential of curcumin as a therapeutic agent for colorectal cancer is highlighted (Howells et al., 2019). Green tea's catechin, EGCG, is being investigated as a potential dietary agent for preventing breast cancer. Laboratory studies show that EGCG can influence cell cycle progression, inhibit signaling pathways, and trigger apoptosis, indicating its potential as a regulator of cell survival. Preclinical data suggests EGCG could be used alongside EGFR-targeting treatments, but extensive epidemiological studies are lacking on the use of EGCG-containing substances in breast cancer prevention and management (Athanasiou et al., 2022).

Quercetin shows anticancer effects in solid tumors and triggers apoptosis in acute myeloid leukemia (AML) cells by suppressing the VEGF/Akt pathway. It relies on caspase activity, reduces mitochondrial membrane potential, and affects Bcl-2 proteins. Quercetin's inhibition of VEGF-Akt signaling and changes in Bcl-2 proteins and mitochondrial membrane depolarization leads to apoptotic cell death in AML cells. It effectively inhibits leukemic cell growth with minimal toxicity to normal monocytes, making it a potential therapy for AML. Overall, quercetin demonstrates promising preclinical anticancer activity in AML cells by influencing VEGF/Akt signaling and inducing apoptosis (Shi et al., 2020). Preclinical studies have shown that pomegranate and its main component, ellagic acid (EA), have potential pharmacological benefits. EA has various therapeutic properties, including anti-inflammatory, antimutagenic, antifibrosis, anticancer, and antiaging effects. Pomegranate and EA also modulate the immune system by regulating T cell function and suppressing humoral immunity in preclinical models (Baradaran Rahimi et al., 2020).

Neelu Batra and colleagues found that genistein combined polysaccharide (GCP) can inhibit the production of intracrine androgens in prostate cancer cells. GCP reduces testosterone levels by three times and suppresses androgen synthesis enzyme expression. Three active fractions of GCP were identified, with genistein being the primary active component (antiproliferative, pro-apoptotic, and anti-androgen receptor). GCP, fraction 40, and genistein all showed a minimum two fold change in biological activities compared to the control group. GCP treatment also influenced the expression of crucial androgen synthesis molecules (Batra et al., 2020). TA showed dose-dependent and time-responsive effects on inhibiting non-small-cell lung cancer (NSCLC) cell growth. It suppressed colony formation and reduced invasiveness and migration of NSCLC cells. TA also displayed antiangiogenic properties by decreasing tube-like network formation and reducing VEGF release (Hatami et al., 2022). TA was also found to inhibit lipogenic signaling, repress lipid metabolic pathways, and induce reactive oxygen species, leading to apoptosis and disruption of lipid signaling and metabolism in prostate cancer cells, suggesting potential therapeutic applications (Nagesh et al., 2020).

Molecular mechanisms of polyphenolic compounds in cancers

Apoptosis plays a crucial and pivotal role in the prevention of various cancers and its loss of function leads to the accumulation of mutations and prolonged survival of cancer cells (Pfeffer & Singh, 2018). In higher concentrations, RSV (phytoalexin and stilbenoid) has been found to impede cell growth and trigger autophagy and apoptosis by activating MAPK signaling in TNBC cells (Farghadani & Naidu, 2023). This impact is attributed to the antioxidant properties of RSV, as evidenced by the increased levels of *SOD3* and *FAM213B*, enhanced catalase activity, and elevated NAD(P)H levels (Cheng et al., 2022). Also, in both MDA-MB-231 and MDA-MB-468 cells that possess constitutively active signal transducer and activator of transcription 3 (STAT3), elevated levels reported in various cancers. RSV's inhibitory effect on Src tyrosine kinase activity leads to the suppression of the STAT3 signaling pathway. As a result, this cascade of events induces apoptosis and causes cell cycle arrest in the G1 and S phases in the respective cell lines (Kotha et al., 2006). Polymerase δ catalytic subunit gene 1(*POLD1*) expression is linked to poor prognosis in breast cancer, with increased lymph node status, histological grade, p53 status, and ki-67 index (Qin et al., 2018). RSV treatment decreases PARP1, PCNA, and Bcl-2 expression while increasing the cleaved-PARP1 and caspase 3 levels, indicating apoptosis. RSV also reduces POLD1 expression, overexpression of POLD1 in MDA-MB-231 cells decreased RSV-induced apoptosis (Liang et al., 2021). In vitro Transwell assays showed that RSV inhibited the migration of MDA-MB-231 breast cancer cells in a concentration-dependent manner. It also reduces the secretion of matrix metalloproteinase-2 (MMP-2) and MMP-9 in these cells. MMP-2 and MMP-9 are reported to induce extracellular matrix (ECM) degradation. RSV reverses the EMT induced by TGF-β1 in MDA-MB-231 cells by increasing E-cadherin expression and decreasing vimentin expression (Sun et al., 2019). RSV also inhibits acetylation, decreases the STAT3 levels, activates the tumor-suppressor genes, enhances the anti-estrogen sensitivity, reduces tumor cell viability, and increases therapeutic sensitivity in melanoma and triple-negative breast cancer (Lee et al., 2012). Jeon et al. found that cis-resveratrol demonstrated greater potency than trans-resveratrol in inhibiting anoctamin 1(ANO1) channel activity in PC-3 cells. Cis-resveratrol exhibited stronger downregulation of ANO1 transcription and translation compared to trans-resveratrol. Both cis-and trans-resveratrol effectively decreased cell proliferation and migration in PC-3 cells through ANO1-dependent mechanisms. The induction of apoptosis was evident in PC-3 cells treated with both cis-and trans-resveratrol, as indicated by increased caspase3 activity, PARP cleavage, and apoptotic sub-G1 phase ratio (Jeon et al., 2023). Researchers synthesized nanoparticles encapsulated with RSV and conjugated it with folic acid to target prostate cancer. These nanoparticles effectively reduced docetaxel levels, increased apoptotic cells, and reduced NF-kB p65, COX-2, and pro-apoptotic genes (Singh et al., 2018). Multi-reservoir nanocarriers showed promise as a substitute for traditional breast cancer treatment by targeting MCF-7 cells, reducing tumor volume, and lowering growth biomarkers in vivo. They can deliver both Monascus yellow pigments (MYPs) and RSV, making them a promising alternative to conventional therapy (El-Far et al., 2018). NF-κB is a key signaling pathway in cancer development and progression (Xia et al., 2018). It regulates the expression of genes like *BCL2, BCL XS, VEGF TNF-α, BCL XL, XIAP,* and *IL-6,* controlling tumor cell growth, survival, and angiogenesis (Baud & Karin, 2009). The antiproliferative activity of curcumin was observed in HCT-116 cell lines. It demonstrated inhibitory effects on the migration and invasion of HCT-116 and LoVo cell lines. Curcumin inhibited the STAT-3 and NF-KB pathways, induced apoptosis through the ubiquitin-proteasome pathway, and increased LDH (Lactate dehydrogenase) release from LoVo cells. It activates caspase3 and caspase9, leading to apoptosis. Curcumin reduces CD24 expression and upregulates E-cadherin, inhibiting epithelial-mesenchymal transition. It also downregulates

Sp-1 and FAK in colorectal cancer and decreases phospho-Erk1/2 and phospho-Akt levels in various cell lines (Calibasi-Kocal et al., 2019). Quercetin showed antitumor potential against various cancer cells via induction of apoptosis. In MDA-MB-231 cells it induces apoptosis via both extrinsic and intrinsic pathways by increasing caspase 3, 8, 9, Bax, and Fas protein levels. Additionally, it also decreased the levels of XIAP and Bcl-2 proteins (Ankit et al., 2022). Quercetin upregulates the expression of p57 and downregulates the expression of cyclin A and B. It inhibits the β-catenin and AKT signaling pathway (Srinivasan et al., 2016). In liver cancer cells, quercetin enhanced the expression of caspase-3, Bax, and p21 proteins, while decreasing the Bcl-2, Akt, cyclin-B1, PLK-1, CDC-2, cyclin-A, and CDK-2 proteins (Sethi et al., 2023). Studies on cervical cancer (HeLa cells) had revealed that Quercetin increased the expression of miR-26b, miR-126, and miR-320a and decreased the β-catenin (Murata et al., 2023). In HepG2 cells, quercetin increased the p53 and Bax levels while decreasing the Bcl-2 and YY1 transcription factors (Guan et al., 2023). It decreased the IL15 levels in HeLa and A549 cells (Guan et al., 2023). It increased the expression of G3BP1 and YWHAZin CAL27 (oral squamous carcinoma cell line) (Hu et al., 2023). In vivo and in vitro studies demonstrated that quercetin increased the level of p53, and p21, and decreased the levels of CDK4, cyclin D1, and PCNA and arrested cell cycle at the G0/G1 phase (Ali et al., 2023) in Ang II-infused C57BL/6 mice. In YD10B and YD38 (oral squamous carcinoma cell line), quercetin arrested the cell cycle at the G1 phase, increased the levels of CDK inhibitor, and activated the p38 MAPK signaling pathway (Son & Kim, 2023). These findings suggested that quercetin showed antitumor potential by modulating different signaling pathways and protein expressions against various cancer cells.

Kaempferol (KAE), a plant-derived flavonoid, inhibited the growth of colorectal cancer cells by decreasing the activation of the JMJD2C/β-catenin signaling pathway, which is directly linked to the progression of colorectal cancer cells. It also reduced the expression of the *circ_0000345* gene involved in the development of colorectal cancer (Pu et al., 2024). Kaempferide, a compound studied for its potential antitumor properties in both in vitro and in vivo, resulted in increased caspase9 activity and decreased TGF-β1, a protein that promotes tumor growth and progression (Chandrababu et al., 2023). KAE-induced apoptosis in HCT-116 cells downregulates the PI3K/Akt signaling pathway (Fig. 8.3), a crucial pathway for cell survival and proliferation (Li, Wei, et al., 2019; Li, Zhao, et al., 2019). The effect of KAE has been screened against various gastric cancer cell lines, including NCI-N87, AGS, MKN-74, and SNU-638, and resulted in activation of the IRE1-JNK-CHOP signaling pathway, involved in cellular stress response. It has been reported that KAE can increase autophagy (a process that helps remove damaged cellular components) and apoptosis in these cancer cells (Kim et al., 2018). These findings highlighted the potential of KAE and Kaempferide as therapeutic agents in treating colorectal and gastric cancers. Further research is needed to fully understand the mechanisms of action against the tumor cells. Antitumor activity of 3,5,7-trihydroxy-2-(4-hydroxyphenyl) − 4 H-1-benzopyran-4-one, has been studied in MDA-MB-231 and MDA-MB-453 cells in vitro. It has resulted in inducing cell cycle arrest at the G2/M phase and inhibited the NF-κB signaling pathway. Additionally, KAE decreased the levels of cyclin A, B, and CDK1 in these cells (Zhu & Xue, 2019). Ellagic acid is a polyphenol synthesized naturally in plants. It has been screened for antitumor properties against various cancer types in vitro conditions. It decreased the phosphorylation of Akt and Erk in MDA-MB-468 and NCI-H358 cells (Ma et al., 2024). Ellagic acid showed antitumor properties against HCT-116 cells via upregulation of AMPK phosphorylation, increased level of apoptosis, and decreased mTOR phosphorylation (Ni et al., 2023). A similar result was seen in A549 cells, it decreased Akt and PI3K phosphorylation, increased apoptosis, and promoted sub-G1 phase retention (Liu et al., 2018). Ellagic acid decreased JAK2, and STAT3 phosphorylation, and increased apoptosis in HeLa, SiHa, and C33A cells (Li et al., 2018). These findings highlight the potential of ellagic acid as an antitumor agent. TA was screened for antitumor properties against different cancer cell lines like MDA-MB-231, MCF-7, SK-BR 3, and T47D. It showed antitumor properties via induction of apoptosis, cell cycle arrest in the G1 phase, and decreased Bcl-2 expression (Darvin et al., 2017). It enhanced the expression of Bim, Bak, cleaved caspase 3, and PARP, while reducing the levels of Bcl-xL, Bcl-2, p18INK4C, cyclin D1, MMP2, and MMP9 in prostate cancer cell line (PC-3, C4−2, and DU 145). It also increased the expression of p21WAF1/CIP1 (Nagesh et al., 2018). TA attenuates TGF-β1 and its receptor in A549 and H1299 cell lines, suggesting its potential therapeutic applications in cancer research and its ability to modulate cellular processes (Pattarayan et al., 2018). Genistein has been found to have significant effects on various cancer-related factors. In pancreatic cancer, it decreased the expression of VEGF, NFκB, and Bcl-2 in vivo studies (Sallam et al., 2024). Genistein decreased the expression of KCNK9 and the Wnt/β-catenin in SW480 and HT29 cells in male nude mice both in vivo and in vitro conditions (Yu et al., 2023). It hinders the expression of paxillin and Focal

Adhesion Kinase (FAK), as well as snail and twist genes in HeLa cells in vitro conditions. Moreover, it can also impede the expression of vimentin and β-catenin proteins (Chen et al., 2023). These findings highlight the potential of genistein as a therapeutic agent in cancer treatment. Apigenin affects CSC metabolism and reduces CSCs by suppressing the Wnt/β-catenin signaling pathway (Fig. 8.3), inhibiting nuclear factor-κB protein expression, and downregulating the cell cycle via p21 and cyclin-dependent kinases (CDK) (Ghanbari-Movahed et al., 2023). Apigenin promotes autophagy in Ishikawa cells by upregulating Beclin 1, ULK1, ATG5, ATG13, and LC3B, and inhibits tumor growth by induction of ferroptosis (Liang et al., 2023).

Conclusion

Polyphenolic compounds, which are easily available and one of the common dietary sources, revealed crucial potential in curing various cancers. These compounds have minimal cytotoxicity and are economically feasible. The major demerit associated with using polyphenols as antitumor and antigenotoxic agents is their low bioavailability in the human body. Another concern is their interaction with other natural compounds present in a diet may alter their therapeutic efficacy. Epidemiological and preclinical evidence suggests that the consumption of polyphenolic vegetables and fruits has antitumor and antigenotoxic properties. Despite antitumor and antigenotoxic characteristics, most of the polyphenolic compounds are unable to cure cancer and have lower antigenotoxic potentials due to poor solubility, poor pharmacokinetic profiles low stability, and poor retention effect. Hence, most of the polyphenolic compounds fail in their initial stage of clinical trials. Yet, in the last few decades, few studies related to polyphenolic compounds showed promising results and raised the hope for the selective antigenotoxic and antitumor potential of polyphenolic compounds for better therapeutic potential in an economic way against various cancers. This chapter spotlights the most recent plant-based compounds that have been showing antigenotoxic and antitumor properties in preclinical and clinical studies. Several studies suggest that polyphenolic compounds may have significant antigenotoxic and antitumor therapeutic potentials soon (Table 8.1).

Table 8.1 In vitro and in vivo antitumor effect of polyphenols.

Polyphenols	Cell line / Model system	Molecular mechanism	References
Resveratrol (RSV), 3,5,4′-trihydroxy-trans-stilbene; Cis- and Trans-Resveratrol Nanoformulation of Resveratrol	MDA-MB-231 and MDA MB-468 cells, PC-3 cells (LNCaP, PC-3 and C4−2B); Breast cancer (MCF-7)	↑ ERK1/2; ↓ DNA polymerase delta-1 (POLD1); ↓ PCNA, ↓ Bcl-2 ↑ PARP1 and -caspase 3; ↓ EGFR and VEGF; inhibited EGFR/PI3k/AKT signaling; ↑ E-cadherin; ↓ MMP-2, MMP-9, SMA, fibronectin, and vimentin; ↓ STAT3; ↓ ANO1, ↑ caspase3; inhibited NF-kB pathway ↓ PI3K/Akt pathway	El-Far et al. (2018); Jeon et al. (2023); Lee et al. (2012); Liang et al. (2021); Qin et al. (2018); Singh et al. (2018); Sun et al. (2019)
Quercetin	MDA-MB-231; Liver cancer; HeLa; HepG2, HeLa, A549 and CAL27 cell lines Ang II-infused C57BL/6 mice; YD10B and YD38 cell line	↑ apoptosis; ↑ caspase 3, 8 and 9; ↑ Bax and Fas; ↓ XIAP ↓ Bcl-2; ↑ p57 ↓ cyclin A and B; inhibited β-catenin and AKT; ↑ caspase3, Bax and p21; ↓ Bcl-2, Akt, cyclin-B1, PLK-1, CDC-2, cyclin-A and CDK-2, ↑ expression of miR-26b, miR-126, and miR-320a; ↓ β-catenin, ↑ p53 and Bax, ↓ Bcl-2, ↓ *YY1*, ↓ IL15, ↑ G3BP1 and YWHAZ; cell cycle, cell cycle arrest G0/G1 phase, ↑ p53 and p21, ↓ CDK4, cyclin D1, ↓ PCNA; cell cycle, G1 phase arrest, ↑ CDK inhibitor, ↑ p38 MAPK	Ali et al. (2023); Guan et al. (2023); Hu et al. (2023); Malviya and Sharma (2022); Murata et al. (2023); Sethi et al. (2023); Srinivasan et al. (2016); Zhang, Aray et al. (2023); Zhang, Lu, et al. (2023); Sun and Kim (2023)

Kaempferol (KAE) (3,5,7-trihydroxy-2-(4-hydroxyphenyl) – 4 H-1-benzopyran-4-one)	Colorectal cancer cell MDA-MB-231; MDA-MB-453 HCT-8 or HCT-116, NCI-N87, AGS, MKN-74 and SNU-638 cell lines	↓ JMJD2C/β-catenin; ↓ circ_0000345 cell cycle arrest G2/M; inhibited NF-κB; ↓ cyclin A, and B, CDK1, ↑ apoptosis, ↓ PI3K/Akt, ↑ IRE1-JNK-CHOP, ↑ autophagy and apoptosis	Kim et al. (2018); Li, Wei, et al. (2019); Li, Zhao, et al. (2019); Pu et al. (2024); Zhu and Xue (2019)
Kaempferide	HepG2 cells	↑ caspase9, ↓ TGF-β1	Chandrababu et al. (2023)
Genistein	Pancreatic cancer cells, SW480 and HT29; male nude mice; HeLa cell line	↓ VEGF, NFκB and Bcl-2 ↓ KCNK9, ↓ Wnt/β-catenin, inhibited paxillin and FAK; inhibited *Snail* and *Twist*; inhibited vimentin and β-catenin	Chen et al. (2023); Sallam et al. (2024); Yu et al. (2023)
Ellagic acid	MDA-MB-468 and NCI-H358, HCT-116, A549, HeLa, SiHa, and C33A cell lines	↓ Akt and Erk phosphorylation ↑ apoptosis, ↓mTOR phosphorylation, ↑AMPK phosphorylation, ↑ autophagy ↓ Akt and PI3K phosphorylation, ↑apoptosis and sub-G1 phase arrest, ↓ JAK2 and STAT3 phosphorylation, ↑ apoptosis	Li et al. (2018); Liu et al. (2018); Ma et al. (2024); Ni et al. (2023)

(*Continued*)

Table 8.1 (Continued)

Polyphenols	Cell line / Model system	Molecular mechanism	References
Tannic acid	MDA-MB 231, MCF-7, SK-BR 3, T47D, PC-3, C4−2 and DU 145, A549 and H1299 cell lines	Arrest cell cycle G1 phase, ↑ apoptosis, ↓ BCl-2, ↑ Bim, Bak, cleaved caspase 3, and cleaved PARP; ↓ Bcl-xL, ↓ Bcl-2, ↓ p18INK4C and ↑ p21WAF1/CIP1, ↓ cyclin D1, ↓ MMP2, ↓ MMP9, inhibited TGF-β1 and TGF-β1R	Darvin et al. (2017); Nagesh et al. (2018); Pattarayan et al. (2018)
Apigenin or isovitexin	Cancer stem cells, human endometrial carcinoma	inhibited Wnt/β-catenin, inhibited NF-κB, ↑ p21 and cdks, ↑ Bax, ↑ caspases, ↓ Bcl-2, ↓ VEGF, ↑ *ULK1*, *Beclin 1*, *LC3B*, *ATG5*, *ATG13* and ↓ mTOR, AMPK, ATG4, P70S6K	Ghanbari-Movahed et al. (2023); Liang et al. (2023)
Curcumin	HCT-116, LoVo, A549, H1299 MDA-MB-231 cell lines	↑ Beclin1 and LC3, ACSL4 ↓ P62, SLC7A11 and GPX4 ↓ PTCH1, Gli1, Gli2, ↓ E-cadherin, ↓ vimentin and Gli1, Gli2, inhibited Gli1	Calibasi-Kocal et al. (2019); Li et al. (2022); Tang et al. (2021)
Coumarin, Tacrine-coumarin hybrid molecules, Scopoletin,6-methoxy-7 hydroxycoumarin	MCF-7 breast and PC-3 prostate cancer cell lines human and mouse cancer cell linesA549, HCT-116 and HepG2 cell line (Human)	inhibited VEGFR-2 kinase, ↑ pre G1 phase, ↑ apoptosis, inhibited G2/M phase, ↑ caspase9, ↓ proliferation, inhibited RAS-RAF-MEK-ERK and PI3K/AKT), ↑ apoptosis, inhibited	Ahmed et al. (2020); Sherapura et al. (2022); Solárová et al. (2018); Yuan et al. (2021)

2-bromobenzoyl-4-methylphenoxy-acetyl hydra acetyl Coumarin (BP-1C)		JAK2, modulation in Akt/Src survival signal	
Gallic acid (GA) + pre-treatment with low-level laser, Gallic acid [Fuzheng Jiedu Xiaoji formulation (FZJDXJ)]	HDF, MCF10A, MDA-MB-231 and A375 cell line Hepatocellular carcinoma (HCC)	↑ ROS production, apoptosis, modulation of the AKT/CyclinD1/p21/p27 pathways, cell growth, invasion, ↑ apoptosis	Khorsandi et al. (2020); Yang et al. (2021)
Myricetin (3, 5, 7, 3', 4', 5' hexahydroxyflavone)	HCT-116 cell line 4T1 mouse lung metastasis mouse model, MDA-MB-231, MCF-7, PC3 and A549	↓ PI3K, AKT, mTOR, ↑ apoptosis, ↓ migration, invasion, and adhesion, cell cycle arrest G0/G1 phase, ↑ apoptosis ↓ autophagy, ↓ Beclin1, AKT, and phospho-AKT proteins	Anwer et al. (2022); Chen et al. (2022); Jiang et al. (2019); Zhang, Aray et al. (2023); Zhang, Lu, et al. (2023)
Fisetin (3,3',4',7-tetrahydroxyflavone)	PANC-1 cell line, HSC3 cell line, SK-Hep1 cells, Ca9−22, CAL-27	↓ PI3K/AKT/mTOR; inhibited PANC-1 cells, ↑ apoptosis, inhibit Bcl-2 ↑ Sestrin2, PARP ↓ mTOR and Mcl-1, ↓ VEGFR, VEGF	Rahmani et al. (2022); Won et al. (2019); Xiao et al. (2021)
Glycitein	AGS cell line	Cell cycle arrest G0/G1 via ROS-related NF-κB, MAPK, STAT3, ↑ MAPK, ↓ STAT3, NF-κB, ↑ AGS cell apoptosis	Zang et al. (2019)

(*Continued*)

Table 8.1 (Continued)

Polyphenols	Cell line / Model system	Molecular mechanism	References
Naringin	TPC-1, SW1736	Inhibited TPC-1 and SW1736 cell proliferation, ↑ apoptosis, caspase3, Bax, ↓ cyclin D1, c-Myc, survivin, and Bcl-2, ↓ PI3K/AKT in TC cells	Zhou et al. (2019)
Daidzein	LA795 cell lines MCF-7, T47D	Inhibited TMEM16A channel, ↓ TMEM16A expression, cell cycle arrest G1/S, ↓ NGB	Montalesi et al. (2020); Wang et al. (2023)
Neohesperidin (NH), propionyl neohesperidin ester (PNHE)	MCF-7	↓ Proliferation, ↑ apoptosis, ↑ ROS in MCF-7 cells, cell arrest in S phase	Xia et al. (2020)
Vanillic acid (VA)	SKBR3 and H1299 cell lines	↑ STING/TBK1/IRF3, ↑ IFN, ↑ cytotoxic T cells	Xia et al. (2020)
Tangeretin	KB cell lines	↑ Apoptosis, inhibited proliferation, and cell cycle arrest	Gv et al. (2023)
Nobiletin	HCT-116, RKO CRC cell lines	↓ CD44v6, cell viability of CRC stem cells (CR-CSCs), ↓ S and G2/M phase, ↑ Wnt	Turdo et al. (2021)
Epigallocatechin Gallate (EGCG)	FaDu, SCC-1 and NHBE, A549, 8505 C cell lines	Inhibited DNA hypermethylation, HNSCC cells; ↓ DNMT3a and DNT3b, ↑ G0/G1 phase arrest, ↓ EMT	Agarwal et al. (2023); Bhardwaj and Mandal (2019); Li, Wei, et al. (2019); Li, Zhao, et al. (2019)

Note: ⊣, inhibition; ↑, up-regulation; ↓, down-regulation.

Abbreviations

ANO1	Anoctamin 1
ATG5	Autophagy related 5
ATG13	Autophagy-related protein 13
AML	Acute myeloid leukemia
CA	Caffeic acid
CDK	Cyclin-dependent kinases
DNA	Deoxyribonucleic acid
DPPH	2,2-Diphenyl-1-picrylhydrazyl
EA	Ellagic acid
EGCG	Epigallocatechin-3-gallate
EGFR	Epidermal growth factor receptors
EMT	Epithelial-mesenchymal transition
FA	Ferulic acid
FAK	Focal adhesion kinase
5-FU	5-Fluorouracil
FOLFOX	Folinic acid fluorouracil and oxaliplatin
GA	Gallic acid
GCP	Genistein combined polysaccharide
GPX	Glutathione peroxidase
H_2O_2	Hydrogen peroxide
IGF-1R	Insulin-like growth factor-1 receptor
.OH	Hydroxyl radical
JMJD2C	Jumonji-C histone demethylase family
KCNK9	Potassium channel subfamily K member 9
IL-6	Interleukin-6
LC3B	Microtubule associated protein 1 light chain 3 beta
LiCl	Lithium chloride
PCNA	Proliferating cell nuclear antigen
2 HMC	(E)-1-(2-hydroxyphenyl)-3-(4-methylphenyl)-prop-2-en-1-one
4-MET	7-methoxy-3-(*E*)-3-(3,4,5-trimethoxyphenyl)acryloyl-2*H*-cromen-2-one MMC methyl methanesulfonate
MMP	Matrix metalloproteinase
MSG	Monosodium glutamate
NO	Nitric oxide
NF-kB	Nuclear factor-kappa B
NH	Neohesperidin
NSCLC	Non-small-cell lung cancer
$.O_2$	Superoxide
PME	P*olygonum maritimum* L.
PNHE	Propionyl neohesperidin ester

RA	Rosmarinic acid
RNS	Reactive nitrogen species
ROS	Reactive oxygen species
RSV	Resveratrol
RSV-c-SLNs	Resveratrol-loaded cationic solid lipid nanoparticles
SOD	Superoxide dismutase
STAT3	Signal transducer and activator of transcription 3
SMART	Somatic mutation and recombination test
TA	Tannic acid
TME	Tumor microenvironment
ULK1	Unc-51-like autophagy-activating kinases 1
VA	Vanillic acid
VEGF	Vascular endothelial growth factor
VPA	Valproic acid
YWHAZ	Tyrosine 3-monooxygenase/tryptophan 5-monooxygenase activation protein zeta
YY1	Yin Yang 1

Acknowledgment

Vipendra Kumar Singh acknowledges the Indian Council of Medical Research—Research Associateship (ICMR-RA) Fellowship, New Delhi, India. Arjun Kumar Mehara acknowledges the UGC-Junior Research Fellowship (NTA Ref. No.-221610202178), New Delhi, India. Vikky Kumar Mahto acknowledges the CSIR-Junior Research Fellowship [NTA Ref. No.-17/06/2018(i) EU-V], New Delhi, India.

Conflict of interest

The authors declare no conflict of interest in this chapter.

References

Abiola, T., John, E. O., Sossou, I. T., & Charles Callistus, B. (2023). Immune boosting and ameliorative properties of aqueous extract of *Vernonia amygdalina* Delile against MSG-induced genotoxicity: An *in silico* and *in vivo* approach. *Heliyon*, *10*(1), e23226. Available from https://doi.org/10.1016/j.heliyon.2023.e23226.

Abraham, S. K., Khandelwal, N., Hintzsche, H., & Stopper, H. (2016). Antigenotoxic effects of resveratrol: Assessment of in vitro and in vivo response. *Mutagenesis*, *31*(1), 27–33. Available from https://doi.org/10.1093/mutage/gev048.

Afzal, S., Abdul Manap, A. S., Attiq, A., Albokhadaim, I., Kandeel, M., & Alhojaily, S. M. (2023). From imbalance to impairment: The central role of reactive oxygen species in oxidative stress-induced disorders and therapeutic exploration. *Frontiers in pharmacology*, *14*, 1269581. Available from https://doi.org/10.3389/fphar.2023.1269581.

Agarwal, A., Kansal, V., Farooqi, H., Prasad, R., & Singh, V. K. (2023). Epigallocatechin Gallate (EGCG), an active phenolic compound of green tea, inhibits tumor growth of head and neck cancer cells by targeting DNA hypermethylation. *Biomedicines*, *11*(3), 789. Available from https://doi.org/10.3390/biomedicines11030789.

Ahmed, E. Y., Abdel Latif, N. A., El-Mansy, M. F., Elserwy, W. S., & Abdelhafez, O. M. (2020). VEGFR-2 inhibiting effect and molecular modeling of newly synthesized coumarin derivatives as anti-breast cancer agents. *Bioorganic & Medicinal Chemistry*, *28* (5), 115328. Available from https://doi.org/10.1016/j.bmc.2020.115328.

Ali, F., Wang, D., Cheng, Y., Wu, M., Saleem, M. Z., Wei, L., Xie, Y., Yan, M., Chu, J., Yang, Y., Shen, A., & Peng, J. (2023). Quercetin attenuates angiotensin II-induced proliferation of vascular smooth muscle cells and p53 pathway activation in vitro and in vivo. *BioFactors (Oxford, England)*, *49*(4), 956–970. Available from https://doi.org/10.1002/biof.1959.

Ankit, A., Malviya, R., & Sharma, A. (2022). Sources, properties, and pharmacological effects of quercetin. *Current Nutrition & Food Science*, *18*(5), 457–465. Available from https://doi.org/10.2174/1573401318666220127140859.

Anwer, S. T., Mobashir, M., Fantoukh, O. I., Khan, B., Imtiyaz, K., Naqvi, I. H., & Rizvi, M. M. A. (2022). Synthesis of silver nano particles using myricetin and the in-vitro assessment of anti-colorectal cancer activity: In-silico integration. *International journal of molecular sciences*, *23*(19), 11024. Available from https://doi.org/10.3390/ijms231911024.

Asif, P. J., Longobardi, C., Hahne, M., & Medema, J. P. (2021). The role of cancer-associated fibroblasts in cancer invasion and metastasis. *Cancers*, *13*(18), 4720. Available from https://doi.org/10.3390/cancers13184720.

Athanasiou, E., Verras, G. I., Papageorgiou, S., Kelesis, I., Gatsis, A., Karaoulani, C., Stouras, I., Kanatas, P., Saitani, E. M., Oikonomou, M. E., Vlassi, D. A., Vasileiou, M., Tsagkaris, C., Alexiou, A., & Kamal, M. A. (2022). The association between the risk of breast cancer and epigallocatechin- 3-gallate intake: A literature review of a potential chemopreventive agent. *Current Medicinal Chemistry*, *29*(40), 6169–6196. Available from https://doi.org/10.2174/0929867329666220726153412.

Azmi, A. S., Bhat, S. H., Hanif, S., & Hadi, S. M. (2006). Plant polyphenols mobilize endogenous copper in human peripheral lymphocytes leading to oxidative DNA breakage: a putative mechanism for anticancer properties. *FEBS Letters*, *580*(2), 533–538. Available from https://doi.org/10.1016/j.febslet.2005.12.059.

Baradaran Rahimi, V., Ghadiri, M., Ramezani, M., & Askari, V. R. (2020). Antiinflammatory and anti-cancer activities of pomegranate and its constituent, ellagic acid: Evidence from cellular, animal, and clinical studies. *Phytotherapy Research: PTR*, *34*(4), 685–720. Available from https://doi.org/10.1002/ptr.6565.

Batra, N., Sam, A., Woldemariam, T., Talbott, G., de Vere White, R. W., Ghosh, P. M., Gaikwad, N. W., Kotchoni, S. O., & Vinall, R. L. (2020). Genistein combined polysaccharide (GCP) can inhibit intracrine androgen synthesis in prostate cancer cells. *Biomedicines*, *8*(8), 282. Available from https://doi.org/10.3390/biomedicines8080282.

Baud, V., & Karin, M. (2009). Is NF-kappaB a good target for cancer therapy? Hopes and pitfalls. *Nature Reviews. Drug Discovery*, *8*(1), 33–40. Available from https://doi.org/10.1038/nrd2781.

Becker, L., Zaiter, A., Petit, J., Karam, M. C., Sudol, M., Baudelaire, E., & Dicko, A. (2017). How do grinding and sieving impact on physicochemical properties, polyphenol content, and antioxidant activity of Hieracium pilosella L. powders? *Journal of Functional Foods*, *35*, 666–672. Available from https://doi.org/10.1016/j.jff.2017.06.043.

Ben-Abdallah, S., Sefi, M., Soudani, N., Hamdi, A., Bejaoui, S., Issaoui, H., El Cafsi, M., & Karray-Bouraoui, N. (2021). Potential antioxidant effects of Narcissus tazetta phenolic compounds against cadmium chloride-induced hepatotoxicity in Swiss albino mice. *Environmental Science and Pollution Research International*, *28*(46), 66193–66205. Available from https://doi.org/10.1007/s11356-021-15497-8.

Berdeja, J. G., Madduri, D., Usmani, S. Z., Jakubowiak, A., Agha, M., Cohen, A. D., Stewart, A. K., Hari, P., Htut, M., Lesokhin, A., Deol, A., Munshi, N. C., O'Donnell, E., Avigan, D., Singh, I., Zudaire, E., Yeh, T. M., Allred, A. J., Olyslager, Y., ... Jagannath, S. (2021). Ciltacabtagene autoleucel, a B-cell maturation antigen-directed chimeric antigen receptor T-cell therapy in patients with relapsed or refractory multiple myeloma (CARTITUDE-1): a phase 1b/2 open-label study. *Lancet, 398* (10297), 314–324. Available from https://doi.org/10.1016/S0140-6736(21)00933-8.

Bhardwaj, V., & Mandal, A. K. A. (2019). Next-generation sequencing reveals the role of epigallocatechin-3-gallate in regulating putative novel and known microRNAs which target the MAPK pathway in non-small-cell lung cancer A549 cells. *Molecules, 24*(2), 368. Available from https://doi.org/10.3390/molecules24020368.

Brimson, J. M., Prasanth, M. I., Malar, D. S., Thitilertdecha, P., Kabra, A., Tencomnao, T., & Prasansuklab, A. (2021). Plant polyphenols for aging health: Implication from their autophagy modulating properties in age-associated diseases. *Pharmaceuticals, 14* (10), 982. Available from https://doi.org/10.3390/ph14100982.

Calibasi-Kocal, G., Pakdemirli, A., Bayrak, S., Ozupek, N. M., Sever, T., Basbinar, Y., Ellidokuz, H., & Yigitbasi, T. (2019). Curcumin effects on cell proliferation, angiogenesis and metastasis in colorectal cancer. *Journal of B.U.ON.: Official Journal of the Balkan Union of Oncology, 24*(4), 1482–1487.

Canbolat, F., Demir, N., Yayıntas, O. T., Pehlivan, M., Eldem, A., Ayna, T. K., & Senel, M. (2024). Chitosan nanoparticles loaded with quercetin and valproic acid: A novel approach for enhancing antioxidant activity against oxidative stress in the SH-SY5Y human neuroblastoma cell line. *Biomedicines, 12*(2), 287. Available from https://doi.org/10.3390/biomedicines12020287.

Carranza-Torres, I. E., Viveros-Valdez, E., Guzmán-Delgado, N. E., García-Davis, S., Morán-Martínez, J., Betancourt-Martínez, N. D., Balderas-Rentería, I., & Carranza-Rosales, P. (2019). Protective effects of phenolic acids on mercury-induced DNA damage in precision-cut kidney slices. *Iranian journal of basic medical sciences, 22*(4), 367–375. Available from https://doi.org/10.22038/ijbms.2019.30056.7242.

Chandrababu, G., Varkey, M., Devan, A. R., Anjaly, M. V., Unni, A. R., & Nath, L. R. (2023). Kaempferide exhibits an anticancer effect against hepatocellular carcinoma in vitro and in vivo. *Naunyn-Schmiedeberg's Archives of Pharmacology, 396*(10), 2461–2467. Available from https://doi.org/10.1007/s00210-023-02468-8.

Chen, S. L., Yu, H., Luo, H. M., Wu, Q., Li, C. F., & Steinmetz, A. (2016). Conservation and sustainable use of medicinal plants: problems, progress, and prospects. *Chinese Medicine, 11*, 37. Available from https://doi.org/10.1186/s13020-016-0108-7.

Chen, T., Wang, J., Li, M., Wu, Q., & Cui, S. (2023). Genistein inhibits proliferation and metastasis in human cervical cancer cells through the focal adhesion kinase signaling pathway: A network pharmacology-based in vitro study in HeLa cells. *Molecules, 28* (4), 1919. Available from https://doi.org/10.3390/molecules28041919.

Chen, Y. C., He, X. L., Qi, L., Shi, W., Yuan, L. W., Huang, M. Y., Xu, Y. L., Chen, X., Gu, L., Zhang, L. L., & Lu, J. J. (2022). Myricetin inhibits interferon-γ-induced PD-L1 and IDO1 expression in lung cancer cells. *Biochemical Pharmacology, 197*, 114940. Available from https://doi.org/10.1016/j.bcp.2022.114940.

Cheng, T., Wang, C., Lu, Q., Cao, Y., Yu, W., Li, W., Liu, B., Gao, X., Lü, J., & Pan, X. (2022). Metformin inhibits the tumor-promoting effect of low-dose resveratrol, and enhances the anti-tumor activity of high-dose resveratrol by increasing its reducibility in triple negative breast cancer. *Free Radical Biology & Medicine, 180*, 108–120. Available from https://doi.org/10.1016/j.freeradbiomed.2022.01.010.

Chethna, P., Iyer, S. S., Gandhi, V. V., Kunwar, A., Singh, B. G., Barik, A., Balaji, N. V., Ramani, M. V., Subbaraju, G. V., & Priyadarsini, K. I. (2018). Toxicity and antigenotoxic effect of hispolon derivatives: Role of structure in modulating cellular redox state and thioredoxin reductase. *ACS Omega*, *3*(6), 5958–5970. Available from https://doi.org/10.1021/acsomega.8b00415.

Dai, J., & Mumper, R. J. (2010). Plant phenolics: Extraction, analysis and their antioxidant and anticancer properties. *Molecules*, *15*(10), 7313–7352. Available from https://doi.org/10.3390/molecules15107313.

Darvin, P., Joung, Y. H., Kang, D. Y., Sp, N., Byun, H. J., Hwang, T. S., Sasidharakurup, H., Lee, C. H., Cho, K. H., Park, K. D., Lee, H. K., & Yang, Y. M. (2017). Tannic acid inhibits EGFR/STAT1/3 and enhances p38/STAT1 signalling axis in breast cancer cells. *Journal of Cellular and Molecular Medicine*, *21*(4), 720–734. Available from https://doi.org/10.1111/jcmm.13015.

Darvin, P, Joung, Y. H., Kang, D. Y., Sp, N., Byun, H. J., Hwang, T. S., Sasidharakurup, H., Lee, C. H., Cho, K. H., Park, K. D., Lee, H. K., & Yang, Y. M. (2017). Tannic acid inhibits EGFR/STAT 1/3 and enhances p38/STAT 1 signalling axis in breast cancer cells. *Journal of cellular and molecular medicine*, *21*(4), 720–734. Available from https://doi.org/10.1111/jcmm.13015.

Dereli Can, G., Taner, G., & İnci Aydemir, Ç. (2023). Evaluation of the cytotoxic and genotoxic/antigenotoxic effects of resveratrol in human limbal explant cultures. *International Ophthalmology*, *43*(6), 1977–1985. Available from https://doi.org/10.1007/s10792-022-02597-8.

Dini, I., & Grumetto, L. (2022). Recent advances in natural polyphenol research. *Molecules*, *27*(24), 8777. Available from https://doi.org/10.3390/molecules27248777.

Dini, I., Seccia, S., Senatore, A., Coppola, D., & Morelli, E. (2020). Development and validation of an analytical method for total polyphenols quantification in extra virgin olive oils. *Food Anal. Methods*, *13*, 457–464. Available from https://doi.org/10.1007/s12161-019-01657-7.

Dormousoglou, M., Efthimiou, I., Antonopoulou, M., Dailianis, S., Herbst, G., & Vlastos, D. (2023). Phytochemical analysis and genotoxicological evaluation of prickly pear peel extracts. *Plants*, *12*(7), 1537. Available from https://doi.org/10.3390/plants12071537.

Ekor, M. (2014). The growing use of herbal medicines: Issues relating to adverse reactions and challenges in monitoring safety. *Frontiers in Pharmacology*, *4*, 177. Available from https://doi.org/10.3389/fphar.2013.00177.

El-Far, S. W., Helmy, M. W., Khattab, S. N., Bekhit, A. A., Hussein, A. A., & Elzoghby, A. O. (2018). Phytosomal bilayer-enveloped casein micelles for codelivery of monascus yellow pigments and resveratrol to breast cancer. *Nanomedicine*, *13*(5), 481–499. Available from https://doi.org/10.2217/nnm-2017-0301.

Farghadani, R., & Naidu, R. (2023). The anticancer mechanism of action of selected polyphenols in triple-negative breast cancer (TNBC). *Biomedicine & Pharmacotherapy*, *165*, 115170. Available from https://doi.org/10.1016/j.biopha.2023.115170.

Ferreira-Santos, P., Badim, H., Salvador, Â. C., Silvestre, A. J. D., Santos, S. A. O., Rocha, S. M., Sousa, A. M., Pereira, M. O., Wilson, C. P., Rocha, C. M. R., Teixeira, J. A., & Botelho, C. M. (2021). Chemical characterization of *Sambucus nigra* L. flowers aqueous extract and its biological implications. *Biomolecules*, *11*(8), 1222. Available from https://doi.org/10.3390/biom11081222.

Figat, R., Śliwińska, A., Stochmal, A., Soluch, A., Sobczak, M., Zgadzaj, A., Sykłowska-Baranek, K., & Pietrosiuk, A. (2020). Antigenotoxic, anti-photogenotoxic, and antioxidant properties of *Polyscias filicifolia* shoots cultivated in vitro. *Molecules*, *25*(5), 1090. Available from https://doi.org/10.3390/molecules25051090.

García-Rodríguez, Md. C., Carvente-Juárez, M. M., & Altamirano-Lozano, M. A. (2013). Antigenotoxic and apoptotic activity of green tea polyphenol extracts on hexavalent chromium-induced DNA damage in peripheral blood of CD-1 mice: analysis with differential acridine orange/ethidium bromide staining. *Oxidative Medicine and Cellular Longevity*, *2013*, 486419. Available from https://doi.org/10.1155/2013/486419.

Ghanbari-Movahed, M., Shafiee, S., Burcher, J. T., Lagoa, R., Farzaei, M. H., & Bishayee, A. (2023). Anticancer potential of apigenin and isovitexin with focus on oncogenic metabolism in cancer stem cells. *Metabolites*, *13*(3), 404. Available from https://doi.org/10.3390/metabo13030404.

Guan, H., Zhang, W., Liu, H., Jiang, Y., Li, F., Wu, M., Waterhouse, G. I. N., Sun-Waterhouse, D., & Li, D. (2023). Quercetin induces apoptosis in HepG2 Cells *via* directly interacting with YY1 to disrupt YY1-p53 interaction. *Metabolites*, *13*(2), 229. Available from https://doi.org/10.3390/metabo13020229.

Gv, V., Ranganathan, P., Jr, & Palati, S. (2023). Tangeretin's anti-apoptotic signaling mechanisms in oral cancer cells: In vitro anti-cancer activity. *Cureus*, *15*(10), e47452. Available from https://doi.org/10.7759/cureus.47452.

Hartwig, A., Arand, M., Epe, B., Guth, S., Jahnke, G., Lampen, A., Martus, H. J., Monien, B., Rietjens, I. M. C. M., Schmitz-Spanke, S., Schriever-Schwemmer, G., Steinberg, P., & Eisenbrand, G. (2020). Mode of action-based risk assessment of genotoxic carcinogens. *Archives of Toxicology*, *94*(6), 1787–1877. Available from https://doi.org/10.1007/s00204-020-02733-2.

Hatami, E., Nagesh, P. K. B., Sikander, M., Dhasmana, A., Chauhan, S. C., Jaggi, M., & Yallapu, M. M. (2022). Tannic acid exhibits antiangiogenesis activity in nonsmall-cell lung cancer cells. *ACS omega*, 7(27), 23939–23949. Available from https://doi.org/10.1021/acsomega.2c02727.

Heo, M. Y., Sohn, S. J., & Au, W. W. (2001). Anti-genotoxicity of galangin as a cancer chemopreventive agent candidate. *Mutation Research*, *488*(2), 135–150. Available from https://doi.org/10.1016/s1383-5742(01)00054-0.

Howells, L. M., Iwuji, C. O. O., Irving, G. R. B., Barber, S., Walter, H., Sidat, Z., Griffin-Teall, N., Singh, R., Foreman, N., Patel, S. R., Morgan, B., Steward, W. P., Gescher, A., Thomas, A. L., & Brown, K. (2019). Curcumin combined with FOLFOX chemotherapy is safe and tolerable in patients with metastatic colorectal cancer in a randomized Phase IIa trial. *The Journal of Nutrition*, *149*(7), 1133–1139. Available from https://doi.org/10.1093/jn/nxz029.

Hu, M., Song, H. Y., & Chen, L. (2023). Quercetin acts via the G3BP1/YWHAZ axis to inhibit glycolysis and proliferation in oral squamous cell carcinoma. *Toxicology Mechanisms and Methods*, *33*(2), 141–150. Available from https://doi.org/10.1080/15376516.2022.2103480.

Izquierdo-Vega, J. A., Morales-González, J. A., SánchezGutiérrez, M., Betanzos-Cabrera, G., Sosa-Delgado, S. M., Sumaya-Martínez, M. T., Morales-González, Á., Paniagua-Pérez, R., Madrigal-Bujaidar, E., & Madrigal-Santillán, E. (2017). Evidence of some natural products with antigenotoxic effects. Part 1: Fruits and polysaccharides. *Nutrients*, *9*(2), 102. Available from https://doi.org/10.3390/nu9020102.

Jeon, D., Jo, M., Lee, Y., Park, S. H., Phan, H. T. L., Nam, J. H., & Namkung, W. (2023). Inhibition of ANO1 by *Cis*- and *Trans*-resveratrol and their anticancer activity in human prostate cancer PC-3 cells. *International Journal of Molecular Sciences*, *24*(2), 1186. Available from https://doi.org/10.3390/ijms24021186.

Jiang, M., Zhu, M., Wang, L., & Yu, S. (2019). Anti-tumor effects and associated molecular mechanisms of myricetin. *Biomedicine & Pharmacotherapy*, *120*, 109506. Available from https://doi.org/10.1016/j.biopha.2019.109506.

Juan, C. A., Pérez de la Lastra, J. M., Plou, F. J., & Pérez-Lebeña, E. (2021). The chemistry of reactive oxygen species (ROS) revisited: Outlining their role in biological macromolecules (DNA, lipids and proteins) and induced pathologies. *International Journal of Molecular Sciences*, *22*(9), 4642. Available from https://doi.org/10.3390/ijms22094642.

Kasote, D. M., Katyare, S. S., Hegde, M. V., & Bae, H. (2015). Significance of antioxidant potential of plants and its relevance to therapeutic applications. *International Journal of Biological Sciences*, *11*(8), 982–991. Available from https://doi.org/10.7150/ijbs.12096.

Khorsandi, K., Kianmehr, Z., Hosseinmardi, Z., & Hosseinzadeh, R. (2020). Anti-cancer effect of gallic acid in presence of low level laser irradiation: ROS production and induction of apoptosis and ferroptosis. *Cancer Cell International*, *20*, 18. Available from https://doi.org/10.1186/s12935-020-1100-y.

Kim, T. W., Lee, S. Y., Kim, M., Cheon, C., & Ko, S. G. (2018). Kaempferol induces autophagic cell death via IRE1-JNK-CHOP pathway and inhibition of G9a in gastric cancer cells. *Cell Death & Disease*, *9*(9), 875. Available from https://doi.org/10.1038/s41419-018-0930-1.

Kotha, A., Sekharam, M., Cilenti, L., Siddiquee, K., Khaled, A. R., Zervos, A. S., Carter, B., Turkson, J., & Jove, R. (2006). Resveratrol inhibits Src and Stat3 signaling and induces the apoptosis of malignant cells containing activated Stat3 protein. *Molecular Cancer Therapeutics*, *5*(3), 621–629. Available from https://doi.org/10.1158/1535-7163.mct-05-0268.

Kumar, H., Choudhary, N., Varsha, K. N., & Suman, S. R. (2014). Phenolic compounds and their health benefits: A review. *Journal of Food Research and Technology*, *2*(2), 46–59.

Lee, H., Zhang, P., Herrmann, A., Yang, C., Xin, H., Wang, Z., Hoon, D. S., Forman, S. J., Jove, R., Riggs, A. D., & Yu, H. (2012). Acetylated STAT3 is crucial for methylation of tumor-suppressor gene promoters and inhibition by resveratrol results in demethylation. *Proceedings of the National Academy of Sciences of the United States of America*, *109*(20), 7765–7769. Available from https://doi.org/10.1073/pnas.1205132109.

Li, L. W., Na, C., Tian, S. Y., Chen, J., Ma, R., Gao, Y., & Lou, G. (2018). Ellagic acid induces HeLa cell apoptosis via regulating signal transducer and activator of transcription 3 signaling. *Experimental and Therapeutic Medicine*, *16*(1), 29–36. Available from https://doi.org/10.3892/etm.2018.6182.

Li, M., Guo, T., Lin, J., Huang, X., Ke, Q., Wu, Y., Fang, C., & Hu, C. (2022). Curcumin inhibits the invasion and metastasis of triple negative breast cancer via Hedgehog/Gli1 signaling pathway. *Journal of Ethnopharmacology*, *283*, 114689. Available from https://doi.org/10.1016/j.jep.2021.114689.

Li, Q., Wei, L., Lin, S., Chen, Y., Lin, J., & Peng, J. (2019). Synergistic effect of kaempferol and 5-fluorouracil on the growth of colorectal cancer cells by regulating the PI3K/Akt signaling pathway. *Molecular Medicine Reports*, *20*(1), 728–734. Available from https://doi.org/10.3892/mmr.2019.10296.

Li, T., Zhao, N., Lu, J., Zhu, Q., Liu, X., Hao, F., & Jiao, X. (2019). Epigallocatechin gallate (EGCG) suppresses epithelial-Mesenchymal transition (EMT) and invasion in anaplastic thyroid carcinoma cells through blocking of TGF-β1/Smad signaling pathways. *Bioengineered*, *10*(1), 282–291. Available from https://doi.org/10.1080/21655979.2019.1632669.

Li, Y., & Chemat, F. (2019). *Plant based "Green Chemistry 2.0.". Moving from evolutionary to revolutionary*. Singapore: Springer, Singapore. Available from https://doi.org/10.1007/978-981-13-3810-6.

Liang, Y., Zhong, Q., Ma, R., Ni, Z., Thakur, K., Zhang, J., & Wei, Z. (2023). Apigenin, a natural flavonoid, promotes autophagy and ferroptosis in human endometrial carcinoma Ishikawa cells in vitro and in vivo. *Food Science and Human Wellness*, *12*(6), 2242–2251. Available from https://doi.org/10.1016/j.fshw.2023.03.044.

Liang, Z. J., Wan, Y., Zhu, D. D., Wang, M. X., Jiang, H. M., Huang, D. L., Luo, L. F., Chen, M. J., Yang, W. P., Li, H. M., & Wei, C. Y. (2021). Resveratrol mediates the apoptosis of triple negative breast cancer cells by reducing POLD1 expression. *Frontiers in Oncology*, *11*, 569295. Available from https://doi.org/10.3389/fonc.2021.569295.

Liu, Q., Liang, X., Niu, C., & Wang, X. (2018). Ellagic acid promotes A549 cell apoptosis via regulating the phosphoinositide 3-kinase/protein kinase B pathway. *Experimental and Therapeutic Medicine*, *16*(1), 347–352. Available from https://doi.org/10.3892/etm.2018.6193.

Lu, Y., Zhang, R., Zhang, X., Zhang, B., & Yao, Q. (2020). Curcumin may reverse 5-fluorouracil resistance on colonic cancer cells by regulating TET1-NKD-Wnt signal pathway to inhibit the EMT progress. *Biomedicine & Pharmacotherapy = Biomedecine & Pharmacotherapie*, *129*, 110381. Available from https://doi.org/10.1016/j.biopha.2020.110381.

Ma, C. H., Zhao, J. F., Zhang, X. G., Ding, C. H., Hao, H. H., Ji, Y. H., Li, L. P., Guo, Z. T., & Liu, W. S. (2024). Discovery of ellagic acid as a competitive inhibitor of Src homology phosphotyrosyl phosphatase 2 (SHP2) for cancer treatment: In vitro and in silico study. *International Journal of Biological Macromolecules*, *254*(Pt 2)), 127845. Available from https://doi.org/10.1016/j.ijbiomac.2023.127845.

Majrashi, T. A., Alshehri, S. A., Alsayari, A., Muhsinah, A. B., Alrouji, M., Alshahrani, A. M., Shamsi, A., & Atiya, A. (2023). Insight into the biological roles and mechanisms of phytochemicals in different types of cancer: Targeting cancer therapeutics. *Nutrients*, *15*(7), 1704. Available from https://doi.org/10.3390/nu15071704.

Malviya, R., & Sharma, A. (2022). Sources, properties, and pharmacological effects of quercetin. *Current Nutrition & Food Science*, *18*(5), 457–465. Available from https://doi.org/10.2174/1573401318666220127140859.

Marino, A., Battaglini, M., Moles, N., & Ciofani, G. (2022). Natural antioxidant compounds as potential pharmaceutical tools against neurodegenerative diseases. *ACS Omega*, 7(30), 25974–25990. Available from https://doi.org/10.1021/acsomega.2c03291.

Montalesi, E., Cipolletti, M., Cracco, P., Fiocchetti, M., & Marino, M. (2020). Divergent effects of Daidzein and its metabolites on estrogen-induced survival of breast cancer cells. *Cancers*, *12*(1), 167. Available from https://doi.org/10.3390/cancers12010167.

Murata, M., Komatsu, S., Miyamoto, E., Oka, C., Lin, I., Kumazoe, M., Yamashita, S., Fujimura, Y., & Tachibana, H. (2023). Quercetin up-regulates the expression of tumor-suppressive microRNAs in human cervical cancer. *Bioscience of Microbiota, Food and Health*, *42*(1), 87–93. Available from https://doi.org/10.12938/bmfh.2022-056.

Nagesh, P. K. B., Chowdhury, P., Hatami, E., Jain, S., Dan, N., Kashyap, V. K., Chauhan, S. C., Jaggi, M., & Yallapu, M. M. (2020). Tannic acid inhibits lipid metabolism and induce ROS in prostate cancer cells. *Scientific Reports*, *10*(1), 980. Available from https://doi.org/10.1038/s41598-020-57932-9.

Nagesh, P. K. B., Hatami, E., Chowdhury, P., Kashyap, V. K., Khan, S., Hafeez, B. B., Chauhan, S. C., Jaggi, M., & Yallapu, M. M. (2018). Tannic acid induces endoplasmic reticulum stress-mediated apoptosis in prostate cancer. *Cancers*, *10*(3), 68. Available from https://doi.org/10.3390/cancers10030068.

Ni, X., Shang, F. S., Wang, T. F., Wu, D. J., Chen, D. G., & Zhuang, B. (2023). Ellagic acid induces apoptosis and autophagy in colon cancer through the AMPK/mTOR pathway. *Tissue & Cell*, *81*, 102032. Available from https://doi.org/10.1016/j.tice.2023.102032.

Oliveira, D., Dias, M. I., Barros, L., Custódio, L., & Oliveira, R. (2023). Antigenotoxic properties of the halophyte *Polygonum maritimum* L. highlight its potential to mitigate oxidative stress-related damage. *Scientific Reports*, *13*(1), 3727. Available from https://doi.org/10.1038/s41598-022-20402-5.

Pattarayan, D., Sivanantham, A., Krishnaswami, V., Loganathan, L., Palanichamy, R., Natesan, S., Muthusamy, K., & Rajasekaran, S. (2018). Tannic acid attenuates TGF-β1-induced epithelial-to-mesenchymal transition by effectively intervening TGF-β signaling in lung epithelial cells. *Journal of Cellular Physiology*, *233*(3), 2513–2525. Available from https://doi.org/10.1002/jcp.26127.

Pattarayan, D., Sivanantham, A., Krishnaswami, V., Loganathan, L., Palanichamy, R., Natesan, S., Muthusamy, K., & Rajasekaran, S. (2018). Tannic acid attenuates TGF-β1-induced epithelial-to-mesenchymal transition by effectively intervening TGF-β signaling in lung epithelial cells. *Journal of Cellular Physiology*, *233*(3), 2513–2525. Available from https://doi.org/10.1002/jcp.26127.

Pfeffer, C. M., & Singh, A. T. K. (2018). Apoptosis: A target for anticancer therapy. *International Journal of Molecular Sciences*, *19*(2), 448. Available from https://doi.org/10.3390/ijms19020448.

Phillips, D. H., & Arlt, V. M. (2009). Genotoxicity: Damage to DNA and its consequences. *EXS*, *99*, 87–110. Available from https://doi.org/10.1007/978-3-7643-8336-7_4.

Pramesh, C. S., Badwe, R. A., Bhoo-Pathy, N., Booth, C. M., Chinnaswamy, G., Dare, A. J., de Andrade, V. P., Hunter, D. J., Gopal, S., Gospodarowicz, M., Gunasekera, S., Ilbawi, A., Kapambwe, S., Kingham, P., Kutluk, T., Lamichhane, N., Mutebi, M., Orem, J., Parham, G., Ranganathan, P., … Weiderpass, E. (2022). Priorities for cancer research in low- and middle-income countries: A global perspective. *Nature Medicine*, *28*(4), 649–657. Available from https://doi.org/10.1038/s41591-022-01738-x.

Pu, Y., Han, Y., Ouyang, Y., Li, H., Li, L., Wu, X., Yang, L., Gao, J., Zhang, L., Zhou, J., Ji, Q., & Song, Q. (2024). Kaempferol inhibits colorectal cancer metastasis through circ_0000345 mediated JMJD2C/β-catenin signalling pathway. *Phytomedicine: International Journal of Phytotherapy and Phytopharmacology*, 155261. Available from https://doi.org/10.1016/j.phymed.2023.155261.

Qin, Q., Tan, Q., Li, J., Yang, W., Lian, B., Mo, Q., & Wei, C. (2018). Elevated expression of POLD1 is associated with poor prognosis in breast cancer. *Oncology Letters*, *16*(5), 5591–5598. Available from https://doi.org/10.3892/ol.2018.9392.

Rahman, M., Almalki, W. H., Afzal, O., Alfawaz Altamimi, A. S., Kazmi, I., Al-Abbasi, F. A., Choudhry, H., Alenezi, S. K., Barkat, M. A., Beg, S., Kumar, V., & Alhalmi, A. (2020). Cationic Solid lipid nanoparticles of resveratrol for hepatocellular carcinoma treatment: Systematic optimization, in vitro characterization and preclinical investigation. *International Journal of Nanomedicine*, *15*, 9283–9299. Available from https://doi.org/10.2147/IJN.S277545.

Rahman, M. M., Rahaman, M. S., Islam, M. R., Rahman, F., Mithi, F. M., Alqahtani, T., Almikhlafi, M. A., Alghamdi, S. Q., Alruwaili, A. S., Hossain, M. S., Ahmed, M., Das, R., Emran, T. B., & Uddin, M. S. (2021). Role of phenolic compounds in human disease: Current knowledge and future prospects. *Molecules*, *27*(1), 233. Available from https://doi.org/10.3390/molecules27010233.

Rahmani, A. H., Almatroudi, A., Allemailem, K. S., Khan, A. A., & Almatroodi, S. A. (2022). The potential role of fisetin, a flavonoid in cancer prevention and treatment. *Molecules*, *27*(24), 9009. Available from https://doi.org/10.3390/molecules27249009.

Roszkowski, S. (2023). Application of polyphenols and flavonoids in oncological therapy. *Molecules*, *28*(10), 4080. Available from https://doi.org/10.3390/molecules28104080.

Rudrapal, M., Khairnar, S. J., Khan, J., Dukhyil, A. B., Ansari, M. A., Alomary, M. N., Alshabrmi, F. M., Palai, S., Deb, P. K., & Devi, R. (2022). Dietary polyphenols and their role in oxidative stress-induced human diseases: Insights into protective effects, antioxidant potentials and mechanism(s) of action. *Frontiers in Pharmacology*, *13*, 806470. Available from https://doi.org/10.3389/fphar.2022.806470.

Sallam, N. G., Boraie, N. A., Sheta, E., & El-Habashy, S. E. (2024). Targeted delivery of genistein for pancreatic cancer treatment using hyaluronic-coated cubosomes bioactivated with frankincense oil. *International Journal of Pharmaceutics*, *649*, 123637. Available from https://doi.org/10.1016/j.ijpharm.2023.123637.

Sethi, G., Rath, P., Chauhan, A., Ranjan, A., Choudhary, R., Ramniwas, S., Sak, K., Aggarwal, D., Rani, I., & Tuli, H. S. (2023). Apoptotic mechanisms of quercetin in liver cancer: recent trends and advancements. *Pharmaceutics*, *15*(2), 712. Available from https://doi.org/10.3390/pharmaceutics15020712.

Sharifi-Rad, J., Quispe, C., Butnariu, M., Rotariu, L. S., Sytar, O., Sestito, S., Rapposelli, S., Akram, M., Iqbal, M., Krishna, A., Kumar, N. V. A., Braga, S. S., Cardoso, S. M., Jafernik, K., Ekiert, H., Cruz-Martins, N., Szopa, A., Villagran, M., Mardones, L., Martorell, M., ... Calina, D. (2021). Chitosan nanoparticles as a promising tool in nanomedicine with particular emphasis on oncological treatment. *Cancer Cell International*, *21*(1), 318. Available from https://doi.org/10.1186/s12935-021-02025-4.

Sharma, A., Kaur, M., Katnoria, J. K., & Nagpal, A. K. (2018). Polyphenols in food: Cancer prevention and apoptosis induction. *Current Medicinal Chemistry*, *25*(36), 4740–4757. Available from https://doi.org/10.2174/0929867324666171006144208.

Sharma, E., Attri, D. C., Sati, P., Dhyani, P., Szopa, A., Sharifi-Rad, J., Hano, C., Calina, D., & Cho, W. C. (2022). Recent updates on anticancer mechanisms of polyphenols. *Frontiers in Cell and Developmental Biology*, *10*, 1005910. Available from https://doi.org/10.3389/fcell.2022.1005910.

Sherapura, A., Malojirao, V. H., Thirusangu, P., Sharath, B. S., Kandagalla, S., Vigneshwaran, V., Novak, J., Ranganatha, L., Ramachandra, Y. L., Baliga, S. M., Khanum, S. A., & Prabhakar, B. T. (2022). Anti-neoplastic pharmacophore benzophenone-1 coumarin (BP-1C) targets JAK2 to induce apoptosis in lung cancer. *Apoptosis: An International Journal on Programmed Cell Death*, *27*(1-2), 49–69. Available from https://doi.org/10.1007/s10495-021-01699-5.

Shi, H., Li, X. Y., Chen, Y., Zhang, X., Wu, Y., Wang, Z. X., Chen, P. H., Dai, H. Q., Feng, J., Chatterjee, S., Li, Z. J., Huang, X. W., Wei, H. Q., Wang, J., Lu, G. D., & Zhou, J. (2020). Quercetin induces apoptosis via downregulation of vascular endothelial growth factor/Akt signaling pathway in acute myeloid leukemia cells. *Frontiers in Pharmacology*, *11*, 534171. Available from https://doi.org/10.3389/fphar.2020.534171.

Singh, S. K., Lillard, J. W., Jr, & Singh, R. (2018). Reversal of drug resistance by planetary ball milled (PBM) nanoparticle loaded with resveratrol and docetaxel in prostate cancer. *Cancer Letters*, *427*, 49–62. Available from https://doi.org/10.1016/j.canlet.2018.04.017.

Singh, V. K., & Sharma, P. K. (2020). Oxidative stress and lifestyle-based changes in breast cancer progression. *Current Advances in Breast Cancer Research: A Molecular Approach*, 208–235.

Singh, N., & Yadav, S. S. (2022). A review on health benefits of phenolics derived from dietary spices. *Current Research in Food Science*, *5*, 1508–1523. Available from https://doi.org/10.1016/j.crfs.2022.09.009.

Solárová, Z., Kello, M., Hamuľaková, S., Mirossay, L., & Solár, P. (2018). Anti-cancer effect of tacrine-coumarin derivatives on diverse human and mouse cancer cell lines. *Acta chimica Slovenica*, *65*(4), 875–881. Available from https://doi.org/10.17344/acsi.2018.4519.

Son, H. K., & Kim, D. (2023). Quercetin induces cell cycle arrest and apoptosis in YD10B and YD38 oral squamous cell carcinoma cells. *Asian Pacific Journal of Cancer Prevention: APJCP*, *24*(1), 283–289. Available from https://doi.org/10.31557/APJCP.2023.24.1.283.

Srinivasan, A., Thangavel, C., Liu, Y., Shoyele, S., Den, R. B., Selvakumar, P., & Lakshmikuttyamma, A (2016). Quercetin regulates β-catenin signaling and reduces the migration of triple negative breast cancer. *Molecular carcinogenesis*, *55*(5), 743–756. Available from https://doi.org/10.1002/mc.22318.

Srinivasan, A., Thangavel, C., Liu, Y., Shoyele, S., Den, R. B., Selvakumar, P., & Lakshmikuttyamma, A. (2016). Quercetin regulates β-catenin signaling and reduces the migration of triple negative breast cancer. *Molecular Carcinogenesis*, *55*(5), 743–756. Available from https://doi.org/10.1002/mc.22318.

Sun, Y., Zhou, Q. M., Lu, Y. Y., Zhang, H., Chen, Q. L., Zhao, M., & Su, S. B. (2019). Resveratrol inhibits the migration and metastasis of MDA-MB-231 human breast cancer by reversing TGF-β1-induced epithelial-mesenchymal transition. *Molecules*, *24*(6), 1131. Available from https://doi.org/10.3390/molecules24061131.

Tang, X., Ding, H., Liang, M., Chen, X., Yan, Y., Wan, N., Chen, Q., Zhang, J., & Cao, J. (2021). Curcumin induces ferroptosis in non-small-cell lung cancer via activating autophagy. *Thoracic Cancer*, *12*(8), 1219–1230. Available from https://doi.org/10.1111/1759-7714.13904.

Tong, Q., & Wu, Z. (2024). Curcumin inhibits colon cancer malignant progression and promotes T cell killing by regulating miR-206 expression. *Clinical Anatomy*, *37*(1), 2–11. Available from https://doi.org/10.1002/ca.24057.

Turdo, A., Glaviano, A., Pepe, G., Calapà, F., Raimondo, S., Fiori, M. E., Carbone, D., Basilicata, M. G., Di Sarno, V., Ostacolo, C., Parrino, B., Cascioferro, S., Pecoraro, C., Di Franco, S., Bellavia, D., Gaggianesi, M., Veschi, V., Lo Iacono, M., Ganduscio, G., Pantina, V. D., ... Stassi, G. (2021). Nobiletin and xanthohumol sensitize colorectal cancer stem cells to standard chemotherapy. *Cancers*, *13*(16), 3927. Available from https://doi.org/10.3390/cancers13163927.

Véras, J. H., do Vale, C. R., da Silva Lima, D. C., Dos Anjos, M. M., Bernardes, A., de Moraes Filho, A. V., E Silva, C. R., de Oliveira, G. R., Pérez, C. N., & Chen-Chen, L. (2022). Modulating effect of a hydroxychalcone and a novel coumarin-chalcone hybrid against mitomycin-induced genotoxicity in somatic cells of *Drosophila melanogaster*. *Drug and Chemical Toxicology*, *45*(2), 775–784. Available from https://doi.org/10.1080/01480545.2020.1776314.

Vladimir-Knežević, S., Blažeković, B., Bival Štefan, M., & Babac, M. (2012). *Plant polyphenols as antioxidants influencing the human health* (pp. 155–180). IntechOpen. Available from https://doi.org/10.5772/27843.

Wang, X., Hao, A., Song, G., Elena, V., Sun, Y., Zhang, H., Zhan, Y., An, H., & Chen, Y. (2023). Inhibitory effect of daidzein on the calcium-activated chloride channel TMEM16A and its anti-lung adenocarcinoma activity. *International Journal of Biological Macromolecules*, *253* (Pt 6)), 127261. Available from https://doi.org/10.1016/j.ijbiomac.2023.127261.

Won, D. H., Chung, S. H., Shin, J. A., Hong, K. O., Yang, I. H., Yun, J. W., & Cho, S. D. (2019). Induction of sestrin 2 is associated with fisetin-mediated apoptosis in human head and neck cancer cell lines. *Journal of Clinical Biochemistry and Nutrition*, *64* (2), 97–105. Available from https://doi.org/10.3164/jcbn.18-63.

Xia, L., Tan, S., Zhou, Y., Lin, J., Wang, H., Oyang, L., Tian, Y., Liu, L., Su, M., Wang, H., Cao, D., & Liao, Q. (2018). Role of the NFκB-signaling pathway in cancer. *OncoTargets and Therapy*, *11*, 2063–2073. Available from https://doi.org/10.2147/OTT.S161109.

Xia, N., Wan, W., Zhu, S., & Liu, Q. (2020). Synthesis of hydrophobic propionyl neohesperidin ester using an immobilied enzyme and description of its anti-proliferative and pro-apoptotic effects on MCF-7 human breast cancer cells. *Frontiers in Bioengineering and Biotechnology*, *8*, 1025. Available from https://doi.org/10.3389/fbioe.2020.01025.

Xiao, Y., Liu, Y., Gao, Z., Li, X., Weng, M., Shi, C., Wang, C., & Sun, L. (2021). Fisetin inhibits the proliferation, migration and invasion of pancreatic cancer by targeting PI3K/AKT/mTOR signaling. *Aging*, *13*(22), 24753–24767. Available from https://doi.org/10.18632/aging.203713.

Xie, C., Liang, C., Wang, R., Yi, K., Zhou, X., Li, X., Chen, Y., Miao, D., Zhong, C., & Zhu, J. (2023). Resveratrol suppresses lung cancer by targeting cancer stem-like cells and regulating tumor microenvironment. *The Journal of Nutritional Biochemistry*, *112*, 109211. Available from https://doi.org/10.1016/j.jnutbio.2022.109211.

Yang, X., Feng, Y., Liu, Y., Ye, X., Ji, X., Sun, L., Gao, F., Zhang, Q., Li, Y., Zhu, B., & Wang, X. (2021). Fuzheng Jiedu Xiaoji formulation inhibits hepatocellular carcinoma progression in patients by targeting the AKT/CyclinD1/p21/p27 pathway. *Phytomedicine: International Journal of Phytotherapy and Phytopharmacology*, *87*, 153575. Available from https://doi.org/10.1016/j.phymed.2021.153575.

Yin, L., Duan, J. J., Bian, X. W., & Yu, S. C. (2020). Triple-negative breast cancer molecular subtyping and treatment progress. *Breast Cancer Research: BCR*, *22*(1), 61. Available from https://doi.org/10.1186/s13058-020-01296-5.

Yu, X., Yan, J., Li, Y., Cheng, J., Zheng, L., Fu, T., & Zhu, Y. (2023). Inhibition of castration-resistant prostate cancer growth by genistein through suppression of AKR1C3. *Food & Nutrition Research*, *67*. Available from https://doi.org/10.29219/fnr.v67.9024, https://doi.org/10.29219/fnr.v67.9024.

Yuan, C., Wang, M. H., Wang, F., Chen, P. Y., Ke, X. G., Yu, B., Yang, Y. F., You, P. T., & Wu, H. Z. (2021). Network pharmacology and molecular docking reveal the mechanism of Scopoletin against non-small cell lung cancer. *Life Sciences*, *270*, 119105. Available from https://doi.org/10.1016/j.lfs.2021.119105.

Zang, Y. Q., Feng, Y. Y., Luo, Y. H., Zhai, Y. Q., Ju, X. Y., Feng, Y. C., Wang, J. R., Yu, C. Q., & Jin, C. H. (2019). Glycitein induces reactive oxygen species-dependent apoptosis and G0/G1 cell cycle arrest through the MAPK/STAT3/NF-κB pathway in human gastric cancer cells. *Drug Development Research*, *80*(5), 573–584. Available from https://doi.org/10.1002/ddr.21534.

Zhanataev, A. K., Pigarev, S. E., Fedoros, E. I., Panchenko, A. V., Anisina, E. A., Chayka, Z. V., Durnev, A. D., & Anisimov, V. N. (2022). Antigenotoxic and antimutagenic effects of lignin derivative BP-C2 against dioxidine and cyclophosphamide in vivo in murine cells. *Toxicology Reports*, *9*, 743–749. Available from https://doi.org/10.1016/j.toxrep.2022.03.041.

Zhang, J., Aray, B., Zhang, Y., Bai, Y., Yuan, T., Ding, S., Xue, Y., Huang, X., & Li, Z. (2023). Synergistic effect of cucurbitacin E and myricetin on Anti-Non-Small cell lung cancer: Molecular mechanism and therapeutic potential. *Phytomedicine: International Journal of Phytotherapy and Phytopharmacology*, *111*, 154619. Available from https://doi.org/10.1016/j.phymed.2022.154619.

Zhang, M., Lu, A., Wang, H., & Yang, J. (2023). Quercetin downregulates the expression of IL15 in cancer cells through DNA methylation. *European Review for Medical and Pharmacological Sciences*, *27*(6), 2580–2590. Available from https://doi.org/10.26355/eurrev_202303_31795.

Zhou, J., Xia, L., & Zhang, Y. (2019). Naringin inhibits thyroid cancer cell proliferation and induces cell apoptosis through repressing PI3K/AKT pathway. *Pathology, Research and Practice*, *215*(12), 152707. Available from https://doi.org/10.1016/j.prp.2019.152707.

Zhu, L., & Xue, L. (2019). Kaempferol suppresses proliferation and induces cell cycle arrest, apoptosis, and DNA damage in breast cancer cells. *Oncology Research*, *27*(6), 629–634. Available from https://doi.org/10.3727/096504018X15228018559434.

Zor, M., Aydin, S., Güner, N. D., Başaran, N., & Başaran, A. A. (2017). Antigenotoxic properties of Paliurus spina-christi Mill fruits and their active compounds. *BMC Complementary and Alternative Medicine*, *17*(1), 229. Available from https://doi.org/10.1186/s12906-017-1732-1.

SECTION 5

Polyphenols and cutaneous-related conditions

CHAPTER NINE

Oral health and polyphenols

Laurence James Walsh and Chaminda Jaya Seneviratne
The University of Queensland, School of Dentistry, Brisbane, QLD, Australia

Introduction

As discussed in earlier chapters of this book, many types of polyphenols have been described, including phenolic acids, flavonoids, stilbenes, lignans, and others. Interest in the use of polyphenols to improve oral health has grown steadily over the past decade. A literature scan undertaken in late 2023 revealed that in the PUBMED database there were over 800 publications on dental applications of polyphenols (Fig. 9.1). There was a slow and steady growth in this literature from the late 1980s until 2013, after which time a much more rapid growth occurred, as indicated by a steeper gradient for the cumulative total. This is not surprising

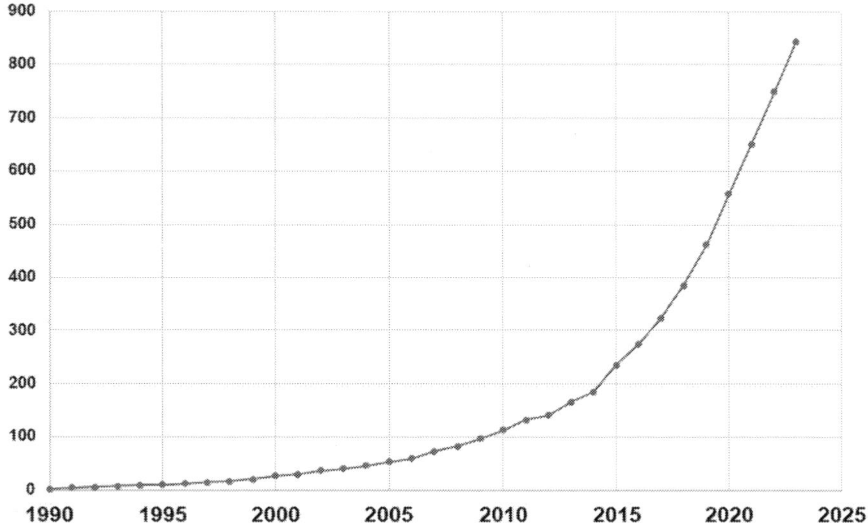

Figure 9.1 Cumulative publications.

Bioactive Polyphenols for Health and Pathology Treatment.
DOI: https://doi.org/10.1016/B978-0-443-13821-8.00009-3

given growing awareness over time of the range of potential uses of polyphenols to prevent or treat oral disease, and the exponential increase in publications identifying health benefits of polyphenols (Rana et al., 2022). It is also possible that another driver for rising interest in polyphenols was greater awareness internationally of the emerging problems of antimicrobial resistance around the globe, and the need to find new ways of tackling diseases with a microbial etiology. Some of the most common polyphenols that suppress the formation of dental plaque and have other useful features for dental caries prevention are summarized in Table 9.1, while those attracting considerable interest for future applications in periodontal disease treatments are summarized in Table 9.2.

An analysis of publications reveals that the major themes in the literature each year for the past decade have been the prevention of oral diseases, particularly dental caries (tooth decay) and periodontal disease,

Table 9.1 Examples of polyphenols used in oral care products.

Name	Source	Properties	Benefits	Applications
Epigallocatechin gallate	Green tea	Antimicrobial, antioxidant	Reduced inflammation	Toothpastes, mouthrinses, chewing gums
Tannins	Tea, grapes, and berries	Antimicrobial, anti-inflammatory	Reduced inflammation	Toothpastes, chewing gums
Grape seed extract (procyanidins and anthocyanins)	Grape seeds	Antimicrobial	Suppressed growth of cariogenic bacteria	Toothpastes, mouthrinses
Cranberry extracts (proanthocyanidins)	Cranberry	Antimicrobial	Prebiotic action	Toothpastes

Table 9.2 Polyphenols of interest for future applications in dentistry.

Name	Properties	Benefits	Potential applications
Resveratrol	Antioxidant, anti-inflammatory	Reduced plaque formation and reduced inflammation	Toothpastes, mouthrinses, chewing gums
Quercetin	Antimicrobial, anti-inflammatory, antioxidant	Reduced plaque formation and reduced inflammation	Toothpastes, mouthrinses

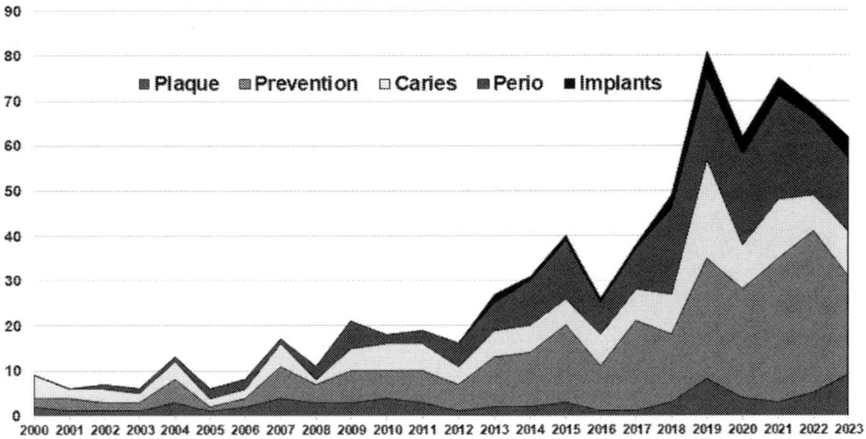

Figure 9.2 Distribution of publications.

and inhibiting or modifying dental plaque biofilms that form on teeth or dental implants (Fig. 9.2) These major themes will be explored in this chapter.

Numerous detailed reviews of individual polyphenol molecules to support oral health have been published (Flemming et al., 2021; Gill et al., 2016; Kováč et al., 2023; Li et al., 2019; Petti & Scully, 2009; Sakagami et al., 2019; Sakagami, 2014; Varoni et al., 2012). These emphasize the possible uses of various polyphenols as primary or additional ingredients in toothpastes, mouthwashes, and other oral care products, and build upon knowledge of traditional medicines where various plant extracts were used for the topical treatment of oral lesions. Such reviews have typically identified situations where polyphenols could serve as alternatives to conventional treatments.

A challenge when assessing the existing literature is that some studies have used one or more purified polyphenols, whereas others have used plant extracts containing polyphenols with multiple other molecules that could also be active. A further issue is that the common oral diseases of dental caries and periodontal disease typically have slow rates of progression. Clinical studies with short timeframes tend to use surrogate measures to infer what long-term benefits may accrue, since they cannot measure long-term outcomes such as tooth loss. This highlights the need for long-term clinical studies on humans to confirm the optimism that comes from the many short-term studies of polyphenols that show promising results.

Issues with formulating polyphenols into oral care products

When any potential plant-based polyphenol is proposed for use in the prevention of oral disease, there are important considerations around ensuring (Rana et al., 2022) bioavailability in the oral cavity (Petti & Scully, 2009), safety as a topical agent (Varoni et al., 2012), and safety when ingested accidentally or deliberately. This includes a requirement for rigorous testing of the product for any minor components or contaminants that may be allergenic or toxic. The presence of impurities and the cost of production both reflect the ease of extraction from natural sources, which in turn raises questions around variations in composition due to seasonal variations as well as from climate change. The latter is then linked to the sustainability of the source. The choice of techniques used for extraction will influence the extent of such undesirable molecules that may be present in extracts, as well as how much of the total polyphenol content is not extracted. Such non-extractable polyphenols (NEP) are generally larger polymeric polyphenols, or small phenolics linked to polysaccharides, proteins, or other macromolecules. In some cases, NEP can contain molecules of considerable interest for therapeutic purposes, such as proanthocyanidins (PACs). A range of extraction methods have been developed to recover NEP from waste material and release them from the matrix that they are bound to (Domínguez-Rodríguez et al., 2017). The efficiency of the extraction and purification processes used for recovering such NEP influences the cost of the material that will be used in commercial products.

Using polyphenols in pharmaceutical applications targeting oral health has been challenging because of frequently encountered issues with irritancy, or with taste (such as astringency, or a tart or bitter taste). Particularly challenging materials in this regard are polyphenols with a marked astringent taste, such as proanthocyanidins, anthocyanins, flavonoids, quercetin, and resveratrol. Astringency is the sensation of dryness, puckering, and tightening in the mouth from the polyphenols as they bind to and interact with salivary proteins, particularly proline-rich proteins (Charlton et al., 2002; García-Estévez et al., 2017). Astringency concerns will limit the concentrations of polyphenols that can be provided in an oral care product (such as a toothpaste, mouth rinse, gel, wafer, spray, crème, chewing gum, etc.).

Taste

Taste issues with polyphenols arise because of the way that they can bind to the bitter taste receptors on the tongue. Including appropriate flavors such as intense sweeteners can help achieve a positive consumer experience and improved palatability. Altering the pH can also reduce the bitter taste of some polyphenols. In the case of quercetin, the bitter taste is more pronounced at lower pH. Under these acidic conditions, the quercetin is more soluble. Raising the pH to the neutral range will reduce the bitter taste. Another advantage of changing the pH is it may prevent side reactions, such as quercetin reacting with acids to form quinones, which themselves have a bitter taste. There are limits to what can be achieved by changing the pH, since the water solubility of some polyphenols declines rapidly at alkaline conditions, and alkaline oral care products have a brackish taste and are tolerated very poorly.

With regard to the bitter taste of polyphenols, there is significant genetic variability between individuals in taste sensitivity for bitterness. This influences what individuals will tolerate, especially when little or no intense sweetener is included in the product. Some people will find the bitter taste of polyphenols to be a major problem in terms of their compliance with using oral care products, while others will find the bitter taste tolerable or perhaps even enjoyable, as it can stimulate saliva production. Conveniently, individuals who are less sensitive to bitter tastes tend to have more dental caries than those who are highly sensitive to bitter-tasting polyphenols. (Melis et al., 2017; Yousaf et al., 2020, 2022) This is advantageous when considering polyphenols in oral care products for those with a high dental caries risk.

Solubility

Physicochemical properties create challenges when formulating oral care products with polyphenols. The molecule of interest may be hydrophobic and have poor solubility in water, or be prone to degradation. In some cases, it is necessary to adopt strategies such as changing from the normal water-based or glycerin-based vehicle, to include propylene glycol or polyethylene glycol as solvents, or to use encapsulation. Propylene glycol is already used widely in liquid sweeteners and many food products, while polyethylene glycol is used in some food and beverage products as a viscosity modifier and emulsifier. Swallowing small amounts of propylene glycol or polyethylene glycol from oral care products does not pose a concern to general health.

Fortunately, some polyphenols are sufficiently water-soluble that they can be added directly to conventional toothpastes, mouthrinses, and other oral care products, to provide antimicrobial, anti-inflammatory, antioxidant, and other beneficial activities. An example of this is the water-soluble polyphenols from cranberry such as PACs, as well anthocyanins and flavonoids that are present in lower quantities. Fortunately, most of the oral health benefits come from the PACs, and these can be obtained commercially in high purity, and readily incorporated into existing oral care products such as toothpastes, without causing problems such as clumping or phase separation within the toothpaste.

Prevention

The dental profession around the world is looking constantly for better ways to address major oral health problems. The common oral diseases (dental caries, periodontal diseases, and oral cancer) represent sizeable public health concerns as noncommunicable diseases, in all countries. These conditions share common risk factors with other diseases and have complex interactions with systemic health. For certain conditions, such as diabetes mellitus, there are a well-known two-way interactions between periodontal disease and diabetic control. On the other hand, many systemic diseases can impair oral health by depressing or altering the normal host inflammatory responses to the accumulation of dental plaque biofilms around teeth, dental implants, and prostheses. Likewise, many medical conditions can impair the production of saliva at rest between meals, which then drives the development of dysbiosis in the oral microbiome. As a consequence of these complex interactions, many of the applications of polyphenols to prevent or manage systemic disease that are discussed in earlier chapters of this book can have positive effects on oral health. Such positive downstream effects have not yet attracted the attention that they deserve.

Given that dental plaque biofilms are the major etiological factor in both dental caries and periodontal disease, much of the literature on polyphenols for the prevention of oral diseases has focused on how the formation and maturation of those biofilms can be prevented, and how the composition of the dental plaque biofilms can be altered by polyphenols, both when used in oral care products, and when consumed as part of a normal diet. For the latter, the underlying concept is that a healthy diet that is rich in polyphenols should reduce the accumulation of dental plaque biofilm, and alter its composition, to lower the levels of disease-associated microorganisms. (Kurek-Górecka et al., 2022)

It is not surprising that polyphenols from plants exert a broad range of antimicrobial actions, since their primary purposes are to protect the plant from predation and from adverse external environmental influences, as well as from internal stresses such as from free radicals. In line with this, polyphenols have antibacterial and antifungal actions relevant to the microflora of the oral cavity, where there are over 700 known species of bacteria and numerous fungi. (Ghannoum et al., 2010; Kilian et al., 2016)

Caries prevention by suppression of *Streptococcus mutans*

Polyphenols such as catechin, apigenin, luteolin, morin, myricetin, naringin, quercetin, and rutin can exert potent antimicrobial actions on *Streptococcus mutans*. This is a key bacterial species involved in producing organic acids and building a resilient biofilm on teeth by laying down exopolysaccharides (EPS) that resist disruption during normal chewing motions. As well as these direct effects on the growth and viability of *S. mutans* (Gutiérrez-Venegas et al., 2019; Melok et al., 2018; Schneider-Rayman et al., 2021; Smullen et al., 2007; Wang et al., 2022; Wu et al., 2021; Xu et al., 2012), polyphenols can also impair the adhesion of these and other bacteria to the salivary pellicle layer that forms on tooth surfaces (Wang et al., 2022), and interfere with the proper expression and function of the three types of the glucosyl transferase enzymes secreted by *S. mutans* which builds EPS by polymerizing glucosyl moieties from sucrose and starch carbohydrates into water-insoluble α1,3- and α1,6-linked glucans (Ferrazzano et al., 2011; Souissi et al., 2021; Xu et al., 2012).

By interacting with the membranes of *S. mutans* and other oral bacteria, polyphenols can increase microbial cell permeability, leading to leakage of the internal contents and impaired metabolism. Such effects have been seen at low doses of tea polyphenols and theaflavins, with reduced lactic acid production and impaired adhesion of *S. mutans* as well as *Streptococcus sobrinus*, a related bacterial species that also plays a key role in dental caries (Cui et al., 2023).

Numerous sources of polyphenols with potential use for suppressing key cariogenic bacteria have been described in the literature, including tea, cranberry, grapes, pomegranate, propolis, and magnolia bark. (Cheng et al., 2015; Philip et al., 2019b; Yoo et al., 2011) Clinical trials have shown that epigallocatechin gallate (EGCG) reduced saliva levels of *S. mutans* by an average of 80% after a 1 minute rinse, (Vilela et al., 2020) while a cranberry mouthwash (0.6% w/v) used twice daily for 2 weeks gave a 68% reduction in the *S. mutans* count, which was no different from the chlorhexidine gluconate positive control (69%). (Khairnar et al., 2015) Such positive results indicate the viability of regular delivery of EGCG (e.g., by mouth rinsing or using a lozenge) for patients with a high dental caries risk.

While the largest body of evidence exists for polyphenols from cranberry and tea, many others have been tested with positive results, including from propolis (Choudhary et al., 2023; Curuțiu et al., 2020; Kurek-Górecka et al., 2023; Skaba et al., 2013; Veloz et al., 2019), blueberry (*Vaccinium corymbosum* and *Vaccinium angustifolium*) and from tart cherry (*Prunus cerasus* L.). (Lagha et al., 2020) The polyphenols from cranberry have the unique ability to serve as prebiotic agents, even at very low concentrations. Moreover, they can disrupt cariogenic virulence properties of bacteria without being bactericidal. This avoids concerns regarding the emergence of resistance species (Philip & Walsh, 2019).

Prebiotic actions of polyphenols

Ingested polyphenols influence the composition of the gut microbiota, and this has been linked to several health benefits. Polyphenol intake alters the composition of the gut microflora through a prebiotic action, increasing the levels of health-promoting bacteria (e.g., *Bifidobacterium* and *Lactobacillus spp.*) while decreasing the concentration of harmful bacteria (e.g., *Clostridium spp.*) (Lavefve et al., 2020; Zheng et al., 2017).

There is an important parallel with the oral microbiota, in terms of polyphenol effects on the composition of supragingival dental plaque biofilms. While a range of polyphenol-rich extracts can suppress the growth of dental plaque bacteria (Ferrazzano et al., 2011), PACs from American cranberry (*Vaccinium macrocarpon*) have a unique A-type interflavan linkage in their structure. These PACs exert a powerful prebiotic action that greatly facilitates the growth of health-associated oral streptococci, such as *Streptococcus sanguinis*, which utilize the arginine deiminase system and metabolize proteins to generate ammonia. This ammonia alkalinizes the plaque fluid, and impairs the growth of aciduric bacteria and fungi that are key pathogens in dental caries and oral fungal infections (Bradshaw & Marsh, 1998; Huang et al., 2016; Slomka et al., 2017).

Randomized double-blinded placebo-controlled clinical trials have shown that cranberry PACs delivered in a toothpaste to patients at a high risk of dental caries (because of wearing fixed orthodontic appliances) alters the composition of dental plaque, with greatly reduced levels of key dental caries-associated bacteria, and greatly increased levels of health-associated bacteria (Philip et al., 2020). These effects mirror those seen in complex multi-species oral biofilm laboratory models (Philip et al., 2019a).

By following an ecological approach and using cranberry-derived polyphenols as prebiotic agents, the development of dental caries can be prevented, without meeting the unrealistic requirement of having no visible deposits of dental plaque. This represents an important progression in how ecological approaches to caries prevention are used (Marsh et al., 2015; Philip et al., 2018a, 2018b).

As with the situation in the gut, there is likely to be a two-directional component to this interaction, as the oral microbiome metabolizes PACs. When the microbiome shifts from a state of dysbiosis to one that is associated with health, the metabolism of PACs and other polyphenols will also change. At the present time, relatively little is known about the biotransformation of PACs and other polyphenols by the oral microbiome.

A major difference between the situation in the gut and in the oral cavity is that in the oral cavity there is the opportunity for direct application of the polyphenol agent to the site, without concerns of absorption or chemical transformation that may lower the bioavailability of polyphenols when ingested (Di Lorenzo et al., 2021). An example of this issue with ingested polyphenols is seen with anthocyanins, the common polyphenols in colored plants that often are glycosylated. When ingested, the low pH of the stomach and intestines alters their chemical form, and only about 1%–2% retain their original structure (Tena et al., 2020). Said another way, when an oral care product is used to directly influence the oral tissues and oral biofilms, they can reach the target unchanged. As a consequence, either the polyphenol itself or its metabolites can then exert actions on the target. Oral delivery enhances bioavailability by preventing issues from numerous modifications and removing variations caused by limited absorption. A final key point is that the presence of saliva may enhance the availability of polyphenols that are not soluble in water but are only soluble in lipids, and enhance their retention in the oral cavity (Ginsburg et al., 2012).

Antifungal actions of polyphenols against *Candida albicans*

Candida albicans is the most prevalent and most important opportunistic pathogenic fungus in humans and accounts for a large burden of systemic disease as well as oral disease (Samaranayake et al., 2009). In the oral cavity, it is an important contributor to oral mucosal disease and dental caries. It has a symbiotic relationship with *S. mutans*, a major pathogen in dental caries, and can ferment dietary carbohydrates to produce organic acids that cause mineral loss from teeth (Koo & Bowen, 2014). *C. albicans* can also contribute to oral cancer, and to periodontal diseases, especially those associated with dental implants.

In light of its importance in oral diseases, attention has been directed to using polyphenols of various types to inhibit the growth of *C. albicans* (Bagieh & Bageah, 2014; Shahzad et al., 2014; Sitheeque et al., 2009), with special interest in EGCG, the major polyphenol from tea leaves (from *Camellia sinensis*). EGCG inhibits the growth and hyphal formation of *C. albicans* in a pH-dependent manner, being far more potent at neutral pH than under acidic pH conditions (Hirasawa & Takada, 2004). EGCG blocks multiple metabolic pathways including fungal cell membrane synthesis (Ning et al., 2015) and causes extensive cell wall damage (Sitheeque et al., 2009).

Likewise, studies have been conducted into how polyphenols can impair the ability of *C. albicans* to form dual-species biofilms with *S. mutans*. Positive effects have been shown in biofilm inhibition assays for polyphenol-rich berry extracts, cranberry PACs, and green tea extracts that are rich in EGCG. These agents reduce biofilm biomass, lower the production of EPS, and the suppress the metabolism of sucrose to lactic acid (Farkash et al., 2019; Koo et al., 2010).

More recent work has shown that some polyphenols, such as extracts from *Punica granatum* (pomegranate), have not only anti-fungal actions but may also exert additional benefits, including anti-viral actions against HIV by blocking binding of the virus to CD4 and CXCR4/CCR5 (Huerta-Reyes et al., 2022; Neurath et al., 2004).

Prevention of dental erosion

Flavonoids, tannins, and other polyphenols can bind to and modify the salivary pellicle that forms on the surface of teeth and prostheses. By binding to salivary mucins and proline-rich proteins, polyphenols cause the pellicle layer to thicken. This reduces its permeability to ions, and thus lessens the likelihood that endogenous or exogenous acids will cause the enamel to dissolve away chemically. The complex of the polyphenol with the salivary proteins acts as a protective layer when a person ingests acidic beverages that cause dental erosion, such as a cola drink containing phosphoric acid, or a fruit juice containing citric acid. Slow sipping of a beverage that contains polyphenols could help prevent dental erosion (Carvalho et al., 2021; Joiner et al., 2004; Weber et al., 2015). Such anti-erosive effects have been shown for green tea, black tea, grape seed extract, and grapefruit seed extracts (Niemeyer et al., 2021; Schestakow et al., 2022). Polyphenols can also enhance the actions of fluoride ions in preventing erosion (Baumann et al., 2023).

Dentine biomodification

Cranberry PACs reduce degradation of type I collagen in the dentine by matrix metalloproteinase enzymes once the collagen becomes exposed, e.g., by loss of the overlying enamel from erosion or dental caries. In fact, cranberry PAC promotes the formation of cross-links in dentine collagen (Bedran-Russo et al., 2008; Boteon et al., 2017). This makes them useful for dentine biomodification, especially when bonding to the dentine is undertaken in restorative dental procedures. Oligomeric PACs form a tight interface with hydrophobic methacrylate adhesives on damp surfaces. This concept opens the possibility of developing a new mechanism for adhesion of restorative materials to dentine and to cementum, both of which are rich in type I collagen. The PACs act as a type of bridging agent between the dentine surface and the methacrylate resin (Leme-Kraus et al., 2017).

For bonding to dentine, multiple polyphenols can enhance the stability of collagen fibrils in the hybrid layer, and improve the bond strength for resin adhesives. PACs, quercetin, resveratrol, myricetin, and kaempferol have all been used successfully for conserving unprotected collagen fibrils within the hybrid layer, and thereby maintaining or enhancing the bond strength of the dentine to resin interface (Hass et al., 2021; Porto et al., 2018; Wang et al., 2023).

Periodontal diseases

Antimicrobial actions against key periodontal pathogens

Periodontitis is a chronic inflammatory disease associated with dysbiotic plaque biofilms. Hence, polyphenols can play key roles in altering the biofilm and in modulating host immune responses. *Porphyromonas gingivalis* is associated with the initiation and progression of periodontitis. It possesses several key virulence factors, including proteolytic enzymes like gingipains, collagenases, lipopolysaccharides (LPS) or endotoxin, fimbriae, and capsules. EGCG inhibits the growth of *P. gingivalis* (Sakanaka et al., 1996). EGCG prevents biofilm formation and eradicates pre-formed *P. gingivalis* biofilms (Asahi et al., 2014), and also boosts the synthesis of protective beta-defensins (hBDs) by gingival epithelial cells, whilst also preventing their bacterial degradation.

Theaflavins inhibit *P. gingivalis* growth and biofilm formation and impair proteinase activities of its secreted collagenases and gingipains, while lowering secretion of matrix metalloproteinases (MMPs) and pro-inflammatory cytokines by host cells, and boosting secretion of protective hBDs (Lagha & Grenier, 2017). Cranberry polyphenols impair *P. gingivalis* biofilm formation and gingipain protease activity (Yamanaka et al., 2007) and can inhibit the secretion of pro-inflammatory cytokines and chemokines in response to *P. gingivalis* (Bodet et al., 2006). Resveratrol lowers *P. gingivalis* bacterial adherence to host matrix proteins, prevents biofilm formation, and is effective against pre-formed biofilms (Lagha et al., 2019). It also downregulates P. *gingivalis* LPS-mediated activation of the nuclear factor-kappaB (NF-κB) signaling pathway (Andreou et al., 2004), and attenuates *P. gingivalis* virulence factors related to fimbriae and proteinases (Kugaji et al., 2019).

Polyphenols from natural sources are used in some commercial products for their effects on periodontal bacteria, such as OroPhenol, which contains polyphenols from blueberries, cranberries, and strawberries. This attenuates the growth of *P. gingivalis*, reduces its adherence to basement membranes, and impairs its proteinase activities (Vaillancourt, et al., 2022). Its efficacy needs to be confirmed by randomized clinical trials.

Antiinflammatory properties of polyphenols in the context of periodontitis

A primary aim of periodontal therapy is to resolve inflammation, and in this context polyphenols with anti-inflammatory and anti-proteolytic activity are of interest. Resveratrol inhibits *P. gingivalis* LPS-induced adhesion of leukocytes to endothelial cells by down-regulating the cell adhesion molecules ICAM-1 and VCAM-1(Park et al., 2009), by blocking the NF-kB pathway and the synthesis of interleukin-1β(Ma et al., 2023). Animal studies have shown that administration of resveratrol reduces periodontal inflammation (Bhattarai et al., 2016; Casati et al., 2013; Tamaki et al., 2014). Antiinflammatory actions in periodontitis have also been seen for cranberry PACs and green tea EGCG in vitro (Lagha & Grenier, 2019; Lombardo Bedran et al., 2015) and in animal models (Fan et al., 2023), by modulating NF-κB signaling. This provides a strong basis for polyphenol-based therapies to progress into clinical studies.

Clinical trials of polyphenols on periodontitis

Positive clinical outcomes have been shown for polyphenols in randomized, double-blind clinical trials conducted on chronic periodontitis patients. Resveratrol as a dietary supplement (500 mg/day) has been shown to improve clinical periodontal parameters, including the bleeding index and attachment levels (Nikniaz et al., 2023), and to lower systemic endotoxin levels, and inflammatory markers in the serum and in crevicular fluid (Zhang et al., 2022). These benefits occur in diabetic and nondiabetic patients; and the former gain by lowered insulin resistance (Javid et al., 2019).

EGCG has been added into the water coolant flowing through ultrasonic scalers, and this enhanced pocket reduction and clinical attachment gain (Wang et al., 2021), or lowered the bleeding index (Zeng et al.,

2022). Positive results have also been seen for a polyphenol-containing grape seed extract delivered in the same manner; however, more clinical evidence is required to confirm the clinical efficacy of polyphenols as adjunctive agents. The promise in polyphenols has substantial evidence in laboratory studies. More clinical studies are warranted to fine-tune the pharmacokinetics and pharmacodynamics aspects of promising polyphenols to bring them into clinical practice for treating periodontal disease.

Future perspectives for polyphenols in oral health care

There are four major areas where future studies into polyphenols in dentistry should be directed.

Of the over 10,000 known polyphenol compounds, relatively few have been closely examined for oral care applications. Consequently, there is great potential for further work to explore the landscape of naturally occurring polyphenols in terms of their potential benefits to oral health. In the same way, more attention should be paid to oral health benefits that may occur when a polyphenol is used regularly for systemic health reasons, such as dietary supplementation with quercetin or resveratrol. Likewise, the effects of regular dietary consumption of foods and food extracts that are rich in polyphenols on oral health deserve closer inspection. The large-scale studies that have been undertaken exploring the oral health of tea drinkers provide a platform for future studies of dental caries and periodontal diseases in those who are ingesting polyphenols on a regular basis. One of the most interesting of these future topics will be the interactions between dietary polyphenol intake and type II diabetes, since there are direct implications for improving diabetic status to reduce its impact as a modifying factor in periodontal diseases.

Further work is needed to explore how metabolic products of polyphenols may alter oral diseases. Since the oral microbiome varies dramatically in its composition between individuals, these differences will affect the metabolism of polyphenols. Adding to this, certain polyphenols such as PACs exert potent influences on the composition of the oral microbiome, and thus two-way interactions need to be contemplated. There is also the opportunity to explore further how known polyphenols exert their oral health benefits and use this information to inform the development of future synthetic compounds that may have even greater biological activity and bioavailability. This moves the field of polyphenols in dentistry beyond the concept of healthy foods and food supplements, to the idea of targeted therapy based on being "inspired by nature." This workflow is well-known in the pharmaceutical industry.

A further area of potential development is the deployment of polyphenols into other areas of dental clinical practice where microbial diseases are problematic. A particular example is endodontics, where bacteria with antibiotic resistance are commonly found when there has been persisting infection of the root canal system of a nonvital tooth. One of the most important pathogens in this context is *Enterococcus faecalis*. Polyphenols from green tea such as EGCG have been shown to be highly effective at eradicating *E. faecalis*, both in the planktonic state and in biofilms. This indicates the potential value of including EGCG in solutions that are used during root canal treatment to irrigate the root canal (Lee & Tan, 2015).

Lastly, there are considerable opportunities to explore altered delivery systems for polyphenols, moving beyond the traditional oral care products of toothpastes and mouth rinses. One way to achieve this is to deliver the polyphenol of interest using a chewing gum. As the active agent is diluted in the saliva, this provides a topical effect on the teeth and the oral soft tissues, as well as a systemic effect when the saliva is eventually swallowed. Application using a chewing gum also means that the polyphenol of interest will be released into an oral environment at a time when the salivary pH is elevated into the neutral or slightly alkaline range because of chewing action. As mentioned earlier, keeping the pH in the neutral or alkaline range can mitigate the bitter taste of some polyphenols. As an example, conventional chewing gums are a suitable vehicle for releasing quercetin. An effective antimicrobial concentration can be achieved in the saliva from the first few minutes of chewing, which causes a reduction in the levels of *S. mutans*(Ferrazzano et al., 2016).

Other novel delivery systems for polyphenols should also be explored, ranging from micro-encapsulation approaches through to nano-scale delivery concepts, such as nano vesicles (Xu et al., 2021), and sustained-release technologies. Examples of the latter include slow release from reservoirs in dentures and orthodontic appliances, or from adhesive wafers bonded to teeth that dissolve slowly over prolonged periods of time. Existing adhesive technologies such as tooth whitening strips could be adapted to give short intense bursts of polyphenols release to achieve particular oral care objectives. A similar approach could be used employing green bio-based nano-carriers (Zhang et al., 2022) to encapsulate, protect, and deliver the relevant polyphenols into the oral environment. Such delivery systems could be based on lipids, proteins, or polysaccharides.

Conclusions

Based on the above considerations, there are numerous ways that polyphenols could be incorporated into oral care products or used as part of various sequences of dental treatment, to improve outcomes of existing treatments, and to prevent further disease development. The challenge lies in translating the strong body of existing work into a range of successful and well-proven products that can be adopted into everyday clinical practice. Many of the issues around formulating polyphenols into oral care products and dental materials can be resolved with appropriate formulation expertise, and there is a well-developed rationale around which polyphenol to choose for which applications. It is likely that the amount of research efforts devoted to using polyphenols in dentistry is going to increase in the future, especially given their unique and multifunctional capabilities as therapeutic agents.

References

Andreou, V., D'Addario, M., Zohar, R., Sukhu, B., Casper, R. F., Ellen, R. P., & Tenenbaum, H. C. (2004). Inhibition of osteogenesis in vitro by a cigarette smoke-associated hydrocarbon combined with *Porphyromonas gingivalis* lipopolysaccharide: Reversal by resveratrol. *Journal of Periodontology, 75*(7), 939–948.

Asahi, Y., Noiri, Y., Miura, J., Maezono, H., Yamaguchi, M., Yamamoto, R., Azakami, H., Hayashi, M., & Ebisu, S. (2014). Effects of the tea catechin epigallocatechin gallate on *Porphyromonas gingivalis* biofilms. *Journal of Applied Microbiology, 116*(5), 1164–1171. Available from https://doi.org/10.1111/jam.12458.

Bagieh, N., & Bageah. (2014). Effect of tannic acid on growth and acid production of *Candida albicans*. *Pakistan Oral & Dental Journal, 34*(3), 405–408, 2014.

Baumann, T., Niemeyer, S. H., Lussi, A., Scaramucci, T., & Carvalho, T. S. (2023). Rinsing solutions containing natural extracts and fluoride prevent enamel erosion in vitro. *Journal of Applied Oral Science, 31*. Available from https://doi.org/10.1590/1678-7757-2023-0108.

Bedran-Russo, A. K. B., Pashley, D. H., Agee, K., Drummond, J. L., & Miescke, K. J. (2008). Changes in stiffness of demineralized dentin following application of collagen crosslinkers. *Journal of Biomedical Materials Research — Part B Applied Biomaterials, 86*(2), 330–334. Available from https://doi.org/10.1002/jbm.b.31022.

Bhattarai, G., Poudel, S. B., Kook, S.-H., & Lee, J.-C. (2016). Resveratrol prevents alveolar bone loss in an experimental rat model of periodontitis. *Acta Biomaterialia, 29*, 398–408. Available from https://doi.org/10.1016/j.actbio.2015.10.031.

Bodet, C., Piché, M., Chandad, F., & Grenier, D. (2006). Inhibition of periodontopathogen-derived proteolytic enzymes by a high-molecular-weight fraction isolated from cranberry. *Journal of Antimicrobial Chemotherapy*, *57*(4), 685–690.

Boteon, A. P., Kato, M. T., Buzalaf, M. A. R., Prakki, A., Wang, L., Rios, D., & Honório, H. M. (2017). Effect of Proanthocyanidin-enriched extracts on the inhibition of wear and degradation of dentin demineralized organic matrix. *Archives of Oral Biology*, *84*, 118–124. Available from https://doi.org/10.1016/j.archoralbio.2017.09.027, http://www.journals.elsevier.com/archives-of-oral-biology/.

Bradshaw, D. J., & Marsh, P. D. (1998). Analysis of pH–driven disruption of oral microbial communities in vitro. *Caries Research*, *32*(6), 456–462. Available from https://doi.org/10.1159/000016487.

Carvalho, T. S., Pham, K. N., Niemeyer, S. H., & Baumann, T. (2021). The effect of red wine in modifying the salivary pellicle and modulating dental erosion kinetics. *European Journal of Oral Sciences*, *129*(1). Available from https://doi.org/10.1111/eos.12749, http://onlinelibrary.wiley.com/journal/10.1111/(ISSN)1600-0722.

Casati, M. Z., Algayer, C., Cardoso da Cruz, G., Ribeiro, F. V., Casarin, R. C. V., Pimentel, S. P., & Cirano, F. R. (2013). Resveratrol decreases periodontal breakdown and modulates local levels of cytokines during periodontitis in rats. *Journal of Periodontology*, *84*(10). Available from https://doi.org/10.1902/jop.2013.120746.

Charlton, A. J., Baxter, N. J., Khan, M. L., Moir, A. J. G., Haslam, E., Davies, A. P., & Williamson, M. P. (2002). Polyphenol/peptide binding and precipitation. *Journal of Agricultural and Food Chemistry*, *50*(6), 1593–1601. Available from https://doi.org/10.1021/jf010897z.

Cheng, L., Li, J., He, L., & Zhou, X. (2015). Natural products and caries prevention. *Caries Research*, *49*(Suppl. 1), 38–45. Available from https://doi.org/10.1159/000377734.

Choudhary, P., Tushir, S., Bala, M., Sharma, S., Sangha, M. K., Rani, H., Yewle, N. R., Kumar, P., Singla, D., Chandran, D., Kumar, M., & Mekhemar, M. (2023). Exploring the potential of bee-derived antioxidants for maintaining oral hygiene and dental health: A comprehensive review. *Antioxidants*, *12*(7), 1452. Available from https://doi.org/10.3390/antiox12071452.

Cui, X., Xu, L., Qi, K., & Lan, H. (2023). Effects of tea polyphenols and theaflavins on three oral cariogenic bacteria. *Molecules*, *28*(16), 6034. Available from https://doi.org/10.3390/molecules28166034.

Curuțiu, C., Dițu, L. M., Grumezescu, A. M., & Maria Holban, A. (2020). Polyphenols of honeybee origin with applications in dental medicine. *Antibiotics*, *9*(12), 856. Available from https://doi.org/10.3390/antibiotics9120856.

Di Lorenzo, C., Colombo, F., Biella, S., Stockley, C., & Restani, P. (2021). Polyphenols and human health: The role of bioavailability. *Nutrients*, *13*(1), 273. Available from https://doi.org/10.3390/nu13010273.

Domínguez-Rodríguez, G., Marina, M. L., & Plaza, M. (2017). Strategies for the extraction and analysis of non-extractable polyphenols from plants. *Journal of Chromatography A*, *1514*, 1–15. Available from https://doi.org/10.1016/j.chroma.2017.07.066.

Fan, Q., Zhou, X.-H., Wang, T.-F., Zeng, F.-J., Liu, X., Gu, Y., Chen, B., Yang, J., Pang, Z.-Y., Liu, J.-G., & Bai, G.-H. (2023). Effects of epigallocatechin-3-gallate on oxidative stress, inflammation, and bone loss in a rat periodontitis model. *Journal of Dental Sciences*, *18*(4), 1567–1575. Available from https://doi.org/10.1016/j.jds.2023.02.019.

Farkash, Y., Feldman, M., Ginsburg, I., Steinberg, D., & Shalish, M. (2019). Polyphenols inhibit *Candida albicans* and *Streptococcus mutans* biofilm formation. *Dentistry Journal*, 7(2), 42. Available from https://doi.org/10.3390/dj7020042.

Ferrazzano, G. F., Amato, I., Ingenito, A., Zarrelli, A., Pinto, G., & Pollio, A. (2011). Plant polyphenols and their anti-cariogenic properties: A review. *Molecules*, *16*(2), 1486–1507. Available from https://doi.org/10.3390/molecules16021486Italy, http://www.mdpi.com/1420-3049/16/2/1486/pdf.

Ferrazzano, G., Cantile, T., Coda, M., Alcidi, B., Sangianantoni, G., Ingenito, A., Di Stasio, M., & Volpe, M. (2016). In vivo release kinetics and antibacterial activity of novel polyphenols-enriched chewing gums. *Molecules*, *21*(8), 1008. Available from https://doi.org/10.3390/molecules21081008.

Flemming, J., Meyer-Probst, C. T., Speer, K., Kölling-Speer, I., Hannig, C., & Hannig, M. (2021). Preventive applications of polyphenols in dentistry—A review. *International Journal of Molecular Sciences*, *22*(9), 4892. Available from https://doi.org/10.3390/ijms22094892.

García-Estévez, I., Cruz, L., Oliveira, J., Mateus, N., de Freitas, V., & Soares, S. (2017). First evidences of interaction between pyranoanthocyanins and salivary proline-rich proteins. *Food Chemistry*, *228*, 574–581. Available from https://doi.org/10.1016/j.foodchem.2017.02.030.

Ghannoum, M. A., Jurevic, R. J., Mukherjee, P. K., Cui, F., Sikaroodi, M., Naqvi, A., Gillevet, P. M., & May, R. C. (2010). Characterization of the oral fungal microbiome (mycobiome) in healthy individuals. *PLoS Pathogens*, *6*(1), e1000713. Available from https://doi.org/10.1371/journal.ppat.1000713.

Gill, S., Kaur, A., Kapoor, D., Goyal, J., & Duhan, H. (2016). Cranberry polyphenols: Beneficial effects for prevention of periodontal disease and dental caries. *The Saint's International Dental Journal*, *2*(2), 38. Available from https://doi.org/10.4103/2454-3160.202222.

Ginsburg, I., Koren, E., Shalish, M., Kanner, J., & Kohen, R. (2012). Saliva increases the availability of lipophilic polyphenols as antioxidants and enhances their retention in the oral cavity. *Archives of Oral Biology*, *57*(10), 1327–1334. Available from https://doi.org/10.1016/j.archoralbio.2012.04.019.

Gutiérrez-Venegas, G., Gómez-Mora, J. A., Meraz-Rodríguez, M. A., Flores-Sánchez, M. A., & Ortiz-Miranda, L. F. (2019). Effect of flavonoids on antimicrobial activity of microorganisms present in dental plaque. *Heliyon*, *5*(12), e03013. Available from https://doi.org/10.1016/j.heliyon.2019.e03013.

Hass, V., Liu, H., Cook, W., Walker, M. P., & Wang, Y. (2021). Distinct effects of polyphenols and solvents on dentin collagen crosslinking interactions and biostability. *Dental Materials*, *37*(12), 1794–1805. Available from https://doi.org/10.1016/j.dental.2021.09.009.

Hirasawa, M., & Takada, K. (2004). Multiple effects of green tea catechin on the antifungal activity of antimycotics against *Candida albicans*. *Journal of Antimicrobial Chemotherapy*, *53*(2), 225–229. Available from https://doi.org/10.1093/jac/dkh046.

Huang, X., Palmer, S. R., Ahn, S. J., Richards, V. P., Williams, M. L., Nascimento, M. M., & Burne, R. A. (2016). A highly arginolytic Streptococcus species that potently antagonizes *Streptococcus mutans*. *Applied and Environmental Microbiology*, *82*(7), 2187–2201. Available from https://doi.org/10.1128/AEM.03887-15, http://aem.asm.org/content/82/7/2187.full.pdf.

Huerta-Reyes, M., Gaitán-Cepeda, L. A., & Sánchez-Vargas, L. O. (2022). Punica granatum as anticandidal and anti-HIV Agent: An HIV oral cavity potential drug. *Plants*, *11*(19), 2622. Available from https://doi.org/10.3390/plants11192622.

Javid, A. Z., Hormoznejad, R., Yousefimanesh, H. A., Haghighi-Zadeh, M. H., & Zakerkish, M. (2019). Impact of resveratrol supplementation on inflammatory, antioxidant, and periodontal markers in type 2 diabetic patients with chronic periodontitis. *Diabetes & Metabolic Syndrome*, *13*(4), 2769–2774. Available from https://doi.org/10.1016/j.dsx.2019.07.042.

Joiner, A., Muller, D., Elofsson, U. M., & Arnebrant, T. (2004). Ellipsometry analysis of the in vitro adsorption of tea polyphenols onto salivary pellicles. *European Journal of Oral Sciences*, *112*(6), 510–515. Available from https://doi.org/10.1111/j.1600-0722.2004.00166.x.

Khairnar, M. R., Karibasappa, G. N., Dodamani, A. S., Vishwakarma, P., Naik, R. G., & Deshmukh, M. A. (2015). Comparative assessment of Cranberry and Chlorhexidine mouthwash on streptococcal colonization among dental students: A randomized parallel clinical trial. *Contemporary Clinical Dentistry*, *6*(1), 35–39. Available from https://doi.org/10.4103/0976-237X.149289, http://www.contempclindent.org/.

Kilian, M., Chapple, I. L. C., Hannig, M., Marsh, P. D., Meuric, V., Pedersen, A. M. L., Tonetti, M. S., Wade, W. G., & Zaura, E. (2016). The oral microbiome — An update for oral healthcare professionals. *British Dental Journal*, *221*(10), 657–666. Available from https://doi.org/10.1038/sj.bdj.2016.865, http://www.bdj.co.uk.

Koo, H., & Bowen, W. H. (2014). *Candida albicans* and *Streptococcus mutans*: A potential synergistic alliance to cause virulent tooth decay in children. *Future Microbiology*, *9*(12), 1295–1297. Available from https://doi.org/10.2217/fmb.14.92, http://www.futuremedicine.com/loi/fmb.

Koo, H., Duarte, S., Murata, R. M., Scott-Anne, K., Gregoire, S., Watson, G. E., Singh, A. P., & Vorsa, N. (2010). Influence of cranberry proanthocyanidins on formation of biofilms by *Streptococcus mutans* on Saliva-coated apatitic surface and on dental caries development in vivo. *Caries Research*, *44*(2), 116–126. Available from https://doi.org/10.1159/000296306.

Kováč, J., Slobodníková, L., Trajčíková, E., Rendeková, K., Mučaji, P., Sychrová, A., & Bittner Fialová, S. (2023). Therapeutic potential of flavonoids and tannins in management of oral infectious diseases—A review. *Molecules*, *28*(1), 158. Available from https://doi.org/10.3390/molecules28010158.

Kugaji, M. S., Kumbar, V. M., Peram, M. R., Patil, S., Bhat, K. G., & Diwan, P. V. (2019). Effect of Resveratrol on biofilm formation and virulence factor gene expression of *Porphyromonas gingivalis* in periodontal disease. *APMIS: Acta Pathologica, Microbiologica, et Immunologica Scandinavica*, *127*(4), 187–195. Available from https://doi.org/10.1111/apm.12930.

Kurek-Górecka, A., Ramos, P., Kłósek, M., Bobela, E., Czuba, Z. P., Balwierz, R., & Olczyk, P. (2023). Propolis as a cariostatic agent in lozenges and impact of storage conditions on the stability of propolis. *Pharmaceutics*, *15*(6), 1768. Available from https://doi.org/10.3390/pharmaceutics15061768.

Kurek-Górecka, A., Walczyńska-Dragon, K., Felitti, R., Baron, S., & Olczyk, P. (2022). Propolis and diet rich in polyphenols as cariostatic agents reducing accumulation of dental plaque. *Molecules*, *27*(1), 271. Available from https://doi.org/10.3390/molecules27010271.

Lagha, A. B., & Grenier, D. (2019). Tea polyphenols protect gingival keratinocytes against TNF-α-induced tight junction barrier dysfunction and attenuate the inflammatory response of monocytes/macrophages. *Cytokine*, *115*, 64–75. Available from https://doi.org/10.1016/j.cyto.2018.12.009.

Lagha, A. B., Andrian, E., & Grenier, D. (2019). Resveratrol attenuates the pathogenic and inflammatory properties of *Porphyromonas gingivalis*. *Molecular Oral Microbiology*, *34*(3), 118–130. Available from https://doi.org/10.1111/omi.12260.

Lagha, A. B., & Grenier, D. (2017). Black tea theaflavins attenuate *Porphyromonas gingivalis* virulence properties, modulate gingival keratinocyte tight junction integrity and exert anti-inflammatory activity. *Journal of Periodontal Research*, *52*(3), 458–470. Available from https://doi.org/10.1111/jre.12411.

Lagha, A. B., LeBel, G., & Daniel, G. (2020). Tart cherry (*Prunus cerasus* L.) fractions inhibit biofilm formation and adherence properties of oral pathogens and enhance oral epithelial barrier function. *Phytotherapy Research, 34*(4), 886–895. Available from https://doi.org/10.1002/ptr.6574.

Lavefve, L., Howard, L. R., & Carbonero, F. (2020). Berry polyphenols metabolism and impact on human gut microbiota and health. *Food and Function, 11*(1), 45–65. Available from https://doi.org/10.1039/c9fo01634a, http://pubs.rsc.org/en/journals/journal/fo.

Lee, P., & Tan, K. S. (2015). Effects of Epigallocatechin gallate against *Enterococcus faecalis* biofilm and virulence. *Archives of Oral Biology, 60*(3), 393–399. Available from https://doi.org/10.1016/j.archoralbio.2014.11.014, http://www.journals.elsevier.com/archives-of-oral-biology/.

Leme-Kraus, A. A., Aydin, B., Vidal, C. M. P., Phansalkar, R. M., Nam, J. W., McAlpine, J., Pauli, G. F., Chen, S., & Bedran-Russo, A. K. (2017). Biostability of the proanthocyanidins-dentin complex and adhesion studies. *Journal of Dental Research, 96*(4), 406–412. Available from https://doi.org/10.1177/0022034516680586.

Li, Y., Jiang, X., Hao, J., Zhang, Y., & Huang, R. (2019). Tea polyphenols: Application in the control of oral microorganism infectious diseases. *Archives of Oral Biology, 102*, 74–82. Available from https://doi.org/10.1016/j.archoralbio.2019.03.027.

Lombardo Bedran, T. B., Palomari Spolidorio, D., & Grenier, D. (2015). Green tea polyphenol epigallocatechin-3-gallate and cranberry proanthocyanidins act in synergy with cathelicidin (LL-37) to reduce the LPS-induced inflammatory response in a three-dimensional co-culture model of gingival epithelial cells and fibroblasts. *Archives of Oral Biology, 60*(6), 845–853. Available from https://doi.org/10.1016/j.archoralbio.2015.02.021.

Marsh, P. D., Head, D. A., & Devine, D. A. (2015). Ecological approaches to oral biofilms: Control without killing. *Caries Research, 49*, 46–54. Available from https://doi.org/10.1159/000377732, http://content.karger.com/ProdukteDB/produkte.asp?Aktion = JournalHome&ProduktNr = 224219.

Ma, Y., Qian, Y., Chen, Y., Ruan, X., Peng, X., Sun, Y., Zhang, J., Luo, J., Zhou, S., & Deng, C. (2023). Resveratrol modulates the inflammatory response in hPDLSCs via the NRF2/HO-1 and NF-κB pathways and promotes osteogenic differentiation. *Journal of Periodontal Research*. Available from https://doi.org/10.1111/jre.13200.

Melis, M., Yousaf, N. Y., Mattes, M. Z., Cabras, T., Messana, I., Crnjar, R., Tomassini Barbarossa, I., & Tepper, B. J. (2017). Sensory perception of and salivary protein response to astringency as a function of the 6-n-propylthioural (PROP) bitter-taste phenotype. *Physiology and Behavior, 173*, 163–173. Available from https://doi.org/10.1016/j.physbeh.2017.01.031, http://www.elsevier.com/locate/physbeh.

Melok, A. L., Lee, L. H., Yussof, S. A. M., & Chu, T. (2018). Green tea polyphenol epigallocatechin-3-gallate-stearate inhibits the growth of *Streptococcus mutans*: A promising new approach in caries prevention. *Dentistry Journal, 6*(3). Available from https://doi.org/10.3390/dj6030038, https://www.mdpi.com/2304-6767/6/3/38/pdf.

Neurath, A. R., Strick, N., Li, Y. Y., & Debnath, A. K. (2004). Punica granatum (Pomegranate) juice provides an HIV-1 entry inhibitor and candidate topical microbicide. *BMC Infectious Diseases, 4*. Available from https://doi.org/10.1186/1471-2334-4-41, http://www.biomedcentral.com/1471-2334/4/41, United States.

Niemeyer, S. H., Baumann, T., Lussi, A., Meyer-Lueckel, H., Scaramucci, T., & Carvalho, T. S. (2021). Salivary pellicle modification with polyphenol-rich teas and natural extracts to improve protection against dental erosion. *Journal of Dentistry, 105*, 103567. Available from https://doi.org/10.1016/j.jdent.2020.103567.

Nikniaz, S., Vaziri, F., & Mansouri, R. (2023). Impact of resveratrol supplementation on clinical parameters and inflammatory markers in patients with chronic periodontitis: a randomized clinical trail. *BMC Oral Health, 23*(1). Available from https://doi.org/10.1186/s12903-023-02877-4.

Ning, Y., Ling, J., & Wu, C. D. (2015). Synergistic effects of tea catechin epigallocatechin gallate and antimycotics against oral Candida species. *Archives of Oral Biology*, *60*(10), 1565–1570. Available from https://doi.org/10.1016/j.archoralbio.2015.07.001, http://www.journals.elsevier.com/archives-of-oral-biology/.

Park, H.-J., Jeong, S.-K., Kim, S.-R., Bae, S.-K., Kim, W.-S., Jin, S.-D., Koo, T. H., Jang, H.-O., Yun, I., Kim, K.-W., & Bae, M.-K. (2009). Resveratrol inhibits *Porphyromonas gingivalis* lipopolysaccharide-induced endothelial adhesion molecule expression by suppressing NF-kappaB activation. *Archives of Pharmacal Research*, *32*(4), 583–591. Available from https://doi.org/10.1007/s12272-009-1415-7.

Petti, S., & Scully, C. (2009). Polyphenols, oral health and disease: A review. *Journal of Dentistry*, *37*(6), 413–423. Available from https://doi.org/10.1016/j.jdent.2009.02.003.

Philip, N., Bandara, H. M. H. N., Leishman, S. J., & Walsh, L. J. (2019a). Effect of polyphenol-rich cranberry extracts on cariogenic biofilm properties and microbial composition of polymicrobial biofilms. *Archives of Oral Biology*, *102*, 1–6. Available from https://doi.org/10.1016/j.archoralbio.2019.03.026, http://www.journals.elsevier.com/archives-of-oral-biology/.

Philip, N., Bandara, H. M. H. N., Leishman, S. J., & Walsh, L. J. (2019b). Inhibitory effects of fruit berry extracts on *Streptococcus mutans* biofilms. *European Journal of Oral Sciences*, *127*(2), 122–129. Available from https://doi.org/10.1111/eos.12602, http://onlinelibrary.wiley.com/journal/10.1111/(ISSN)1600-0722.

Philip, N., Leishman, S. J., Bandara, H. M. H. N., Healey, D. L., & Walsh, L. J. (2020). Randomized controlled study to evaluate microbial ecological effects of CPP-ACP and cranberry on dental plaque. *JDR Clinical & Translational Research*, *5*(2), 118–126. Available from https://doi.org/10.1177/2380084419859871.

Philip, N., Suneja, B., & Walsh, L. (2018a). Beyond *Streptococcus mutans*: Clinical implications of the evolving dental caries aetiological paradigms and its associated microbiome. *British Dental Journal*, *224*(4), 219–225. Available from https://doi.org/10.1038/sj.bdj.2018.81.

Philip, N., Suneja, B., & Walsh, L. J. (2018b). Ecological approaches to dental caries prevention: paradigm shift or shibboleth? *Caries Research*, *52*(1–2), 153–165. Available from https://doi.org/10.1159/000484985, http://content.karger.com/ProdukteDB/produkte.asp?Aktion = JournalHome&ProduktNr = 224219.

Philip, N., & Walsh, L. (2019). Cranberry polyphenols: Natural weapons against dental caries. *Dentistry Journal*, 7(1), 20. Available from https://doi.org/10.3390/dj7010020.

Porto, I. C. C. M., Nascimento, T. G., Oliveira, J. M. S., Freitas, P. H., Haimeur, A., & França, R. (2018). Use of polyphenols as a strategy to prevent bond degradation in the dentin–resin interface. *European Journal of Oral Sciences*, *126*(2), 146–158. Available from https://doi.org/10.1111/eos.12403, http://www.munksgaard.dk/tidsskrifter.nsf/a3b40ef0ca9b8d86c1256a160050049f/46964c5b7087316ac1256a1c003d81e2?OpenDocument.

Rana, A., Samtiya, M., Dhewa, T., Mishra, V., & Aluko, R. E. (2022). Health benefits of polyphenols: A concise review. *Journal of Food Biochemistry*, *46*(10). Available from https://doi.org/10.1111/jfbc.14264.

Sakagami, H. (2014). Biological activities and possible dental application of three major groups of polyphenols. *Journal of Pharmacological Sciences*, *126*(2), 92–106. Available from https://doi.org/10.1254/jphs.14r04cr.

Sakagami, H., Watanabe, T., Hoshino, T., Suda, N., Mori, K., Yasui, T., Yamauchi, N., Kashiwagi, H., Gomi, T., Oizumi, T., Nagai, J., Uesawa, Y., Takao, K., & Sugita, Y. (2019). Recent progress of basic studies of natural products and their dental application. *Medicines*, *6*(1), 4. Available from https://doi.org/10.3390/medicines6010004.

Sakanaka, S., Aizawa, M., Kim, M., & Yamamoto, T. (1996). Inhibitory effects of green tea polyphenols on growth and cellular adherence of an oral bacterium, *Porphyromonas gingivalis*. *Bioscience, Biotechnology, and Biochemistry*, *60*(5), 745–749.

Samaranayake, L. P., Keung Leung, W., & Jin, L. (2009). Oral mucosal fungal infections. *Periodontology 2000*, *49*(1), 39–59. Available from https://doi.org/10.1111/j.1600-0757.2008.00291.x.

Schestakow, A., Nekrashevych, Y., Hoth-Hannig, W., & Hannig, M. (2022). Influence of periodic polyphenol treatment on the anti-erosive potential of the acquired enamel pellicle—A qualitative exploratory study. *Journal of Dentistry*, *124*, 104236. Available from https://doi.org/10.1016/j.jdent.2022.104236.

Schneider-Rayman, M., Steinberg, D., Sionov, R. V., Friedman, M., & Shalish, M. (2021). Effect of epigallocatechin gallate on dental biofilm of *Streptococcus mutans*: An in vitro study. *BMC Oral Health*, *21*(1). Available from https://doi.org/10.1186/s12903-021-01798-4, http://www.biomedcentral.com/bmcoralhealth/.

Shahzad, M., Sherry, L., Rajendran, R., Edwards, C. A., Combet, E., & Ramage, G. (2014). Utilising polyphenols for the clinical management of *Candida albicans* biofilms. *International Journal of Antimicrobial Agents*, *44*(3), 269–273. Available from https://doi.org/10.1016/j.ijantimicag.2014.05.017.

Sitheeque, M. A. M., Panagoda, G. J., Yau, J., Amarakoon, A. M. T., Udagama, U. R. N., & Samaranayake, L. P. (2009). Antifungal activity of black tea polyphenols (Catechins and Theaflavins) against candida species. *Chemotherapy*, *55*(3), 189–196. Available from https://doi.org/10.1159/000216836.

Skaba, D., Morawiec, T., Tanasiewicz, M., Mertas, A., Bobela, E., Szliszka, E., Skucha-Nowak, M., Dawiec, M., Yamamoto, R., Ishiai, S., Makita, Y., Redzynia, M., Janoszka, B., Niedzielska, I., & Król, W. (2013). Influence of the toothpaste with brazilian ethanol extract propolis on the oral cavity health. *Evidence-Based Complementary and Alternative Medicine*, *2013*, 1–12. Available from https://doi.org/10.1155/2013/215391.

Slomka, V., Hernandez-Sanabria, E., Herrero, E. R., Zaidel, L., Bernaerts, K., Boon, N., Quirynen, M., & Teughels, W. (2017). Nutritional stimulation of commensal oral bacteria suppresses pathogens: The prebiotic concept. *Journal of Clinical Periodontology*, *44*(4), 344–352. Available from https://doi.org/10.1111/jcpe.12700.

Smullen, J., Koutsou, G. A., Foster, H. A., Zumbé, A., & Storey, D. M. (2007). The antibacterial activity of plant extracts containing polyphenols against *Streptococcus mutans*. *Caries Research*, *41*(5), 342–349. Available from https://doi.org/10.1159/000104791.

Souissi, M., Ben Lagha, A., Chaieb, K., & Grenier, D. (2021). Effect of a berry polyphenolic fraction on biofilm formation, adherence properties and gene expression of *Streptococcus mutans* and its biocompatibility with oral epithelial cells. *Antibiotics*, *10*(1), 46. Available from https://doi.org/10.3390/antibiotics10010046.

Tamaki, N., Cristina Orihuela-Campos, R., Inagaki, Y., Fukui, M., Nagata, T., & Hiro-O, I. (2014). Resveratrol improves oxidative stress and prevents the progression of periodontitis via the activation of the Sirt1/AMPK and the Nrf2/antioxidant defense pathways in a rat periodontitis model. *Free Radical Biology & Medicine*, *75*, 222–229. Available from https://doi.org/10.1016/j.freeradbiomed.2014.07.034.

Tena, N., Martín, J., & Asuero, A. G. (2020). State of the art of anthocyanins: antioxidant activity, sources, bioavailability, and therapeutic effect in human health. *Antioxidants*, *9* (5), 451. Available from https://doi.org/10.3390/antiox9050451.

Vaillancourt,, K., Lagha, A. B., & Grenier, D. (2022). Effects of a berry polyphenolic fraction on the pathogenic properties of Porphyromonas gingivalis. *Frontiers in Oral Health*, *3*, 923663. Available from https://doi.org/10.3389/froh.2022.923663.

Varoni, E. M., Lodi, G., Sardella, A., Carrassi, A., & Iriti, M. (2012). Plant polyphenols and oral health: Old phytochemicals for new fields. *Current Medicinal Chemistry*, *19* (11), 1706–1720. Available from https://doi.org/10.2174/092986712799945012.

Veloz, J. J., Alvear, M., & Salazar, L. A. (2019). Antimicrobial and antibiofilm activity against *Streptococcus mutans* of individual and mixtures of the main polyphenolic compounds found in Chilean Propolis. *BioMed Research International*, *2019*. Available from https://doi.org/10.1155/2019/7602343, http://www.hindawi.com/journals/biomed/.

Vilela, M. M., Salvador, S. L., Teixeira, I. G. L., Del Arco, M. C. G., & De Rossi, A. (2020). Efficacy of green tea and its extract, epigallocatechin-3-gallate, in the reduction of cariogenic microbiota in children: a randomized clinical trial. *Archives of Oral Biology*, *114*, 104727. Available from https://doi.org/10.1016/j.archoralbio.2020.104727.

Wang, B., Han, F., You, R., Chen, C., & Xie, H. (2023). Polyphenols can improve resin-dentin bond durability by promoting amorphous calcium phosphate nanoparticles to backfill the dentin matrix. *International Journal of Nanomedicine*, *18*, 1491–1505. Available from https://doi.org/10.2147/IJN.S395631, https://www.dovepress.com/getfile.php?fileID = 88438.

Wang, Y., Samaranayake, L. P., & Dykes, G. A. (2022). Tea extracts inhibit the attachment of streptococci to oral/dental substrata by reducing hydrogen bonding energies. *Biofouling*, *38*(1), 42–54. Available from https://doi.org/10.1080/08927014.2021.2013826, http://www.tandf.co.uk/journals/titles/08927014.asp.

Wang, Y., Zeng, J., Yuan, Q., & Luan, Q. (2021). Efficacy of (-)-epigallocatechin gallate delivered by a new-type scaler tip during scaling and root planing on chronic periodontitis: a split-mouth, randomized clinical trial. *BMC Oral Health*, *21*(1). Available from https://doi.org/10.1186/s12903-021-01418-1.

Weber, M. T., Hannig, M., Pötschke, S., Höhne, F., & Hannig, C. (2015). Application of plant extracts for the prevention of dental erosion: An in situ/in vitro study. *Caries Research*, *49*(5), 477–487. Available from https://doi.org/10.1159/000431294, http://content.karger.com/ProdukteDB/produkte.asp?Aktion = JournalHome&ProduktNr = 224219.

Wu, C. D., Huang, E., Li, W., White, M., Jung, S., & Xie, Q. (2021). Beverages containing plant-derived polyphenols inhibit growth and biofilm formation of *Streptococcus mutans* and children's supragingival plaque bacteria. *Beverages*, 7(3), 43. Available from https://doi.org/10.3390/beverages7030043.

Xu, X., Dai, Z., Zhang, Z., Kou, X., You, X., Sun, H., Guo, H., Liu, M., & Zhu, H. (2021). Fabrication of oral nanovesicle in-situ gel based on Epigallocatechin gallate phospholipid complex: Application in dental anti-caries. *European Journal of Pharmacology*, *897*, 173951. Available from https://doi.org/10.1016/j.ejphar.2021.173951.

Xu, X., Zhou, X. D., & Wu, C. D. (2012). Tea catechin epigallocatechin gallate inhibits *Streptococcus mutans* biofilm formation by suppressing gtf genes. *Archives of Oral Biology*, *57*(6), 678–683. Available from https://doi.org/10.1016/j.archoralbio.2011.10.021.

Yamanaka, A., Kouchi, T., Kasai, K., Kato, T., Ishihara, K., & Okuda, K. (2007). Inhibitory effect of cranberry polyphenol on biofilm formation and cysteine proteases of *Porphyromonas gingivalis*. *Journal of Periodontal Research*, *42*(6), 589–592.

Yoo, S., Murata, R. M., & Duarte, S. (2011). Antimicrobial traits of tea- and cranberry-derived polyphenols against *Streptococcus mutans*. *Caries Research*, *45*(4), 327–335. Available from https://doi.org/10.1159/000329181.

Yousaf, N. Y., Melis, M., Mastinu, M., Contini, C., Cabras, T., Tomassini Barbarossa, I., & Tepper, B. J. (2020). Time course of salivary protein responses to cranberry-derived polyphenol exposure as a function of PROP taster status. *Nutrients*, *12*(9), 2878. Available from https://doi.org/10.3390/nu12092878.

Yousaf, N. Y., Wu, G., Melis, M., Mastinu, M., Contini, C., Cabras, T., Tomassini Barbarossa, I., Zhao, L., Lam, Y. Y., & Tepper, B. J. (2022). Daily exposure to a cranberry polyphenol oral rinse alters the oral microbiome but not taste perception in PROP taster status classified individuals. *Nutrients*, *14*(7), 1492. Available from https://doi.org/10.3390/nu14071492.

Zeng, J., Wang, Y., Yuan, Q., & Luan, Q. (2022). The effect of (-)-epigallocatechin gallate as an adjunct to non-surgical periodontal treatment: a randomized clinical trial. *Trials*, *23*(1). Available from https://doi.org/10.1186/s13063-022-06298-6.

Zhang, Z., Li, X., Sang, S., McClements, D. J., Chen, L., Long, J., Jiao, A., Jin, Z., & Qiu, C. (2022). Polyphenols as plant-based nutraceuticals: health effects, encapsulation, nano-delivery, and application. *Foods*, *11*(15), 2189. Available from https://doi.org/10.3390/foods11152189.

Zheng, C.-jin, Liu, R., Xue, B., Luo, J., Gao, L., Wang, Y., Ou, S., Li, S., & Peng, X. (2017). Impact and consequences of polyphenols and fructooligosaccharide interplay on gut microbiota in rats. *Food & Function*, *8*(5), 1925–1932. Available from https://doi.org/10.1039/c6fo01783e.

CHAPTER TEN

Polyphenols in skin and hair health

Ana Margarida Silva[1,*], Filipa Teixeira[1,*], Maria João Bessa[2], Paulo C. Costa[3,4], Cristina Delerue-Matos[1] and Francisca Rodrigues[1]

[1]REQUIMTE/LAQV, ISEP, Polytechnic of Porto, Porto, Portugal
[2]UNIPRO – Oral Pathology and Rehabilitation Research Unit, University Institute of Health Sciences (IUCS), CESPU, Gandra, Portugal
[3]REQUIMTE/UCIBIO, MedTech-Laboratory of Pharmaceutical Technology, Department of Drug Sciences, Faculty of Pharmacy, University of Porto, Porto, Portugal
[4]Associate Laboratory i4HB – Institute for Health and Bioeconomy, Faculty of Pharmacy, University of Porto, Porto, Portugal

List of Abbreviations

DNFB	1-fluoro-2,4-dinitrobenzene
EGCG	epigallocatechin 3-gallate
HMOX-1	Heme oxygenase-1
HRIPT	Human repeat insult patch test
IC_{50}	Half-maximal inhibitory concentration
IGF-1	Insulin-like growth factor-1
IL-1α	Interleukin-1α
IL-6	Interleukin-6
IL-8	Interleukin-8
LPS	Lipopolysaccharide
MHC-II	Major histocompatibility complex II
MIC	Minimum inhibitory concentration
MMPs	Matrix metalloproteinases
MPN	Metal-phenolic networks
NF-κB	Nuclear factor-κB
OMW	Olive mill wastewater
PPD	Phenylenediamine
RNS	Reactive nitrogen species
ROS	Reactive oxygen species
TNF-α	Tumor necrosis factor alpha

* The authors contributed equally to the work.

Bioactive Polyphenols for Health and Pathology Treatment.
DOI: https://doi.org/10.1016/B978-0-443-13821-8.00010-X

TPA	12-O-tetradecanoyl-phorbol-13-acetate
UV	Ultraviolet
VEGF	Vascular endothelial cell growth factor

Introduction

On the last two decades, cosmetics and pharmaceutical industries have boosted the search for natural compounds as new active ingredients for skin and hair products, with the number of new formulations available on the marked and the patents regarding new actives being more than ever (Silva, Costa, et al., 2021). The potential safeness of natural compounds, coupled to their multiple biological effects and costs effectiveness, attracted the consumers attention, leading to the generation of "eco-cosmetics" (Silva, Costa, et al., 2021). Coupled to this new trend, modern society is more aware of the appearance importance, looking to products that may prevent the aging progress associated with the gradual loss of the physiological skin integrity. A growing interest has emerged regarding the use of natural compounds for promoting skin health, with emphasis on polyphenols, mainly due to their multifunctional biological properties, such as antioxidant, antiinflammatory, or antiallergic activities effects (Jesumani et al., 2020). These activities may be used not only in healthy skin, as antiaging ingredients for example, but also in skin disorders. In fact, skin disorders are the most common diseases (Sun et al., 2022), being triggered by genetic factors, infections, endocrine disorders, and mental stress (Singh et al., 2011; Sun et al., 2022). Traditional drugs, such as clotrimazol or topical corticosteroids, have been employed against skin disorders, but most of the times medication is linked to negative side effects (Jesumani et al., 2020; Sun et al., 2022). Several studies demonstrated satisfactory results regarding the use of natural extracts rich in polyphenols against skin diseases and disorders, not only in terms of biological activity, but also concerning skin maintenance and health promotion (Jesumani et al., 2020; Wei et al., 2009). Among cosmetics, the hair field is also a tendency, with hair health and growth cycle being affected by numerous circumstances, including heavy metals, toxins, drugs, genetics, stress, smoking, diet, menopause, and lifestyle (Kesika et al., 2023). A wide range of hair products with traditional drugs (e.g., minoxidil or finasteride) are available to treat hair disorders, but most of them are also associated with undesirable side effects (Shen et al., 2018). To overcome these drawbacks, polyphenols are also an option, once again due to their biological properties, with several studies demonstrating satisfactory results against multiple hair diseases (Sun et al., 2022; Zhang et al., 2022).

Therefore the aim of this chapter is to explore the polyphenols effects on skin and hair health and diseases, highlighting their biological activities, particularly in what concerns to antioxidant, antiaging, antiinflammatory, antimicrobial, and antiallergic effects. Moreover, it will be given a particular emphasis on the polyphenols effects on hair growth related to hair loss symptoms as well as hair coloring. Commercial formulations available on the market and safety concerns will be also overview.

Anatomy and physiology of skin and hair

Skin

Skin is classified as the body biggest and essential organ, representing 16% of the total body weight (Venus et al., 2010). In adult, its covers around 2 m^2 of surface area and weights about 13 kg (Lai-Cheong & McGrath, 2013; Nunes et al., 2017), covering the entire body and containing accessory structures, such as glands, hair, and nails (McKnight et al., 2022), receiving about one-third of the blood circulating through the body (Supe & Takudage, 2021). This complex structure is divided in three main compartments: epidermis or outer layer, dermis or cutis, and hypodermis or subcutis (Nunes et al., 2017). Fig. 10.1 summarizes the skin structure.

Epidermis constitutes the skin outermost layer and present around 0.1−1.5 mm of thickness (Lai-Cheong & McGrath, 2013). Keratinocytes are the main constituents of this layer, characterizing 95% of the cells and being responsible for the keratin production (Lai-Cheong & McGrath, 2013). The remaining 5% are constituted by melanocytes, Langerhans, and Merkel cells (Lai-Cheong & McGrath, 2021; Nunes et al., 2017). Epidermis is divided in five sublayers, namely *stratum germinativum*, *stratum spinosum*, *stratum granulosum*, *stratum lucidum*, and *stratum corneum* (Supe & Takudage, 2021). The *stratum germinativum*, also known as basal layer, is a renewing level, being keratinocytes and melanocytes the main cell types present (Venus et al., 2010). In this sublayer, the cells contain cylindrically shaped and a relatively large nucleus, where column-shaped keratinocytes are attached to the basement membrane zone with their long axis perpendicular to the dermis (Kolarsick et al., 2011). The *stratum spinosum*, also named squamous cell layer, contains a variable number of cell levels, and

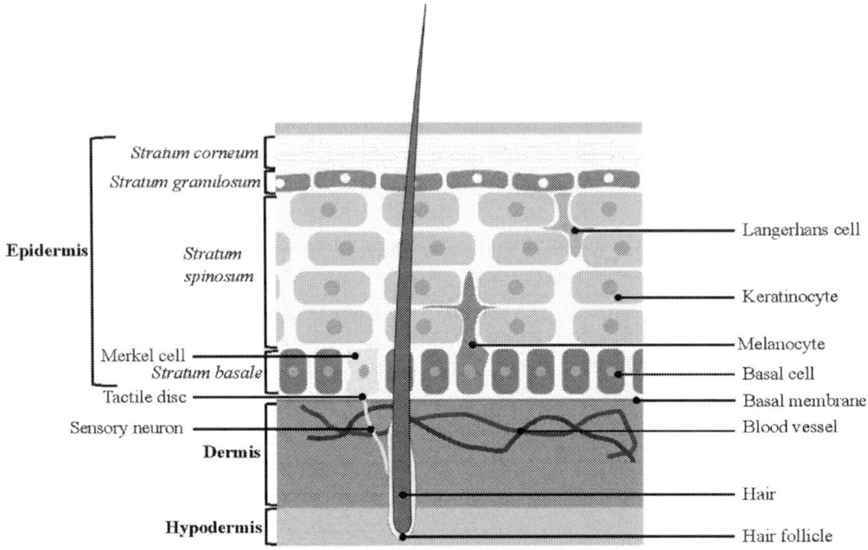

Figure 10.1 Skin structure and composition.

is composed by a variety of cells that differ in shape, structure, and subcellular properties, depending on their location (Kolarsick et al., 2011), being possible to find Langerhans cells (Venus et al., 2010). The *stratum granulosum*, also known as granular layer, is the most superficial layer of epidermis and contains living cells, being composed by intracellular keratohyalin granules (Kolarsick et al., 2011; Venus et al., 2010). The *stratum granulosum* varies in thickness according to its location (Kolarsick et al., 2011). The *stratum lucidum* is not always existent, being present in palmar and plantar skin. Lastly, the *stratum corneum* is the outermost layer of epidermis and corneocytes are the cells present (Lai-Cheong & McGrath, 2021). This thin layer ($\pm$ 15 μm) is composed by several lipids as fatty acids, sterols, and ceramides, which are released from lamellar bodies within the upper (Lai-Cheong & McGrath, 2021). Each cycle in *stratum corneum* takes 28 days from cell division to shedding (Venus et al., 2010). The melanocytes present in epidermis are constituted by dendritic cells and produce melanin, a pigment that is transferred to keratinocytes (Kolarsick et al., 2011). The Langerhans' cells are immunocompetent and derive from the bone marrow, being located above the basal layer (Lai-Cheong & McGrath, 2013), while Merkel's cells are found in the basal layer, and have an important role in sensory perception, once transmit sensory information from skin to sensory nerves (Lai-Cheong & McGrath, 2013).

Dermis is an intermediate layer between epidermis and hypodermis (Nunes et al., 2017). The thickness of this layer varies according to the body site (1–2 mm), being thinnest on the eyelid and thickest on the back (Lai-Cheong & McGrath, 2021; Nunes et al., 2017). The dermis present interstitial and cellular components, such as collagen fibers, elastic tissue, fibroblast, mast cell, lymphocyte, dermal dendritic cell, but also blood vessels, lymphatic channels, and sensory nerves (Lai-Cheong & McGrath, 2021). This layer is divided into papillary dermis and reticular dermis (Mathes et al., 2014). The papillary dermis is superficial, being constituted by a loosely arranged collagen fibers, while the reticular dermis, the deep layer, is formed by a dense collagen fibers arranged in parallel to the skin surface (Mathes et al., 2014). Collagen is the main component present in dermis, being predominant the type I and III (Brown & Krishnamurthy, 2021). The elastic fibers, also present in dermis, have impact in the extensible properties of the skin (Brown & Krishnamurthy, 2021; Lai-Cheong & McGrath, 2021). The collagen and elastic fibers are embedded in a gelatinous substance composed by proteoglycans and glycoproteins, also known as ground substance (Lai-Cheong & McGrath, 2013). This layer is produced by fibroblasts, the main cell type presents in dermis (Lai-Cheong & McGrath, 2013; Nunes et al., 2017). Fibroblasts are mesenchymal cells with embryonic heterogeneity and body site diversity (Lai-Cheong & McGrath, 2021). In contrast to epidermis, dermis present blood and lymph vessels, nerves, sweat, sebaceous glands, and hair follicles (Mathes et al., 2014).

The hypodermis, or subcutis, is the deepest skin layer (Lai-Cheong & McGrath, 2021), being formed by small lobes of fat or lipocytes that are separated from one another by fibrous septa of connective tissue (Lai-Cheong & McGrath, 2021). The fibers formed in dermis are extended into hypodermis, allowing the connection between these two layers (Lai-Cheong & McGrath, 2021).

Skin plays important vital functions, providing a protection barrier and maintaining homeostasis (Kolarsick et al., 2011; McKnight et al., 2022). This barrier constitutes an interface between the human body and the environment, controlling the water loss and the entry of infectious or toxic substances (Abdo et al., 2020; Segre, 2006). Therefore the most important skin function is the protection of the deeper body structures by the presence of a waterproof layer, preventing dehydration and providing protection against physical and chemical traumas (McKnight et al., 2022). Epidermis and *stratum corneum* are responsible for the protection from the environment (Venus et al., 2010). The *stratum corneum* constitutes an impermeable layer, once presents keratinocytes arranged within a multilayer and a histidine-rich protein called involucrin, having also the intercellular spaces filled with a lipid-rich matrix that allow a waterproof barrier (Venus et al., 2010). However, different body sites present distinctive permeation capacities, being face, forehead, and dorsum of the hands the most permeable, while palms of the hands are the least one, for example (Venus et al., 2010). Furthermore, skin is highly permeable to polar molecules such as water, sodium or potassium (Venus et al., 2010). In fact, skin contains almost 20% of the body's total water content (McKnight et al., 2022). In addition, the presence of keratin in epidermis provides protection against physical trauma (McKnight et al., 2022). Nonetheless, the strength, flexibility, and protection of the deeper anatomical structures are provided by dermis (Brown & Krishnamurthy, 2021).

The *stratum corneum* also protects skin from inflammation caused by irritants and allergens (Lai-Cheong & McGrath, 2021), also providing defense against microorganisms, contributing to the innate and adaptive immunity (Lai-Cheong & McGrath, 2021). In this mechanism intervenes keratinocyte-derived endogenous antimicrobial peptides, as defensins and cathelicidins, as well as Langerhans dendritic cells and transient epidermal T cells (Lai-Cheong & McGrath, 2021; Venus et al., 2010). The Langerhans dendritic cells are present in epidermis and act as antigen presenting cells by phagocytosing antigens and initiating immune responses against microbial threats (Lai-Cheong & McGrath, 2021; Venus et al., 2010). This antimicrobial function is also related to the dryness of the outer layer of epidermis, which prevents the growth of skin organisms (Venus et al., 2010).

Thermoregulation is another important role of skin. The maintenance of body temperature is possible by vasodilatation and vasoconstriction, through cutaneous surface, sweat glands, skeletal muscles, and hypothalamus (Lai-Cheong & McGrath, 2021; McKnight et al., 2022). The non-hairy glabrous skin (covers the most distal body parts) is responsible for thermoregulation (Abdo et al., 2020). This part allows to determine the local temperature and tactile signals, once is characterized by the absence of hair, dense vascularization, and a large surface-to-volume ratio (Abdo et al., 2020). Furthermore, skin may synthetize vitamin D, contributing to the calcium metabolism, immune function, hair follicle cycling, photoprotection, upregulation of innate immune defenses, and wound healing (Abdo et al., 2020; Lai-Cheong & McGrath, 2021). Moreover, skin constitutes the first protector barrier against ultraviolet (UV) radiation damages. The *stratum corneum* is able to reflect radiation, reducing the exposure dose (Venus et al., 2010). Melanin is produced and secreted by melanocytes and decreases the absorption of UV radiation, protecting skin from the harmful UV effects (McKnight et al., 2022). This skin function is crucial, once UVA radiation (320−400 nm), for example, may cause damages in DNA molecules and lipid components, triggering skin aging damages (Rodrigues, Nunes, & Oliveira, 2017). UVB radiation (280−320 nm) penetrates into the skin's epidermis layer, causing sunburn and skin injuries (Rodrigues, Nunes, Alves, et al., 2017). However, this radiation can provoke photochemical damages and increase the prevalence of skin cancers, such as malignant melanoma (Venus et al., 2010).

The skin sensory perception is determined by dermis as it is rich in nerve endings (Lai-Cheong & McGrath, 2021). The nerve endings sense the hair movements and act as receptors, triggering sensation to extend beyond the skin's surface (Brown & Krishnamurthy, 2021). Besides the skin functions described above, the facial skin has a particular interest in social interaction, since reflects the identity and mood (Arda et al., 2014).

Hair

Hair plays an important role on appearance (Yadav et al., 2019), despite regulate the body temperature and protect the skin from damages (Yadav et al., 2019). In addition, the stem cells present in hair follicles contribute to the regeneration of the damaged skin (Yadav et al., 2019). Actually, hair is a skin appendage located in dermis that come out from epidermis. One centimeter square area of the skin commonly contains five hairs, with exception of the palms and soles (Yadav et al., 2019). The human body presents around 5 millions of hairs that can be divided in three types, namely lanugo, vellus, and terminal hairs (Lai-Cheong & McGrath, 2021). The lanugo hair is the first one that grows soon after birth, the vellus hair is the fine hair that appear over the body, while the terminal hair is longer and coarser (Lai-Cheong & McGrath, 2021). For this reason, the human body is covered with hair follicles, apart from areas of glabrous skin (Miteva et al., 2019). The hair follicles connect the epidermis to the dermis and are characterized by mesenchymal and epithelial signaling pathways (de Groot et al., 2021). This structure undergoes lifelong cycling characterized by periods of growth (anagen), transformation (catagen), and rest (telogen) (Miteva et al., 2019). Some hairs reside in the anagen phase, maintaining stable the density and the total hair (Natarelli et al., 2023). This phase ranges from 2 to 5 years in women, while in men it takes approximately 1000 days, with the hair growing 0.3 mm per day (Venus et al., 2010).

Nevertheless, the eyebrow hair follicles, for example, have an anagen phase of 2–3 months, producing short hairs (Paus & Cotsarelis, 1999). In the catagen phase, the hair follicles degenerate during 2–3 weeks, and many cells undergo apoptosis (Kolarsick et al., 2011; Venus et al., 2010). Telogen is the resting phase and lasts for 3–4 months on the scalp, being characterized by hair loss (Venus et al., 2010). During this stage, the hair shaft matures into a club hair, which is eventually shed from the follicle with combing or washing (Paus & Cotsarelis, 1999).

Regarding hair fibers, this structure is composed by cuticle, cortex, and medulla (Koch et al., 2020). The cuticle is the outer covering of the hair, while the cortex reinforces imparts strength to the hair shaft and medulla is the central area of the hair (Lai-Cheong & McGrath, 2021). The melanosomes, cortical fusi, and cell membrane complex are presented in hair fibers (Koch et al., 2020), with melanosomes being responsible for the hair color (Kolarsick et al., 2011). The hair bulb presents melanocytes that synthesizes melanosomes and transfers them to the keratinocytes of the bulb matrix (Kolarsick et al., 2011). In the hair bulb occurs the matrix cell proliferation that produces the cortex, a specific intermediate hair composed by filaments and associated to proteins (Paus & Cotsarelis, 1999). The matrix cells are compressed and funneled to enter the rigid inner root sheath, determining the shape, size and curvature of the hair (Kolarsick et al., 2011; Paus & Cotsarelis, 1999). Fig. 10.2 summarizes the

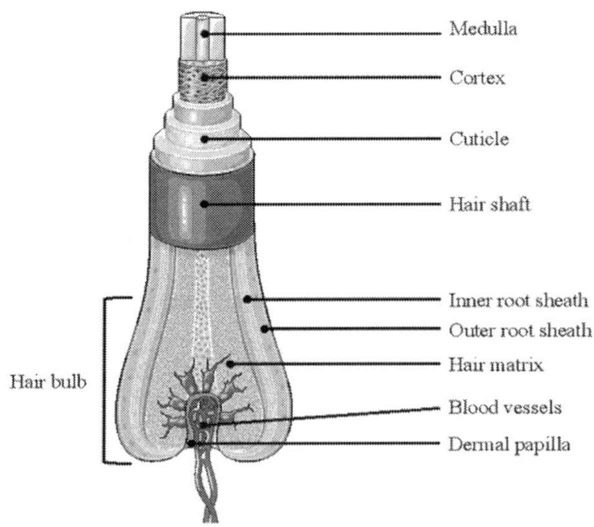

Figure 10.2 Hair structure and composition.

hair structure. However, some factors such as hormonal profile (estrogens, thyroid hormones, glucocorticoids, retinoids, prolactin) or pregnancy can influence the hair growth (Kolarsick et al., 2011). With age, the hair become gray since the melanin and hair follicles activity are reduced, resulting in hair thinning (McKnight et al., 2022).

Polyphenols with potential effects in skin health

Antioxidant and antiaging activities

The skin appearance is a primary feature used in social settings to roughly assess a person age and health. Even though the baseline rate of skin aging is genetically determined, a number of variables, such as mechanical, environmental, or socioeconomic factors, can affect the process (Roh et al., 2017). Still, excessive damage inflicted by UV irradiation is the most significant one (Vayalil et al., 2004). The excessive UV irradiation leads to the generation of free radicals, namely reactive oxygen and nitrogen species (ROS and RNS, respectively) (Korkina et al., 2008), and consequently, to a state of oxidative stress (Schlupp et al., 2019) that is the principal accelerator of the photoaging process (Vayalil et al., 2004). These species are also responsible for the promotion of other skin disorders, such as melasma, freckles, and senile lentigo (Roh et al., 2017), through the enhancement of tyrosinase activity, leading to abnormal skin melanin production and hyperpigmentation (Jesumani et al., 2020). Therefore the inhibition of the skin damage caused by oxidative stress, primarily fueled by reactive species, is one of the most appropriate protective effect of polyphenols in terms of antioxidant and antiaging activity (Jesumani et al., 2020; Roh et al., 2017).

The antioxidant activity of polyphenols are mostly due to their strong capacity for electron-donating, metal ions chelating, and free radicals scavenging abilities (Korkina et al., 2008). These skills are due to polyphenols distinctive physical-chemical features, particularly the presence of aromatic rings in the phenol groups (Jesumani et al., 2020). For example, a polyphenol-enriched extract from olive mill wastewater (OMW) was investigated regarding its antioxidative properties on an immortal keratinocyte cell line (HaCaT cells), achieving a strong reduction of the ROS generated, even more effective than ascorbic acid, the positive control used (Schlupp et al., 2019). Another polyphenol-rich extract of the seaweed *Sargassum vachellianum* demonstrated a marked scavenging ability of hydrogen peroxide and hydroxyl radicals, with half-maximal inhibitory concentration (IC_{50}) values of 0.80 and 0.98 mg/mL, respectively (Jesumani et al., 2020).

The absorption of UV radiation is another capacity of polyphenols (Jesumani et al., 2020), which is due to the structures similarities between polyphenols and chemical UV filters (Carrara et al., 2021). For instance, caffeic acid exhibits a sun protection factor of roughly 7.2 and possesses an absorption maximum at 269 and 360 nm, which are close to the UVB and UVA range (Balupillai et al., 2015). Also, seaweed polyphenols presented capacity to absorb UV radiation at wavelengths of 270 nm (UVC), 312 nm (UVB), and 396 nm (UVA) (Jesumani et al., 2020). Galanakis et al. demonstrated that in the UVB and UVA regions, polyphenols extracted from OMW are even more effective UV filters than ascorbic acid and α-tocopherol, both used as positive controls (Galanakis et al., 2018). These reports are in accordance with other studies that highlighted the strong UV absorbance of natural polyphenols (Abu-Qatouseh et al., 2019; Carrara et al., 2021; Wei et al., 2009), justifying their incorporation in sunscreen products (Scalia et al., 2013). However, even though polyphenols can be applied topically and have extraordinary antioxidant effects, the oral delivery has several advantages, namely the easier intestinal absorption when compared to the cutaneous one, the protection against degradation after contact with atmospheric oxygen or other environmental oxidants, and the avoidance of the skin redox systems (Korkina et al., 2008).

Additionally, polyphenols are chemically similar to several biological entities available in organisms that play important roles toward biological molecules (Korkina et al., 2008). For instance, the antioxidant activity of polyphenols may come from its capacity to enhance the activity or expression of antioxidant enzymes (Korkina et al., 2008; Wei et al., 2009). A propolis extract, rich in hydroxycinnamic acids and derivatives, was capable of increase ($P < .05$) the heme oxygenase-1 (HMOX-1) expression (around 50%), a cytoprotective enzyme with a powerful antioxidant and antiinflammatory action (Conte et al., 2022). Additionally, in the presence of lipopolysaccharide (LPS), which induces an oxidative stress state, propolis extract promoted an even higher expression of HMOX-1 (around 150%). Despite this, polyphenols can inhibit the activity or expression of oxidant enzymes (Vayalil et al., 2004). Most of the abnormalities present in photoaged skin are caused by oxidative damage to proteins present in dermis and produced by fibroblasts (Vayalil et al., 2004), which play a crucial role in maintaining skin health (Miere et al., 2021). Green tea (*Camellia sinensis* L.) is consumed around the world due to its extraordinary panoply of biological effects, particularly the potent radical species scavenging capacity and the exceptionally strong antioxidant activity (Roh et al., 2017; Vayalil et al., 2004). Its polyphenols, such as epigallocatechin 3-gallate (EGCG) (Singh et al., 2011; Wei et al., 2009), have been associated with protective effects against chronic UV irradiation–induced protein oxidation in mouse skin tissue when orally administered (Vayalil et al., 2004). EGCG has demonstrated the ability to reduce the matrix metalloproteinases (MMPs) formation, regardless the UV radiation exposure. When skin is exposed to UV radiation produces high levels of MMP that contribute to the collagen breakdown in the dermal extracellular matrix and accelerate the process of photoaging (Roh et al., 2017; Vayalil et al., 2004). The application of EGCG can potentially mitigate the damaging effects of UV radiation on skin and prevent the collagen degradation. Furthermore, the tyrosinase inhibition, which prevents hyperpigmentation, is another way that polyphenols exert their antioxidant benefits (Roh et al., 2017). Wei et al. evaluated the tyrosinase inhibition power of green tea polyphenols and reported inhibition rates from 57.14% to 75.37% in concentrations from 1 to 100 mg/L (Wei et al., 2009).

Antiinflammatory activity

Persistent inflammation can lead to skin negative effects, such as increased oxidative stress, fibrosis, necrosis, and a variety of reactions, including eczema (Galanakis et al., 2018). Therefore antiinflammatory strategies are crucial and recently polyphenols have been highlighted for this effect, given their ability to inhibit the sustained inflammation driven by ROS and secondary cytokines, inhibiting their generation or release (Korkina et al., 2008). In addition to the other mechanisms of action, polyphenols can also exert antiinflammatory effects through antioxidant properties. Certain polyphenols possess free radical scavenging potential, antioxidant, and transition metal chelating activities that allow the directly reduction of the pro-inflammatory mediators' levels, including free radicals, and inorganic and organic peroxides. By decreasing the oxidative stress and neutralizing the harmful molecules, polyphenols contribute to the alleviation of inflammation and the associated detrimental effects, for example, promoted by the chronic exposure to UV radiation that may conduct to immune suppression.

However, polyphenols predominantly exert their antiinflammatory effects by modulating the essential cellular signaling processes (Korkina et al., 2008). Keratinocytes, which act as sentinels in skin, play a key role in coordinating immune responses and produce various signaling mediators, including interleukin-8 (IL-8), interleukin-1α (IL-1α), interleukin-6 (IL-6), and tumor necrosis factor alpha (TNF-α) (Zhang et al., 2014). The mediators release is normally internally regulated, despite other factors may interfere in the process, namely the UV radiation (Nichols & Katiyar, 2010). If mediators are excessively released, an uncontrolled and intense inflammation occur, contributing to a wide range of skin disorders (Galanakis et al., 2018). Therefore polyphenols capacity to inhibit the release of cellular signaling mediators is crucial (Conte et al., 2022). For example, HaCaT cells released IL-8 in lower levels after being exposed to an OMW extract containing hydroxytyrosol, tyrosol, and verbascoside, possibly through inhibition of the nuclear factor-κB signaling pathway (Schlupp et al., 2019). In another study (Abu-Qatouseh et al., 2019) several polyphenol-rich plant leaves extracts, namely *Eucalyptus globulus* and *Arum palaestinum*, demonstrated a strong inhibition of IL-1α release by more than 60% and 75%, respectively, for concentrations of 0.5 mg/mL. Regarding TNF-α, polyphenols may also interfere in their release, as proven by Bitler et al. According to the authors, the TNF-α levels were reduced by more than 90% in mice stimulated by LPS and orally fed with 500 mg/kg extract of OMW (containing hydroxytyrosol, tyrosol, oleuropein, verbascoside, and caffeic acid) (Bitler et al., 2005). In addition, Zhang et al. investigated the effect of caffeic acid in the production of TNF-α by human keratinocytes stimulated with 12-O-tetradecanoyl-phorbol-13-acetate (TPA, 100 ng/mL) and calcium ionophore A23187 (1 μg/mL). The authors reported a marked inhibition of TPA/A23187-induced TNF-α production at 1 μM and higher concentrations (Zhang et al., 2014). Also, since TNF-α stimulates the production of other mediator, IL-6, the same authors evaluated the IL-6 production by keratinocytes in vitro upon stimulation by 10 ng/mL of TNF-α and reported that caffeic acid was able to markedly attenuated the TNF-α induced IL-6 production in a dose-dependent manner (Zhang et al., 2014).

Antimicrobial activity

Skin is home to millions of bacteria, fungi, and viruses that compose the skin microbiota, ensuring its health (Jesumani et al., 2020; Schlupp et al., 2019). *Staphylococcus aureus*, *Escherichia coli*, and *Pseudomonas aeruginosa* are part of the normal skin flora (Xu et al., 2021). However, in severe skin disorders, due to the loss of the epidermal barrier integrity, microbiota may pass through and lead to exacerbated inflammatory reactions (Miere et al., 2021; Singh et al., 2011). Therefore, the polyphenols antimicrobial activity can be of extreme relevance in such cases, inhibiting the bacteria growth in different ways, such as by interacting with surface proteins to disrupt cell membranes, breaking down necessary metabolites, blocking enzyme, producing ROS stress, and altering the structure of the cell wall (Jesumani et al., 2020; Xu et al., 2021). For example, polyphenols, such as EGCG extracted from green tea, have a great potential to inhibit the growth of *S. aureus*, *P. aeruginosa* and *E. coli* bacteria, displaying minimum inhibitory concentration (MIC) values between 15 and 120 μg/mL (Avila et al., 2021). According to the authors, the inhibitory effect is due to morphological deformations in the bacteria cell wall caused by surface interaction with EGCG, which promotes lysis and cell death. Furthermore, a OMW extract rich in rutin, hydroxytyrosol, chlorogenic and caffeic acid, have also proven to be capable of inhibiting the growth of *S. aureus*, *E. coli*, *P. aeruginosa* bacteria, achieving inhibition zones of 3.0, 4.0 and 3.5, respectively (Schlupp et al., 2019). Worth noting, the extent to which a polyphenol is capable of inhibiting bacteria growth is directly related to the chemical composition, as well as the composition of the bacterial cell wall (Jesumani et al., 2020). For instance, seaweed polyphenols are more efficient against Gram-positive *S. aureus* bacteria (MIC = 0.4 mg/mL; inhibition zone of 12.3 mm) than Gram-negative *E. coli* bacteria (MIC = 0.8 mg/mL; inhibition zone of 7.2 mm), probably due to the special cell wall composition (Jesumani et al., 2020). This information is in line with another study (Miere et al., 2021) that reported a greater antimicrobial activity of *Stellaria media* extract polyphenols, namely chlorogenic, ferulic, and caffeic acid, toward *S. aureus* (inhibition diameter of 15.33 mm) than *E. coli* (inhibition diameter of 9.66 mm).

The Gram-positive bacteria *Propionibacterium acnes* is the basis of inflammatory acne, also known as acne vulgaris, which is considered the most frequent skin disease of sebaceous glands (Abu-Qatouseh et al., 2019). According to Schlupp et al., the OMW extract was able to inhibit the growth of *P. acnes* bacteria, achieving inhibition zones of 9.0 mm (Schlupp et al., 2019). Also, the antimicrobial activity of several polyphenol-rich plant leaves extracts, including *E. globulus*, *A. palaestinum*, *Urtica dioica*, and *Inula viscose*, was evaluated against *P. acne*, and noteworthy growth inhibitory effects (MIC of 0.125, 0.125, 0.25, and 0.5 mg/mL, respectively) and great inhibition zones (28, 26, 22, and 29 mm, respectively) were reported (Abu-Qatouseh et al., 2019). The authors attribute the antimicrobial activity observed to the presence of phenolic acids and flavonoids, namely ascorbic, ferulic, and rosmarinic acid, ferruginol and kaempferol, highlighting the crucial antimicrobial role of such compounds against this bacterium. Additionally, even if polyphenols do not inhibit the bacteria growth, they may inactive the pathway by which they influence an organism. As example, *S. aureus* produce staphylococcus enterotoxins that are responsible for the adverse effects felt in an organism, particularly on skin with inflammation onset (Rasooly et al., 2010). Rasooly et al. evaluated the capacity of polyphenols from commercial apple juices and apple extracts to inhibit the biological activity of staphylococcus enterotoxins (Rasooly et al., 2010). The authors concluded that procyanidins, catechins, epicatechins, and chlorogenic acid, the main polyphenols present in the samples, were able of inhibiting the toxins. These findings offer encouraging evidence for the use of polyphenols as a management strategy for skin inflammatory diseases, even though this study was conducted in splenocyte cells.

Nevertheless, polyphenols may also exert the antimicrobial activity indirectly. For example, polyphenols can promote the fibroblasts viability by maintaining the skin immunity and creating an unfavorable environment for microorganisms by producing acidic sweat (Miere et al., 2021). Wei et al. (2009) also demonstrated the polyphenols' dose-dependent manner effect in fibroblast viability, reaching its maximum (259.9%) at a concentration of 500 mg/L. Moreover, Miere et al. reported that a *S. media* extract can significantly increase ($P < .001$) the viability of human dermal fibroblasts in a concentration of 100 μg/mL (Miere et al., 2021).

Antiallergic activity

Skin allergic disorders result from the sensitization to a specific allergenic that is normally harmless. The re-exposure to the same allergenic triggers an abnormal immune system reaction (Mwakalukwa et al., 2019), which manifests as dryness, itching, inflammation, and erythema (Shen et al.,

2023). Normally, skin allergies manifest the first symptoms at an early age (3–4 months) (Singh et al., 2011), being characterized by periods of relapse and remission during life (Sun et al., 2022). The most common skin allergic disorders are eczema and atopic dermatitis (Korkina et al., 2008; Mwakalukwa et al., 2019), which are also inflammatory events (Shen et al., 2023).

Several drugs, including glucocorticoids and antihistaminics, with a primary focus on symptom reduction, can be used to treat skin allergic diseases (Singh et al., 2011). However, like with any pharmacological therapy, there are several unavoidable adverse effects, such as chronic use and dependence (Sun et al., 2022). As a result, biological therapeutics, particularly polyphenolic compounds, have been highlighted due to their biological activities (with particular focus on antioxidant and antiinflammatory activities) (Singh et al., 2011; Sun et al., 2022). As example, *Prinsepia utilis* leaves, extremely rich in *p*-coumaric acid glucoside, vicenin-2, isoschaftoside, and kaempferol-3-rutinoside, have been pointed for the treatment of allergic contact dermatitis symptoms (Shen et al., 2023). The extract exhibited a strong antihyaluronidase effect ($IC_{50} = 231.93$ μg/mL), also attenuating the allergic contact dermatitis-like symptoms induced by fluorescein isothiocyanate, in a dose-dependent manner, after oral administration.

Polyphenols may intervene at different stages of an allergic reaction, including during sensitization and re-exposure. This ability, along with the other biological properties already reported, contribute to the well-established antiallergic effect (Singh et al., 2011; Sun et al., 2022). In this sense, by suppressing the expression of major histocompatibility complex II or costimulatory molecules (CD80, CD86) on dendritic cell surfaces, polyphenols lead to the ineffective antigen presentation and, therefore, interfere in the sensibilization phase of an allergenic response (Conte et al., 2022; Singh et al., 2011). Conte et al. demonstrated the antiallergic action of polyphenols from propolis, a bee product extremely rich in hydroxycinnamic acids and derivatives, by a significant reduction ($P < .05$) of the CD86 expression in human monocytic cells (THP-1) triggered by the strong skin allergen 1-fluoro-2,4-dinitrobenzene (DNFB) (Conte et al., 2022). Additionally, caffeic, ferulic, and chlorogenic acids were also able to interfere in the sensibilization stage of an allergic response by forming insoluble phenolic–allergen complexes with peanut allergens (Chung & Champagne, 2009), thereby efficiently preventing the antigen presentation. This antiallergic capacity was achieved through the antioxidant capacity (chelating ability) of polyphenolic compounds (Korkina et al., 2008).

During the re-exposure phase of an allergic response, polyphenols may inhibit basophils or mast cells degranulation and, consequently, the release of histamines and other cytokines (Mwakalukwa et al., 2019). For instance, when allergen specific IgE bonds to the FcεRI receptor on the cell surface of mast cells and basophils, the cell degranulation happens. Yano et al. showed that the flavonoids chrysin and apigenin have a strong ability to suppress FcεRI receptor expression and, therefore, prevent the basophil degranulation (Yano et al., 2005). This is also supported by other study that evidenced the capacity of green tea catechin, EGCG, to achieve the same result (Fujimura et al., 2002). Nevertheless, there are other ways to prevent degranulation, for example by preventing the extracellular Ca^{2+} influx, which is a crucial step in the degranulation process (Mwakalukwa et al., 2019). This was demonstrated by Mwakalukwa et al. that assessed the effect of OMW polyphenols on the cells' degranulation (release of β-hexosaminidase), carried out on rat basophil leukemia cells (RBL-2H3) (Mwakalukwa et al., 2019). After ensuring that polyphenols had no impact on the enzymatic activity of β-hexosaminidase, the researchers showed the extract capacity to inhibit the degranulation of RBL-2H3 cells. According to the authors, luteolin reduced degranulation by 68.1%, hydroxytyrosol acetate by 41.4%, and 1-acetoxypinoresinol by 44.6%, which was attributed to the blockage of extracellular Ca^{2+} influx.

Polyphenols with potential effects in hair

Hair growth activity

The market has a wide amplitude of hair care cosmetics, but the major segment of these products focuses on hair growth (Lourith & Kanlayavattanakul, 2021). Hair loss is a common complaint and has multiple origins, such as genetic diseases, stress, local infections, autoimmune, and/or endocrine disorders (Sun et al., 2022). Most of these causes are characterized by oxidative stress and the inflammatory events that contribute to the aging of hair follicles, subsequently resulting in hair loss (Choi et al., 2023). For many years, traditional drugs and synthetic agents such as minoxidil, finasteride, or dutasteride, have been used to treat several diseases related to hair loss (Sattur & Sattur, 2021). Polyphenols have been used for their health-promoting effects within the hair-growth sector since they act as promising antioxidants and antiinflammatory agents (Daniels et al., 2019).

EGCG, a polyphenolic catechin present in green tea (*C. sinensis* L.), has been widely reported for its potential on preventing or treating autoimmune alopecia by inhibiting the activity of 5α-reductase, a NADPH-dependent enzyme responsible for the reduction of testosterone to 5α-dihydrotestosterone, leading to hair follicle elongation (Hamed et al., 2018; Shin, Kim, et al., 2016). The effect of EGCG on inhibiting gamma interferon pathways has also been reported, which is known to reduce the response of human follicles (Chung et al., 2003; Hamed et al., 2018). The potential effect of tea polyphenolic compounds on hair loss has been already demonstrated in a previous in vivo study using mice with spontaneous hair loss, where 33% of the animals receiving polyphenol extract from dehydrated green tea in their drinking water had a significant hair regrowth during six months of treatment (Esfandiari & Kelly, 2005).

Grapes (*Vitis vinifera*) are another great source of polyphenolic compounds, for example in skins (e.g., resveratrol) or seeds (e.g., proanthocyanidins that convert into procyanidins), which possess great antioxidative effects (Lourith & Kanlayavattanakul, 2013). Kubo et al. evidenced that resveratrol and fisetin regulated the genetic expression of several cytokines responsible for maintaining the niche of hair follicle stem cells, as well as induced a shift from telogen to anagen in the hair follicle, which is strongly related to the promotion of hair growth (Kubo et al., 2020). Procyanidins, also present in apple juice (*Malus pumila*), have been found, either in vitro or in vivo studies, to accelerate the hair growth by stimulating the anagen induction of the hair cycle, reducing cell death via 5-α-reductase inhibition and upregulating the antiapoptotic factors (e.g., Bcl-xL) (Kamimura & Takahashi, 2002; Kamimura et al., 2006; Takahashi et al., 1998, 2005). For example, clinical trials using procyanidin B-2 have already shown improvement in hair loss in Japanese men (Takahashi et al., 2001).

Eckol and dieckol are major phlorotannins isolated from *Ecklonia cava*, an edible marine brown alga, that have been demonstrated to increase fibroblast survival by reducing ROS (Thomas & Kim, 2013). Previous in vitro studies with eckol showed that these polyphenols promote radical scavenging, antiplasmin inhibition, antimutagenic, bactericidal, and tyrosinase inhibiting activities in fibroblast cells (Kang et al., 2005). The hair cycle is influenced by dermal papilla cells, which have a critical role in regulating the hair follicle development and regeneration. So, in a pathological state, these cells potentially promote various hair loss disorders (Morgan, 2014). As *E. cava* polyphenols demonstrated to increase fibroblast survival, Shin, Cho, et al. (2016) and Shin, Kim, et al. (2016) attempted to evaluate if polyphenols promote the hair growth by assessing their effects in vitro and ex vivo. The authors observed that purified polyphenols extracted from *E. cava* increased the expression of vascular endothelial cell growth factor (VEGF) and insulin-like growth factor-1 (IGF-1), enhanced proliferation, and reduced the oxidative stress of dermal papilla cells, suggesting that polyphenols could enhance the human hair growth, being a promising therapeutic candidate for hair loss (Shin, Cho, et al., 2016).

OMW extract also demonstrated to be another promising ingredient for hair care to improve hair growth. Sittek et al. (2021) assessed the impact of these polyphenolic compounds in vitro on human follicle dermal papilla cells and observed a positive influence on their proliferation, an increased secretion of IGF-1, and a considerable reduction of the ROS formation. The results strongly suggested that polyphenols from OMW help to maintain a healthy scalp and contribute to the prevention of hair loss (Sittek et al., 2021). Other study demonstrated that the ethanolic extract of *Centella asiatica*, containing polyphenols and flavonoids, is also a potential candidate for hair growth−promoting ingredients since induced the expression of VEGF in human dermal papilla cells, which is responsible for the hair growth (Saansoomchai et al., 2018).

Hair coloring activity

Hair color is strongly affected by aging and exposure to external chemical and physical agents, such as bleaching, coloring, or prolonged exposition to solar radiation (Jeon et al., 2008). Particularly, the solar radiation promotes fading of natural and dyed hair color by the free radicals generation, oxidizing the hair pigments (Dario et al., 2013). Moreover, hair aging is also related to the decrease of melanin, leading to a change of color to gray or even white (Tobin, 2009). When these changes occur, the color can be restored using hair coloring products. However, most of the traditional commercial hair dyeing products commonly contain bleaching agents (e.g., ammonia and hydrogen peroxide) and/or phenylenediamine (PPD) or PPD-derived compounds as key ingredients, which are often associated with hair damage during the dyeing process, being also toxic, allergenic, mutagenic, and potentially carcinogenic (Sun et al., 2022). With the recent advancement in cosmetics, a greater interest has been raised in using natural products for hair dyeing, mostly associated with minimal side effects when compared to conventional chemical products (He et al., 2022). Polyphenols extracted from different plants can either be used for hair coloring, reducing the fading of the natural hair color, or even be incorporated in hair products as posttreatment after hair coloring, to protect and repair damages (Sun et al., 2022). For example, the use of pyrogallol and gallic acid polyphenols has revolutionized the coating methods with safer chemical structure mechanisms than the traditional dying agents (Jin et al., 2023; Nishida et al., 2004). Since polyphenols are powerful antioxidants, it has been proposed the incorporation of these natural biocompounds in hair products to reduce the alterations of hair color. A previous study by Dario et al. (2013) showed reduced photodamage of red dyed hair exposed to UVA radiation after being treated with *Punica granatum* L. hydroalcoholic extracts, which are rich with polyphenols and tannins (Dario et al., 2013). Tannins, a group of polyphenols with high antioxidant activity, can also be applied to reduce fading of natural hair color (Sun et al., 2022).

Recently, polyphenol coatings have been successfully applied for the functional modification of material surfaces (Xu et al., 2018), offering a promising strategy for hair dyeing. For instance, it was reported that the in situ incubation of gray hair with products of laccase (enzyme)-catalyzed oxidation of catechin and catechol (poly)phenolic compounds is effective in hair coloration (Im & Jeon, 2016). Moreover, Jin et al. (2023) produced a biosafe hair dye constituted by oligomers of pyrogallol and gallic acid polyphenols prepared with $CuCl_2$ in an aqueous NaCl solution. This established an ion-channel coating built by Cu (II) in a NaCl aqueous solution with hydroxyl groups on the hair surface, which further protected the hair from damage during the dyeing process, and the hair thickness enhanced after dyeing it up to 10 times (Jin et al., 2023). Notwithstanding, a more simple, fast, and material-independent surface functionalization method using metal-phenolic networks (MPN) has been applied in tannic acid−based MPN and gallic acid−composed MPN, which have reportedly colored natural gray hair without the presence of potentially toxic chemicals, and also protected the dyed hair against repeated shampoo washing (Geng et al., 2020). Jia et al. (2022) also developed MPN with gallic acid and tannic acid polyphenols, where the mutual synergistic effect of both polyphenols chelates with metal ions producing more complex MPN structures and, therefore, more colorful MPN pigments that deposit on the hair surface through covalent and noncovalent interactions. This resulted in a series of hair dyeing effects with high color saturation and excellent stability (Jia et al., 2022).

Antimicrobial activity

The balance between the bacterial population present in the scalp is essential for scalp health, helping to prevent the fungal growth or the appearance of scalp diseases (Clavaud et al., 2013). However, the interaction of these bacterial communities with the hair shaft is not fully elucidated yet (Kerk et al., 2018). There are several microorganisms that play a key role in the pathogenesis and/or predisposition to several scalp disorders, including folliculitis, dandruff, and other fungal diseases (Grimalt, 2007). Fungal infections in hair shaft originated by *Malassezia* species are often found in humans, including trichomycoses (Sentamilselvi et al., 2009) and dandruff conditions (Clavaud et al., 2013).

Natural compounds are becoming increasingly common in dermatology, not only due to the augmented resistance of bacteria to synthetic antibiotics, but also because of their antiseptic and antimicrobial activities (Bittner Fialová et al., 2021). Polyphenols, for instance, have demonstrated to play dynamic antimicrobial (Manso et al., 2021) and antifungal activities (Kaur et al., 2021). In fact, polyphenols have shown to be effective against a variety of bacteria and fungi that can contribute to scalp and hair problems, such as dandruff, scalp irritation, and hair loss, being pointed out as potential alternatives for the treatment of scalp disorders. However, the underlying antimicrobial molecular mechanisms of polyphenols remain poorly understood (Li et al., 2022; Sun et al., 2022). A recent study from Ambarwati et al. (2021) demonstrated the efficacy of a combination of noni fruit (*Morinda citrifolia*) and pandan leaves (*Pandanus amaryllifolius* Roxb.) extracts, rich in tannins and flavonoids polyphenolic compounds, against bacteria isolated from dandruff, highlighting their potentialities as ingredients for an antidandruff shampoo (Ambarwati et al., 2021). Polyphenols from pomegranate (*P. granatum*) extracts have also shown antidandruff effects, simultaneously providing nutrition to the hair (Bhinge et al., 2021). Moreover, tannin, ellagic acid, and phenolic acid extracted from plants possess antidandruff, antifungal, and antimicrobial properties (Vasconcelos et al., 2003; Voravuthikunchai et al., 2005).

A previous study demonstrated the antifungal activity of phlorotannins against *Trichophyton rubrum* fungus (Lopes et al., 2012), which can penetrate the hair follicle and invade the hair shaft causing ring-shaped itchy rashes on the scalp (Kaur et al., 2021).

Safety assessments of polyphenols as skin care and hair care ingredients

The safety assessment of new skin and hair care ingredients is an important aspect to ensure the absence of adverse events when used for consumers in a daily routine (Silva et al., 2023). According to the Regulation (EC) No. 1223/2009, a cosmetic product should be "safe for human health when used under normal or reasonably foreseeable conditions of use" (European Commission, 2009). The safety evaluation should be based on the assessment of each individual ingredient (Nunes et al., 2017). The main guidelines were provided by the Scientific Committee on Consumer Safety (SCCS) of the European Union (EU) that reinforced the requirements for cosmetic ingredients, taking into account adequate studies for the safety evaluation (European Commission, 2016). Some parameters such as (1) acute toxicity, (2) corrosivity and irritation, (3) skin sensitization, (4) dermal/percutaneous absorption, (5) repeated dose toxicity, (6) reproductive toxicity, (7) mutagenicity/genotoxicity, (8) carcinogenicity, (9) toxicokinetic studies, (10) photo-induced toxicity, and (11) human data are required to test and prove the safety of new cosmetic ingredients (European Commission, 2016). According to the Regulation (EC) No. 1223/2009, the animal testing for cosmetic products and cosmetic ingredients is forbidden in the EU since 2009 (European Commission, 2009). In this sense, new methodologies to assess the intrinsic toxicological properties of ingredients are used, namely in vitro, ex vivo, and in silico assays, as well as clinical studies, case reports, epidemiological studies, or a combination of them (European Commission, 2016).

Polyphenols from natural matrixes have gained notoriety by the skin effects, safety, and cost effectiveness (Marto et al., 2016; Sharma et al., 2022; Silva, Costa, et al., 2021). However, skin care products from natural matrixes are absorbed by skin superficial layers and may cause allergic reactions (Mukherjee et al., 2011). Despite the use of herbal products for the skin care purposes is old, it is still imperative to evaluate the safety and effectiveness of these ingredients through validated in vitro studies (Sharma et al., 2022). This is the case of *Actinidia arguta* leaves, a source of polyphenolic compounds (caffeoylquinic acids, proanthocyanidins, quercetin derivatives, and kaempferol) with strong antioxidant activity and great skin health potential (Silva et al., 2022). Silva et al. demonstrated that the extract is nonirritant through in vitro validated skin and ocular irritation models, respectively, the reconstructed human epidermis (Episkin) and the human corneal epithelial models (SkinEthic HCE) (Silva et al., 2022; Silva, Pinto, et al., 2021). Additionally, the authors guaranteed by an in vivo patch test performed in 10 human health volunteers the absence of allergic or irritative reactions provoked by the extract (Silva et al., 2022). Afterward, the extract was incorporated in a topical formulation that demonstrated a good skin compatibility by a patch test performed in 13 healthy human volunteers that confirmed the absence of allergy, irritations, or allergic reactions (edema and erythema) (Silva et al., 2023).

Chestnut (*Castanea sativa*) shells extracts have also been deeply explored and validated according to the European guidelines to be used as cosmetic ingredient. These extracts are rich in polyphenolic compounds, such as pyrogallol, protocatechuic acid, and ellagic acid, presenting high scavenging activity against ROS and RNS (Pinto, Vieira, et al., 2021). Pinto et al. tested the *C. sativa* shells extract obtained by subcritical water extraction in in vitro 3D skin and ocular models (Episkin and SkinEthic HCE, respectively) and reported the absence of irritation effects, classifying the extract as safe (Pinto, Cádiz-Gurrea, et al., 2021). Considering the Regulation (EC) No. 1223/2009, the results were verified by an in vivo patch test performed in human volunteers that attested the absence of irritation (Pinto, Cádiz-Gurrea, et al., 2021). Moreover, the leaves of *C. sativa* also present an interesting polyphenol composition (chlorogenic acid, ellagic acid, rutin, isoquercitrin, and hyperoside) and an in vivo patch test carried out in 20 human volunteers revealed that the extract is safe for topical application (Almeida et al., 2008). In another study, Rodrigues, Pereira, et al. (2015) studied through in vitro EpiSkin and SkinEthics HCE models as well as an in vivo patch test the irritation potential of coffee silverskin extracts. The results demonstrated that the extract is a source of natural antioxidants that does not lead to skin irritation, being classified as safe for topical application (Rodrigues, Palmeira-de-Oliveira, et al., 2015; Rodrigues, Pereira, et al., 2015). Zlabiene et al. (2021) also studied the safety of multiple water-oil-water emulsion containing *Rosmarinus officinalis* leaves extracts, *Avena sativa* and *Linum usitatissimum* extracts by the in vitro reconstituted human skin model EpiDerm and through an in vivo human skin patch test. The results showed the absence of irritation, skin inflammation, or sensitization after the emulsion application (Zlabiene et al., 2021).

Natural extracts obtained from underutilized and residual vegetal and macroalgal biomass processed with food-grade green solvents were also studied for cosmetic application (Balboa et al., 2014). Once again, the authors employed an EpiSkin model to evaluate the potential skin irritation after exposure to the extract proving the absence of irritation (Balboa et al., 2014).

Jasmine rice panicle extract have also been characterized by high levels of phenolic compounds with cutaneous benefits and activity toward skin disorders (Kanlayavattanakul et al., 2016). Based on that, Kanlayavattanakul et al. (2016) developed an oil-in-water formulation and incorporated the rice extract, performing a patch test in 20 volunteers that proved the formulation safety (Kanlayavattanakul et al., 2016). *Cynara scolymus* is another plant with excellent antioxidant properties due to the polyphenols composition, particularly cynarin, cymaroside, chlorogenic acid, caffeoylquinic and dicaffeoylquinic acids (Marques et al., 2017). Marques et al. incorporated the extract in topical formulations (an oil-in-water emulsion and an hydrogel) and evaluated them through human repeat insult patch test in 50 healthy volunteers, reporting very good skin compatibilities without allergic reactions (Marques et al., 2017). Nevertheless, despite being an increasing trend the incorporation of natural extracts rich in polyphenols in cosmetic formulations, few studies evaluated their toxicity and safety as new potential cosmetic ingredients, being a challenge for industry.

Commercial products and patents

As previously mentioned, the use of polyphenols in skin and hair care products is a growing area of interest due to their antioxidant activity, antiinflammatory effects, antimicrobial activity, photoprotection, collagen synthesis, skin brightening, wound healing, and moisturizing properties, among others (Sun et al., 2022). Lately, there are an increasing number of commercial products that incorporate these natural compounds as active ingredients. Table 10.1 depicts some examples of commercial products available in the market containing polyphenols originated from different plant sources. This panoply of products includes serums, creams, moisturizers, toners, masks, and cleansers for skin care cosmetics, whilst shampoos, conditioners, hair serums, and styling products for hair care products.

Table 10.1 Examples of skin and hair care cosmetic products containing polyphenols available on the market.

Type of polyphenol	Plant source (examples)	Skin care product	Hair care product
Flavonoids (e.g., catechins)	Green tea, cocoa, citric fruits, berries	L'Occitane En Provence, Garnier (SkinActive), NoveExpert Paris, Caudalie (Vinopure), Sesderma (C-Vit)	The Body Shop (Fuji Green Tea refreshingly purifying shampoo), OGX (Active Beauty green tea fitness shampoo), iKoo infusions (shampoo for chemically damaged hair), Natural Essence (Hair Shampoo Bar), Pharmaceris (H-Stimuclaris), Garnier Fructis (Cacau Hair food)
Resveratrol	Grapes	Caudalie (Resveratrol Lift), Sesderma (Resveraderm)	L'Oreal (Serie Expert Vitamino Resveratrol, Serioxyl Denser Hair)
Tannins	Pomegranate	Estée Lauder (Nutritious Super-Pomegranate Radiant Energy), Weleda (Pomegranate)	Klorane (pomegranate—color-treated hair), Dr. Organic (pomegranate shampoo)
Caffeic acid	Coffee, rosemary	Cantabria Labs (Heliocare 360° sunscreens), Dr. Alora Organiscs (Coffee Scrub)	Aveda (Rosemary mint shampoo), Mielle (Rosemary mint strengthening hair mask)
Curcumin	Turmeric	Kora Organics (Turmeric 2-in-1 Brightening & Exfoliating Mask), CumarGOLD Gel (Nano Curcumin Skincare)	Aceda (Invati Advanced), OGX (Strength & Length + Golden Turmeric), ALLPA Botanicals (turmeric hair oil)

The scientific information regarding the potential polyphenols effects on skin and hair health cannot only be found in research papers, being also available in patents (Rodrigues et al., 2018). Herein some examples of patents related to polyphenols formulations that can be applied to skin and hair care products can be provided. In fact, many skin and hair care formulations containing polyphenols available on the market have been previously patented. One example is the case of Caudalie that has patented "Novel cosmetic applications of polyphenols and derivatives thereof" (Patent N° EP1519709B1-2011), an invention focused on the cosmetic compositions with skin-lightening and/or spot-removing effects. These applications contain polyphenols selected from the group consisting of ε-viniferin, mixtures of resveratrol and ε-viniferin, and mixtures of polyphenolic, in particular stilbene and/or falvonol, oligomers and/or polymers (Caudalie Paris, 2023). Formulations comprising mangiferin, a bioactive polyphenol found in mango leaves extracts, or its derivates, have also been applied in personal care compositions, due to its moisturizing, antiaging, photoprotective, and antioxidant properties. A patent disclosed a cosmetic formula containing mangiferin or one of its isomers or derivatives to protect the skin, lips, or hair against UV rays, and incorporated in skin care formulations for improving the structural quality of skin or combating skin ageing by claiming to have photoprotective, anticollagenase, antielastase, antiradical, and antityrosinase activity (e.g., Patents N° WO9616632A1-1996, US2006088560A1-2006, WO2009038875A1-2009, KR2011132209-2011) (Telang et al., 2013). Caffeic acid and its derivates can be applied in lotions, sunscreens, and other dermatological formulations due to its skin-whitening and antiaging effects, such as claimed by a patent formulation containing caffeic acid derivate 27 and vitamin C (Patent N° EP0339671B1-2010) (Silva et al., 2014). Several patents of quercetin and its derivates formulations have been reported to be promising agents to be incorporated in cosmeceuticals, with interesting antioxidative/antiaging (Patent N° US20110034427A1-2011) and skin protective properties (JP2016060718 A-2016) (Sharma et al., 2018).

do Carmo et al. (2018) made a patent overviewing the polyphenol products for several purposes and gave examples for skin product applications. One of them was a topical application of nucleic acid inhibitor polyphenols, inducible nitric oxide synthase, and nuclear transcription factor alpha, for the treatment of inflammatory skin diseases (Patent N° PI0401980-6 A2-2005), while the other focused on proanthocyanidin monomers and oligomers, hydroxystilbenes, flavonoid monomers and oligomers, used for combating signs of skin, hair, and nails ageing (Patent N° EP3278798 A1-2018) (do Carmo et al., 2018). The application of *Lilium* plants, a powerful source of polyphenols, in cosmeceuticals products has become quite popular, particularly in skin and hair cosmetics. Therefore Tang et al. (2022) overviewed different patents related to *Lilium* plants in cosmetics and found a broad number of applications in skin and hair care, including masks, emulsions, creams, facial cleansers, makeup-related products, hair restorers, and shampoos. The functions of these *Lilium*-based patent cosmetics were quite diverse, ranging from beautification, slimming, hair-growth promotion, and hair darkening to the treatment of skin diseases (Tang et al., 2022). On the other hand, Rambaran (2022) explored the application of nanotechnology to prepare state-of-the-art polyphenol nano-formulations and their available commercialized forms for the pharmaceutical, nutraceutical, food, and cosmetic industries. The authors reported that nano-curcumin was the most patented and popular commercialized nano-polyphenol, while other commercialized nano-polyphenols were derived from berberine, catechin/EGCG/green tea, quercetin, resveratrol, rutin, silymarin, or crude polyphenol-rich extracts. Herein the authors gave examples of commercial nano-polyphenolic formulations available for skin care products, including CumarGOLD Gel—Nano Curcumin Skincare, Quin Nano Curcumin Clinic Cream, Decurma Gel Anti-Acne Nano Curcumin, Nano Berberine, Elixir Nano Resveratrol, Nano Resveratrol Facial Serum, Nano Resveratrol Serum Lift, Nano Resveratrol Day Skin Antiaging, Nano Resveratrol Night Skin Antiaging, Nano Resveratrol Mask, and Nano Up Factor (Nano Resveratrol) (Rambaran, 2022).

There are also some hair coloring technologies involving polyphenols that have been patented. For instance, Aveda Corp. patented a hair coloring formulation and a method for the oxidative coloring of hair using green tea polyphenols, including epigallocatechin gallate, epigallocatechin, epicatechin gallate, or epicatechin, as coupler-modifiers (Patent N° US6099591A-2000) (Matravers et al., 2000). Another example is a patent from the Natural Medicine Institute of Zhejiang Yangshentang in Hangzhou (Patent N° US9345654 B2-2016), which reports a natural dye product made of plant polyphenols, gardenia blue, cocoa pigment, purple sweet potato pigment, purple cabbage pigment, or algae blue pigment (Lan & Hu, 2016). A circulating hair dyeing method based on a MPN structure is still pendent to be patented by a Chinese group, where it claims to develop a stronger adhesive color-developing coating on the hair surface, resistant after washing, and allowing to prevent the hair from being damaged after dyeing (Patent N° CN110840763A) (Geng et al., 2020). Rose et al. (2018) also reported a technology based on various patents (Patents N° WO2010131049 A2—pendent to be granted, US8361167 B2-2013, and AU2010247136 B2-2016) using anthocyanin colorants extracted from renewable waste blackcurrant (*Ribes nigrum* L.) fruit skins from the fruit pressing industry, to be applied as blue-colored hair dyeing, which is reported to be stable after multiple hair washes (Rose et al., 2018). The previously mentioned study from Sittek et al. (2021) also used a patented formula of polyphenol-rich OMW extracts (Patent N° 8815815B2-2014), and demonstrated its positive effects on the proliferation of the human follicle dermal papilla cells, supporting that this patented formula is a promising ingredient for hair care to improve hair growth and maintain a healthy scalp (Sittek et al., 2021).

Challenges and conclusions

The vast number of studies analyzed through this chapter support the growing knowledge that polyphenols can be used as active ingredients in skin and hair products, not only in healthy stage, but also against disorders. Despite the number of studies supporting these conclusions, a deeply safety assessment should be performed before the incorporation of these ingredients in new formulations. For that purpose, regulation forbidden the animal assays, being already validated in vitro models that allow to conclude about this issue. More importantly, few is known regarding these compounds' stability in final formulations since natural polyphenols are sensitive to environmental factors, such as the presence of light or oxygen, that contribute to their degradation. Therefore novel delivery systems, as encapsulation, should be exploited to improve compounds stability, control their release, and assure active functionality.

Acknowledgments

This research was funded by project EXPL/BAA-GR/0663/2021-Kiwi4Health—Exploring the Eco-Innovative Re-Use of Kiwiberry, supported by national funds by FCT/MCTES and by the projects UIDB/50006/2020 and UIDP/50006/2020 through national funds. This work was also financed by national funds from FCT—Fundação para a Ciência e a Tecnologia, I.P., in the scope of the project UIDP/04378/2020 and UIDB/

04378/2020 of the Research Unit on Applied Molecular Biosciences—UCIBIO and the project LA/P/0140/2020 of the Associate Laboratory Institute for Health and Bioeconomy—i4HB.

Ana Margarida Silva (SFRH/BD/144994/2019) is thankful for the Ph.D. grant financed by POPH-QREN and subsidized by the European Science Foundation and Ministério da Ciência, Tecnologia e Ensino Superior. Filipa Teixeira is thankful for the scholarship from project EXPL/BAA-GR/0663/2021-Kiwi4Health—Exploring the Eco-Innovative Re-Use of Kiwiberry. Maria João Bessa thanks to the FCT Post-Doc grant under the framework of the 4FirHealth project. Francisca Rodrigues (2023.06819. CEECIND) is thankful for her contract financed by FCT/MCTES—CEEC Individual Program Contract.

References

Abdo, J. M., Sopko, N. A., & Milner, S. M. (2020). The applied anatomy of human skin: A model for regeneration. *Wound Medicine, 28*, 100179. Available from https://doi.org/10.1016/j.wndm.2020.100179.

Abu-Qatouseh, L., Mallah, E., & Mansoor, K. (2019). Evaluation of Anti-*Propionibacterium acnes* and anti-inflammatory effects of polyphenolic extracts of medicinal herbs in Jordan. *Biomedical and Pharmacology Journal, 11*, 211–217. Available from https://doi.org/10.13005/bpj/1629.

Almeida, I. F., Valentão, P., Andrade, P. B., Seabra, R. M., Pereira, T. M., Amaral, M. H., Costa, P. C., & Bahia, M. F. (2008). In vivo skin irritation potential of a *Castanea sativa* (chestnut) leaf extract, a putative natural antioxidant for topical application. *Basic and Clinical Pharmacology and Toxicology, 103*(5), 461–467. Available from https://doi.org/10.1111/j.1742-7843.2008.00301.x.

Ambarwati, A., Sujono, T. A., & Sintowati, R. (2021). The potential combination of noni (*Morinda citrifolia*) fruit extract and pandan (*Pandanus amaryllifolius* Roxb.) leaves extract as an anti-dandruff shampoo. *Journal of Pharmacy & Bioallied Sciences, 13*(Suppl 2), S1369.

Arda, O., Göksügür, N., & Tüzün, Y. (2014). Basic histological structure and functions of facial skin. *Clinics in Dermatology, 32*(1), 3–13. Available from https://doi.org/10.1016/j.clindermatol.2013.05.021.

Avila, S. R. R., Schuenck, G. P. D., Silva, L. P. Ce, Keijok, W. J., Xavier, L. M., Endringer, D. C., Oliveira, J. P., Schuenck, R. P., & Guimarães, M. C. C. (2021). High antibacterial in vitro performance of gold nanoparticles synthesized by epigallocatechin 3-gallate. *Journal of Materials Research, 36*(2), 518–532. Available from https://doi.org/10.1557/s43578-020-00012-5, https://www.springer.com/journal/43578.

Balboa, E. M., Soto, M. L., Nogueira, D. R., González-López, N., Conde, E., Moure, A., Vinardell, M. P., Mitjans, M., & Domínguez, H. (2014). Potential of antioxidant extracts produced by aqueous processing of renewable resources for the formulation of cosmetics. *Industrial Crops and Products, 58*, 104–110. Available from https://doi.org/10.1016/j.indcrop.2014.03.041. http://www.elsevier.com/inca/publications/store/5/2/2/8/2/5.

Balupillai, A., Prasad, R. N., Ramasamy, K., Muthusamy, G., Shanmugham, M., Govindasamy, K., & Gunaseelan, S. (2015). Caffeic acid inhibits UVB-induced inflammation and photocarcinogenesis through activation of peroxisome proliferator-activated receptor-γ in mouse skin. *Photochemistry and Photobiology, 91*(6), 1458–1468. Available from https://doi.org/10.1111/php.12522. http://www.blackwellpublishing.com.

Bhinge, S. D., Bhutkar, M. A., Randive, D. S., Wadkar, G. H., Todkar, S. S., Savali, A. S., & Chittapurkar, H. R. (2021). Screening of hair growth promoting activity of *Punica granatum* L. (pomegranate) leaves extracts and its potential to exhibit antidandruff and anti-lice effect. *Heliyon*, 7(4), e06903. Available from https://doi.org/10.1016/j.heliyon.2021.e06903.

Bitler, C. M., Viale, T. M., Damaj, B., & Crea, R. (2005). Hydrolyzed olive vegetation water in mice has anti-inflammatory activity. *Journal of Nutrition*, *135*(6), 1475−1479. Available from https://doi.org/10.1093/jn/135.6.1475. http://jn.nutrition.org.

Bittner Fialová, S., Rendeková, K., Mučaji, P., Nagy, M., & Slobodníková, L. (2021). Antibacterial activity of medicinal plants and their constituents in the context of skin and wound infections, considering European Legislation and Folk Medicine—A review. *International Journal of Molecular Sciences*, *22*(19), 10746. Available from https://doi.org/10.3390/ijms221910746.

Brown, T. M., & Krishnamurthy, K. (2021). *Histology, dermis, StatPearls* (2021). StatPearls Publishing.

Carrara, M., Kelly, M. T., Roso, F., Larroque, M., & Margout, D. (2021). Potential of olive oil mill wastewater as a source of polyphenols for the treatment of skin disorders: A review. *Journal of Agricultural and Food Chemistry*, *69*(26), 7268−7284. Available from https://doi.org/10.1021/acs.jafc.1c00296. http://pubs.acs.org/journal/jafcau.

Caudalie Paris. (2023). *Polyphenols: Anti-oxidant shield*.

Choi, D. I., Choi, J. Y., Lee, J. B., Yun, S. J., Moon, B. K., Ahn, Y. G., Lee, S. Y., & Lee, S. C. (2023). Protective activity against oxidative stress in dermal papillae with extracted herbal essential oils. *Applied Sciences (Switzerland)*, *13*(6). Available from https://doi.org/10.3390/app13063985, http://www.mdpi.com/journal/applsci/.

Chung, J. H., Han, J. H., Hwang, E. J., Seo, J. Y., Cho, K. H., Kim, K. H., Youn, J. I., & Eun, H. C. (2003). Dual mechanisms of green tea extract-induced cell survival in human epidermal keratinocytes. *The FASEB Journal*, *17*(13), 1−21. Available from https://doi.org/10.1096/fj.02-0914fje.

Chung, S. Y., & Champagne, E. T. (2009). Reducing the allergenic capacity of peanut extracts and liquid peanut butter by phenolic compounds. *Food Chemistry*, *115*(4), 1345−1349. Available from https://doi.org/10.1016/j.foodchem.2009.01.052.

Clavaud, C., Jourdain, R., Bar-Hen, A., Tichit, M., Bouchier, C., Pouradier, F., El Rawadi, C., Guillot, J., Ménard-Szczebara, F., Breton, L., Latgé, J.-P., Mouyna, I., & Chaturvedi, V. (2013). Dandruff is associated with disequilibrium in the proportion of the major bacterial and fungal populations colonizing the scalp. *PLoS One*, *8*(3), e58203. Available from https://doi.org/10.1371/journal.pone.0058203.

Conte, F. L., Pereira, A. C., Brites, G., Ferreira, I., Silva, A. C., Sebastião, A. I., Matos, P., Pereira, C., Batista, M. T., Sforcin, J. M., & Cruz, M. T. (2022). Exploring the antioxidant, anti-inflammatory and antiallergic potential of Brazilian propolis in monocytes. *Phytomedicine Plus*, *2*(2). Available from https://doi.org/10.1016/j.phyplu.2022.100231. https://www.journals.elsevier.com/phytomedicine-plus.

Daniels, G., Akram, S., Westgate, G. E., & Tamburic, S. (2019). Can plant-derived phytochemicals provide symptom relief for hair loss? A critical review. *International Journal of Cosmetic Science*, *41*(4), 332−345. Available from https://doi.org/10.1111/ics.12554.

Dario, M. F., Pahl, R., De Castro, J. R., De Lima, F. S., Kaneko, T. M., Pinto, C. A. S. O., Baby, A. R., & Velasco, M. V. R. (2013). Efficacy of *Punica granatum* L. hydroalcoholic extract on properties of dyed hair exposed to UVA radiation. *Journal of Photochemistry and Photobiology B: Biology*, *120*, 142−147. Available from https://doi.org/10.1016/j.jphotobiol.2012.12.011.

de Groot, S. C., Ulrich, M. M. W., Gho, C. G., & Huisman, M. A. (2021). Back to the future: From appendage development toward future human hair follicle neogenesis. *Frontiers in Cell and Developmental Biology*, *9*. Available from https://doi.org/

10.3389/fcell.2021.661787. https://www.frontiersin.org/journals/cell-and-developmental-biology#.

do Carmo, M. A. V., Pressete, C. G., Marques, M. J., Granato, D., & Azevedo, L. (2018). Polyphenols as potential antiproliferative agents: Scientific trends. *Current Opinion in Food Science*, *24*, 26–35. Available from https://doi.org/10.1016/j.cofs.2018.10.013. http://www.journals.elsevier.com/current-opinion-in-food-science/.

Esfandiari, A., & Kelly, A. P. (2005). The effects of tea polyphenolic compounds on hair loss among rodents. *Journal of the National Medical Association*, *97*(8), 1165–1169.

European Commission. (2016). The SCCS notes of guidance for the testing of cosmetic ingredients. *SCCS*, *1564*, 151. Available from http://ec.europa.eu/health//sites/health/files/scientific_committees/consumer_safety/docs/sccs_o_190.pdf.

Fujimura, Y., Tachibana, H., Maeda-Yamamoto, M., Miyase, T., Sano, M., & Yamada, K. (2002). Antiallergic tea catechin, (−)-epigallocatechin-3-O-(3-O-methyl)-gallate, suppresses FcεRI expression in human basophilic KU812 cells. *Journal of Agricultural and Food Chemistry*, *50*(20), 5729–5734. Available from https://doi.org/10.1021/jf025680z.

Galanakis, C. M., Tsatalas, P., & Galanakis, I. M. (2018). Phenols from olive mill wastewater and other natural antioxidants as UV filters in sunscreens. *Environmental Technology & Innovation*, *9*, 160–168. Available from https://doi.org/10.1016/j.eti.2017.12.002.

Geng, H., Zhuang, L., Li, M., Liu, H., Caruso, F., Hao, J., & Cui, J. (2020). Interfacial assembly of metal-phenolic networks for hair dyeing. *ACS Applied Materials and Interfaces*, *12*(26), 29826–29834. Available from https://doi.org/10.1021/acsami.0c06928. http://pubs.acs.org/journal/aamick.

Grimalt, R. (2007). A practical guide to scalp disorders. *Journal of Investigative Dermatology Symposium Proceedings*, *12*(2), 10–14. Available from https://doi.org/10.1038/sj.jidsymp.5650048.

Hamed, F. N., McDonagh, A. J. G., Almaghrabi, S., Bakri, Y., Messenger, A. G., & Tazi-Ahnini, R. (2018). Epigallocatechin-3 gallate inhibits STAT-1/JAK2/IRF-1/HLA-DR/HLA-B and reduces CD8 MKG2D lymphocytes of alopecia areata patients. *International Journal of Environmental Research and Public Health*, *15*(12). Available from https://doi.org/10.3390/ijerph15122882. https://www.mdpi.com/1660-4601/15/12/2882/pdf.

He, L., Michailidou, F., Gahlon, H. L., & Zeng, W. (2022). Hair dye ingredients and potential health risks from exposure to hair dyeing. *Chemical Research in Toxicology*, *35* (6), 901–915. Available from https://doi.org/10.1021/acs.chemrestox.1c00427. http://pubs.acs.org/journal/crtoec.

Im, K. M., & Jeon, J. R. (2016). Synthesis of plant phenol-derived polymeric dyes for direct or mordant-based hair dyeing. *Journal of Visualized Experiments*, *2016*(118). Available from https://doi.org/10.3791/54772. http://www.jove.com/pdf/54772.

Jeon, S.-Y., Pi, L. Q., & Lee, W.-S. (2008). Comparison of hair shaft damage after UVA and UVB irradiation. *Journal of Cosmetic Science*, *59*(2), 151–156.

Jesumani, V., Du, H., Pei, P., Aslam, M., Huang, N., & Achal, V. (2020). Comparative study on skin protection activity of polyphenol-rich extract and polysaccharide-rich extract from *Sargassum vachellianum*. *PLoS One*, *15*(1), e0227308. Available from https://doi.org/10.1371/journal.pone.0227308.

Jia, D., Shen, Y., Zhang, X., Wang, Y., Su, R., & Qi, W. (2022). Colorful pigments based on multicomponent metal-phenol network nanoparticles for hair dyeing. *ChemistrySelect*, 7(48). Available from https://doi.org/10.1002/slct.202203886. http://onlinelibrary.wiley.com/journal/10.1002/(ISSN)2365-6549.

Jin, L., Yun, D., Zhang, W., Lee, J., Shin, H., Kim, D., Kang, T. B., Won, H. S., Jang, H., & Kim, W. (2023). Polyphenols coordinated with Cu (II) in an aqueous system build ion-channel coatings on hair surfaces. *Materials*, *16*(4). Available from https://doi.org/10.3390/ma16041333. http://www.mdpi.com/journal/materials.

Kamimura, A., & Takahashi, T. (2002). Procyanidin B-2, extracted from apples, promotes hair growth: A laboratory study. *British Journal of Dermatology*, *146*(1), 41–51. Available from https://doi.org/10.1046/j.0007-0963.2001.04558.x.

Kamimura, A., Takahashi, T., Morohashi, M., & Takano, Y. (2006). Procyanidin oligomers counteract TGF-β1- and TGF-β2-induced apoptosis in hair epithelial cells: An insight into their mechanisms. *Skin Pharmacology and Physiology*, *19*(5), 259–265. Available from https://doi.org/10.1159/000093981.

Kang, K. A., Lee, K. H., Chae, S., Zhang, R., Jung, M. S., Lee, Y., Kim, S. Y., Kim, H. S., Joo, H. G., Park, J. W., Ham, Y. M., Lee, N. H., & Hyun, J. W. (2005). Eckol isolated from *Ecklonia cava* attenuates oxidative stress induced cell damage in lung fibroblast cells. *FEBS Letters*, *579*(28), 6295–6304. Available from https://doi.org/10.1016/j.febslet.2005.10.008.

Kanlayavattanakul, M., Lourith, N., & Chaikul, P. (2016). Jasmine rice panicle: A safe and efficient natural ingredient for skin aging treatments. *Journal of Ethnopharmacology*, *193*, 607–616. Available from https://doi.org/10.1016/j.jep.2016.10.013.

Kaur, N., Bains, A., Kaushik, R., Dhull, S. B., Melinda, F., & Chawla, P. (2021). A review on antifungal efficiency of plant extracts entrenched polysaccharide-based nanohydrogels. *Nutrients*, *13*(6), 2055. Available from https://doi.org/10.3390/nu13062055.

Kerk, S. K., Lai, H. Y., Sze, S. K., Ng, K. W., Schmidtchen, A., & Adav, S. S. (2018). Bacteria display differential growth and adhesion characteristics on human hair shafts. *Frontiers in Microbiology*, *9*. Available from https://doi.org/10.3389/fmicb.2018.02145. https://www.frontiersin.org/articles/10.3389/fmicb.2018.02145/full.

Kesika, P., Sivamaruthi, B. S., Thangaleela, S., Bharathi, M., & Chaiyasut, C. (2023). Role and mechanisms of phytochemicals in hair growth and health. *Pharmaceuticals*, *16* (2). Available from https://doi.org/10.3390/ph16020206. https://www.mdpi.com/journal/pharmaceuticals.

Koch, S. L., Tridico, S. R., Bernard, B. A., Shriver, M. D., & Jablonski, N. G. (2020). The biology of human hair: A multidisciplinary review. *American Journal of Human Biology*, *32*(2). Available from https://doi.org/10.1002/ajhb.23316. http://onlinelibrary.wiley.com/journal/10.1002/(ISSN)1520-6300.

Kolarsick, P. A. J., Kolarsick, M. A., & Goodwin, C. (2011). Anatomy and physiology of the skin. *Journal of the Dermatology Nurses' Association*, *3*(4), 203–213. Available from https://doi.org/10.1097/JDN.0b013e3182274a98.

Korkina, L. G., Pastore, S., De Luca, C., & Kostyuk, V. A. (2008). Metabolism of plant polyphenols in the skin: Beneficial versus deleterious effects. *Current Drug Metabolism*, *9*(8), 710–729. Available from https://doi.org/10.2174/138920008786049267. http://docstore.ingenta.com/cgi-bin/ds_deliver/1/u/d/ISIS/48971918.1/ben/cdm/2008/00000009/00000008/art00004/43AA06A49298AA44123538129839D241C19E341160.pdf?link = http://www.ingentaconnect.com/error/delivery&format = pdf Italy.

Kubo, C., Ogawa, M., Uehara, N., & Katakura, Y. (2020). Fisetin promotes hair growth by augmenting TERT expression. *Frontiers in Cell and Developmental Biology*, *8*. Available from https://doi.org/10.3389/fcell.2020.566617.

Lai-Cheong, J. E., & McGrath, J. A. (2013). Structure and function of skin, hair and nails. *Medicine (United Kingdom)*, *41*(6), 317–320. Available from https://doi.org/10.1016/j.mpmed.2013.04.017. http://www.elsevier.com/wps/find/journaldescription.cws_home/709606/description#description.

Lai-Cheong, J. E., & McGrath, J. A. (2021). Structure and function of skin, hair and nails. *Medicine (United Kingdom)*, *49*(6), 337–342. Available from https://doi.org/10.1016/

j.mpmed.2021.03.001. http://www.elsevier.com/wps/find/journaldescription.cws_home/709606/description#description.

Lan, H., & Hu, L. (2016). *Hair dye product*. Google Patents.

Li, Y., Miao, Y., Yang, L., Zhao, Y., Wu, K., Lu, Z., Hu, Z., & Guo, J. (2022). Recent advances in the development and antimicrobial applications of metal–phenolic networks. *Advanced Science*, *9*(27), 2202684. Available from https://doi.org/10.1002/advs.202202684.

Lopes, G., Sousa, C., Silva, L. R., Pinto, E., Andrade, P. B., Bernardo, J., Mouga, T., & Valentão, P. (2012). Can phlorotannins purified extracts constitute a novel pharmacological alternative for microbial infections with associated inflammatory conditions? *PLoS One*, 7(2). Available from https://doi.org/10.1371/journal.pone.0031145. http://www.plosone.org/article/fetchObjectAttachment.action?uri = info%3Adoi%2F10.1371%2Fjournal.pone.0031145&representation = PDF.

Lourith, N., & Kanlayavattanakul, M. (2013). Hair loss and herbs for treatment. *Journal of Cosmetic Dermatology*, *12*(3), 210–222. Available from https://doi.org/10.1111/jocd.12051.

Lourith, N., & Kanlayavattanakul, M. (2021). *Phytocosmetics for hair care. Phytocosmetics and cosmetic science* (pp. 57–74). Informa UK Limited. Available from http://doi.org/10.1201/9781003107385-5.

Manso, T., Lores, M., & de Miguel, T. (2021). Antimicrobial activity of polyphenols and natural polyphenolic extracts on clinical isolates. *Antibiotics (Basel)*, *11*(1). Available from https://doi.org/10.3390/antibiotics11010046.

Marques, P., Marto, J., Gonçalves, L. M., Pacheco, R., Fitas, M., Pinto, P., Serralheiro, M. L. M., & Ribeiro, H. (2017). *Cynara scolymus* L.: A promising Mediterranean extract for topical anti-aging prevention. *Industrial Crops and Products*, *109*, 699–706. Available from https://doi.org/10.1016/j.indcrop.2017.09.033.

Marto, J., Gouveia, L. F., Chiari, B. G., Paiva, A., Isaac, V., Pinto, P., Simões, P., Almeida, A. J., & Ribeiro, H. M. (2016). The green generation of sunscreens: Using coffee industrial sub-products. *Industrial Crops and Products*, *80*, 93–100. Available from https://doi.org/10.1016/j.indcrop.2015.11.033.

Mathes, S. H., Ruffner, H., & Graf-Hausner, U. (2014). *The use of skin models in drug development, Advanced Drug Delivery Reviews* (Vol. 69–70, pp. 81–102).

Matravers, P., Milius, U., & Cornuelle, T.L. (2000). *Method and composition for hair coloring using green tea polyphenols*. Google Patents.

McKnight, G., Shah, J., & Hargest, R. (2022). Physiology of the skin. *Surgery (Oxford)*, *40*(1), 8–12. Available from https://doi.org/10.1016/j.mpsur.2021.11.005.

Miere, F., Teuşdea, A. C., Laslo, V., Cavalu, S., Fritea, L., Dobjanschi, L., Zdrinca, M., Zdrinca, M., Ganea, M., Paşc, P., Memete, A. R., Antonescu, A., Vlad, A. M., & Vicas, S. I. (2021). Evaluation of in vitro wound-healing potential, antioxidant capacity, and antimicrobial activity of *Stellaria media* (L.) vill. *Applied Sciences (Switzerland)*, *11*(23). Available from https://doi.org/10.3390/app112311526. https://www.mdpi.com/2076-3417/11/23/11526/pdf.

Miteva, M., Hirt, P., & Paus, R. (2019). Healthy hair (anatomy, biology, morphogenesis, cycling, and function). In M. Miteva (Ed.), *Alopecia*. Elsevier.

Morgan, B. A. (2014). The dermal papilla: An instructive niche for epithelial stem and progenitor cells in development and regeneration of the hair follicle. *Cold Spring Harbor Perspectives in Medicine*, *4*(7), a015180. Available from https://doi.org/10.1101/cshperspect.a015180.

Mukherjee, P. K., Maity, N., Nema, N. K., & Sarkar, B. K. (2011). Bioactive compounds from natural resources against skin aging. *Phytomedicine: International Journal of Phytotherapy and Phytopharmacology*, *19*(1), 64–73. Available from https://doi.org/10.1016/j.phymed.2011.10.003. http://www.urbanfischer.de/journals/phytomed.

Mwakalukwa, R., Ashour, A., Amen, Y., Niwa, Y., Tamrakar, S., Miyamoto, T., & Shimizu, K. (2019). Anti-allergic activity of polyphenolic compounds isolated from olive mill wastes. *Journal of Functional Foods*, *58*, 207–217. Available from https://doi.org/10.1016/j.jff.2019.04.058.

Natarelli, N., Gahoonia, N., & Sivamani, R. K. (2023). Integrative and mechanistic approach to the hair growth cycle and hair loss. *Journal of Clinical Medicine*, *12*(3), 893. Available from https://www.mdpi.com/2077-0383/12/3/893.

Nichols, J. A., & Katiyar, S. K. (2010). Skin photoprotection by natural polyphenols: Anti-inflammatory, antioxidant and DNA repair mechanisms. *Archives of Dermatological Research*, *302*(2), 71–83. Available from https://doi.org/10.1007/s00403-009-1001-3.

Nishida, Y., Ito, T., Hosokawa, M., Aono, M., Yokomaku, A., Konta, H., Iimura, K. I., Kato, T., & Sugiyama, K. (2004). Repairing effects of diglucosyl gallic acid on coloring-damaged hair. *Journal of Oleo Science*, *53*(6), 295–304. Available from https://doi.org/10.5650/jos.53.295.

Nunes, M. A., Rodrigues, F., & Oliveira, M. B. P. P. (2017). Grape processing by-products as active ingredients for cosmetic proposes. In C. M. Galanakis (Ed.), *Handbook of grape processing by-products: Sustainable solutions* (pp. 267–292). Elsevier Inc. https://doi.org/10.1016/B978-0-12-809870-7.00011-9.

Paus, R., & Cotsarelis, G. (1999). The biology of hair follicles. *New England Journal of Medicine*, *341*(7), 491–497.

Pinto, D., Cádiz-Gurrea, MdlL., Garcia, J., Saavedra, M. J., Freitas, V., Costa, P., Sarmento, B., Delerue-Matos, C., & Rodrigues, F. (2021). From soil to cosmetic industry: Validation of a new cosmetic ingredient extracted from chestnut shells. *Sustainable Materials and Technologies*, *29*. Available from https://doi.org/10.1016/j.susmat.2021.e00309. https://www.journals.elsevier.com/sustainable-materials-and-technologies.

Pinto, D., Vieira, E. F., Peixoto, A. F., Freire, C., Freitas, V., Costa, P., Delerue-Matos, C., & Rodrigues, F. (2021). Optimizing the extraction of phenolic antioxidants from chestnut shells by subcritical water extraction using response surface methodology. *Food Chemistry*, *334*. Available from https://doi.org/10.1016/j.foodchem.2020.127521, http://www.elsevier.com/locate/foodchem.

Rambaran, T. F. (2022). A patent review of polyphenol nano-formulations and their commercialization. *Trends in Food Science and Technology*, *120*, 111–122. Available from https://doi.org/10.1016/j.tifs.2022.01.011. http://www.elsevier.com/wps/find/journaldescription.cws_home/601278/description#description.

Rasooly, R., Do, P. M., & Friedman, M. (2010). Inhibition of biological activity of staphylococcal enterotoxin A (SEA) by apple juice and apple polyphenols. *Journal of Agricultural and Food Chemistry*, *58*(9), 5421–5426. Available from https://doi.org/10.1021/jf904021b.

Regulation (EC) No 1223/2009 of the European Parliament and of the Council of 30 November 2009 on cosmetic products (pp. 59–209). (2009).

Rodrigues, F., Cádiz-Gurrea, M. D. L. L., Nunes, M. A., Pinto, D., Vinha, A. F., Linares, I. B., Oliveira, M. B. P. P., & Carretero, A. S. (2018). *Cosmetics. Polyphenols: Properties, recovery, and applications* (pp. 393–427). Elsevier. Available from: https://doi.org/10.1016/B978-0-12-813572-3.00012-9.

Rodrigues, F., Nunes, M. A., Alves, R. C., & Oliveira, M. B. P. (2017). Applications of recovered bioactive compounds in cosmetics and other products. In C. M. Galanakis (Ed.), *Handbook of coffee processing by-products: Sustainable applications* (pp. 195–220). Elsevier.

Rodrigues, F., Nunes, M. A. d M., & Oliveira, M. B. P. P. (2017). Applications of recovered bioactive compounds in cosmetics and health care products. In C. M. Galanakis

(Ed.), *Olive mill waste: Recent advances for sustainable management* (pp. 255–274). Elsevier Inc. https://doi.org/10.1016/B978-0-12-805314-0.00012-1.

Rodrigues, F., Palmeira-de-Oliveira, A., Das Neves, J., Sarmento, B., Amaral, M. H., & Oliveira, M. B. P. P. (2015). Coffee silverskin: A possible valuable cosmetic ingredient. *Pharmaceutical Biology, 53*(3), 386–394. Available from https://doi.org/10.3109/13880209.2014.922589. https://www.tandfonline.com/loi/iphb20.

Rodrigues, F., Pereira, C., Pimentel, F. B., Alves, R. C., Ferreira, M., Sarmento, B., Amaral, M. H., & Oliveira, M. B. P. P. (2015). Are coffee silverskin extracts safe for topical use? An in vitro and in vivo approach. *Industrial Crops and Products, 63*, 167–174. Available from https://doi.org/10.1016/j.indcrop.2014.10.014. http://www.elsevier.com/inca/publications/store/5/2/2/8/2/5.

Roh, E., Kim, J. E., Kwon, J. Y., Park, J. S., Bode, A. M., Dong, Z., & Lee, K. W. (2017). Molecular mechanisms of green tea polyphenols with protective effects against skin photoaging. *Critical Reviews in Food Science and Nutrition, 57*(8), 1631–1637. Available from https://doi.org/10.1080/10408398.2014.1003365. http://www.tandf.co.uk/journals/titles/10408398.asp.

Rose, P. M., Cantrill, V., Benohoud, M., Tidder, A., Rayner, C. M., & Blackburn, R. S. (2018). Application of anthocyanins from blackcurrant (*Ribes nigrum* L.) fruit waste as renewable hair dyes. *Journal of Agricultural and Food Chemistry, 66*(26), 6790–6798. Available from https://doi.org/10.1021/acs.jafc.8b01044. http://pubs.acs.org/journal/jafcau.

Saansoomchai, P., Limmongkon, A., Surangkul, D., Chewonarin, T., & Srikummool, M. (2018). Enhanced VEGF expression in hair follicle dermal papilla cells by *Centella asiatica* linn. *Chiang Mai University Journal of Natural Sciences, 17*(1), 25–37. Available from https://doi.org/10.12982/CMUJNS.2018.0003. http://cmuj.cmu.ac.th/ns/full/2020180129142308.pdf.

Sattur, S. S., & Sattur, I. S. (2021). Pharmacological management of pattern hair loss. *Indian Journal of Plastic Surgery, 54*(4), 422–434. Available from https://doi.org/10.1055/s-0041-1739254. http://www.ijps.org.

Scalia, S., Marchetti, N., & Bianchi, A. (2013). Comparative evaluation of different co-antioxidants on the photochemical- and functional-stability of epigallocatechin-3-gallate in topical creams exposed to simulated sunlight. *Molecules (Basel, Switzerland), 18*(1), 574–587. Available from https://doi.org/10.3390/molecules18010574.

Schlupp, P., Schmidts, T. M., Pössl, A., Wildenhain, S., Lo Franco, G., Lo Franco, A., & Lo Franco, B. (2019). Effects of a phenol-enriched purified extract from olive mill wastewater on skin cells. *Cosmetics, 6*(2). Available from https://doi.org/10.3390/COSMETICS6020030. https://res.mdpi.com/cosmetics/cosmetics-06-00030/article_deploy/cosmetics-06-00030.pdf?filename = &attachment = 1.

Segre, J. A. (2006). Epidermal barrier formation and recovery in skin disorders. *Journal of Clinical Investigation, 116*(5), 1150–1158. Available from https://doi.org/10.1172/JCI28521.

Sentamilselvi, G., Janaki, C., & Murugusundram, S. (2009). Trichomycoses. *International Journal of Trichology, 1*(2), 100.

Sharma, A., Kashyap, D., Sak, K., Tuli, H. S., & Sharma, A. K. (2018). Therapeutic charm of quercetin and its derivatives: A review of research and patents. *Pharmaceutical Patent Analyst, 7*(1), 15–32. Available from https://doi.org/10.4155/ppa-2017-0030. http://www.future-science.com/loi/ppa.

Sharma, R. R., Deep, A., & Abdullah, S. T. (2022). Herbal products as skincare therapeutic agents against ultraviolet radiation-induced skin disorders. *Journal of Ayurveda and Integrative Medicine, 13*(1), 100500. Available from https://doi.org/10.1016/j.jaim.2021.07.016.

Shen, W., Li, S. Y., Pan, Y. Q., Liu, H., Dong, X. W., Zhang, X. Q., Ye, W. C., Hu, X. L., & Wang, H. (2023). *Prinsepia utilis* Royle leaf extract: Ameliorative effects on allergic inflammation and skin lesions in allergic contact dermatitis and polyphenolic profiling through UPLC–MS/MS coupled to chemometric analysis. *Journal of Ethnopharmacology, 305.* Available from https://doi.org/10.1016/j.jep.2022.116093. http://www.elsevier.com/locate/jethpharm.

Shen, Y. L., Li, X. Q., Pan, R. R., Yue, W., Zhang, L. J., & Zhang, H. (2018). Medicinal plants for the treatment of hair loss and the suggested mechanisms. *Current Pharmaceutical Design, 24*(26), 3090–3100. Available from https://doi.org/10.2174/1381612824666180911114810. http://www.eurekaselect.com/165286/article.

Shin, H., Cho, A. R., Kim, D. Y., Munkhbayer, S., Choi, S. J., Jang, S., Kim, S. H., Shin, H. C., & Kwon, O. (2016). Enhancement of human hair growth using *Ecklonia cava* polyphenols. *Annals of Dermatology, 28*(1), 15–21. Available from https://doi.org/10.5021/ad.2016.28.1.15. http://anndermatol.org/Synapse/Data/PDFData/0140AD/ad-28-15.pdf.

Shin, S., Kim, K., Lee, M. J., Lee, J., Choi, S., Kim, K. S., Ko, J. M., Han, H., Kim, S. Y., Youn, H. J., Ahn, K. J., An, I. S., An, S., & Cha, H. J. (2016). Epigallocatechin gallate-mediated alteration of the microRNA expression profile in 5α-dihydrotestosterone-treated human dermal papilla cells. *Annals of Dermatology, 28*(3), 327–334. Available from https://doi.org/10.5021/ad.2016.28.3.327. http://anndermatol.org/Synapse/Data/PDFData/0140AD/ad-28-327.pdf.

Silva, A. M., Costa, P. C., Delerue-Matos, C., Latocha, P., & Rodrigues, F. (2021). Extraordinary composition of *Actinidia arguta* by-products as skin ingredients: A new challenge for cosmetic and medical skincare industries. *Trends in Food Science and Technology, 116*, 842–853. Available from https://doi.org/10.1016/j.tifs.2021.08.031. http://www.elsevier.com/wps/find/journaldescription.cws_home/601278/description#description.

Silva, A. M., Costa, P. C., Delerue-Matos, C., & Rodrigues, F. (2023). Mathematical development and in vivo safety and efficacy studies of a topical formulation containing *Actinidia arguta* leaves extract. *Journal of Drug Delivery Science and Technology, 84.* Available from https://doi.org/10.1016/j.jddst.2023.104547. http://www.editionsde-sante.fr/category.php?id_category = 48.

Silva, A. M., Garcia, J., Dall'Acqua, S., Costa, P., Delerue-Matos, C., & Rodrigues, F. (2022). Eco-friendly insights on kiwiberry leaves valorization through in-vitro and in-vivo studies. *Industrial Crops and Products, 184*, 115090. Available from https://doi.org/10.1016/j.indcrop.2022.115090.

Silva, A. M., Pinto, D., Fernandes, I., de Freitas, V., Cádiz-Gurrea, MdlL., Costa, P., Delerue-Matos, C., & Rodrigues, F. (2021). An insight into kiwiberry leaf valorization: Phenolic composition, bioactivity and health benefits. *Molecules (Basel, Switzerland), 26*(8). Available from https://doi.org/10.3390/molecules26082314. https://www.mdpi.com/1420-3049/26/8/2314/pdf.

Silva, T., Oliveira, C., & Borges, F. (2014). Caffeic acid derivatives, analogs and applications: A patent review (2009–2013). *Expert Opinion on Therapeutic Patents, 24*(11), 1257–1270.

Singh, A., Holvoet, S., & Mercenier, A. (2011). Dietary polyphenols in the prevention and treatment of allergic diseases. *Clinical & Experimental Allergy, 41*(10), 1346–1359. Available from https://doi.org/10.1111/j.1365-2222.2011.03773.x.

Sittek, L.-M., Schmidts, T. M., & Schlupp, P. (2021). Polyphenol-rich olive mill wastewater extract and its potential use in hair care products. *Journal of Cosmetics, Dermatological Sciences and Applications, 11*(4), 356–370.

Sun, M., Deng, Y., Cao, X., Xiao, L., Ding, Q., Luo, F., Huang, P., Gao, Y., Liu, M., & Zhao, H. (2022). Effects of natural polyphenols on skin and hair health: A review. *Molecules (Basel, Switzerland), 27*(22), 7832. Available from https://doi.org/10.3390/molecules27227832.

Supe, S., & Takudage, P. (2021). Methods for evaluating penetration of drug into the skin: A review. *Skin Research and Technology*, *27*(3), 299–308. Available from https://doi.org/10.1111/srt.12968.

Takahashi, T., Kamimura, A., Kagoura, M., Toyoda, M., & Morohashi, M. (2005). Investigation of the topical application of procyanidin oligomers from apples to identify their potential use as a hair-growing agent. *Journal of Cosmetic Dermatology*, *4*(4), 245–249. Available from https://doi.org/10.1111/j.1473-2165.2005.00199.x.

Takahashi, T., Kamimura, A., Yokoo, Y., Honda, S., & Watanabe, Y. (2001). The first clinical trial of topical application of procyanidin B-2 to investigate its potential as a hair growing agent. *Phytotherapy Research*, *15*(4), 331–336. Available from https://doi.org/10.1002/ptr.800.

Takahashi, T., Kamiya, T., & Yokoo, Y. (1998). Proanthocyanidins from grape seeds promote proliferation of mouse hair follicle cells in vitro and convert hair cycle in vivo. *Acta Dermato-Venereologica*, *78*(6).

Tang, Y., Liu, Y., Luo, K., Xu, L., Yang, P., & Ming, J. (2022). Potential applications of *Lilium* plants in cosmetics: A comprehensive review based on research papers and patents. *Antioxidants*, *11*(8), 1458. Available from https://doi.org/10.3390/antiox11081458.

Telang, M., Dhulap, S., Mandhare, A., & Hirwani, R. (2013). Therapeutic and cosmetic applications of mangiferin: A patent review. *Expert Opinion on Therapeutic Patents*, *23*(12), 1561–1580. Available from https://doi.org/10.1517/13543776.2013.836182.

Thomas, N. V., & Kim, S. K. (2013). Beneficial effects of marine algal compounds in cosmeceuticals. *Marine Drugs*, *11*(1), 146–164. Available from https://doi.org/10.3390/md11010146. http://www.mdpi.com/1660-3397/11/1/146/pdf.

Tobin, D. J. (2009). Aging of the hair follicle pigmentation system. *International Journal of Trichology*, *1*(2), 83–93. Available from https://doi.org/10.4103/0974-7753.58550.

Vasconcelos, L. C., Sampaio, M. C., Sampaio, F. C., & Higino, J. S. (2003). Use of *Punica granatum* as an antifungal agent against candidosis associated with denture stomatitis. *Mycoses*, *46*(5–6), 192–196. Available from https://doi.org/10.1046/j.1439-0507.2003.00884.x.

Vayalil, P. K., Mittal, A., Hara, Y., Elmets, C. A., & Katiyar, S. K. (2004). Green tea polyphenols prevent ultraviolet light-induced oxidative damage and matrix metalloproteinases expression in mouse skin. *Journal of Investigative Dermatology*, *122*(6), 1480–1487. Available from https://doi.org/10.1111/j.0022-202X.2004.22622.x. http://www.nature.com/jid/index.html.

Venus, M., Waterman, J., & McNab, I. (2010). Basic physiology of the skin. *Surgery (Oxford)*, *28*(10), 469–472. Available from https://doi.org/10.1016/j.mpsur.2010.07.011.

Voravuthikunchai, S. P., Sririrak, T., Limsuwan, S., Supawita, T., Iida, T., & Honda, T. (2005). Inhibitory effects of active compounds from *Punica granatum* pericarp on verocytotoxin production by enterohemorrhagic *Escherichia coli* O157:H7. *Journal of Health Science*, *51*(5), 590–596. Available from https://doi.org/10.1248/jhs.51.590. http://jhs.pharm.or.jp/51(5)/51_590.pdf Thailand.

Wei, X., Liu, Y., Xiao, J., & Wang, Y. (2009). Protective effects of tea polysaccharides and polyphenols on skin. *Journal of Agricultural and Food Chemistry*, *57*(17), 7757–7762. Available from https://doi.org/10.1021/jf901340f.

Xu, F. W., Lv, Y. L., Zhong, Y. F., Xue, Y. N., Wang, Y., Zhang, L. Y., Hu, X., & Tan, W. Q. (2021). Beneficial effects of green tea EGCG on skin wound healing: A comprehensive review. *Molecules (Basel, Switzerland)*, *26*(20). Available from https://doi.org/10.3390/molecules26206123. https://www.mdpi.com/1420-3049/26/20/6123/pdf.

Xu, L. Q., Neoh, K.-G., & Kang, E.-T. (2018). Natural polyphenols as versatile platforms for material engineering and surface functionalization. *Progress in Polymer Science*, *87*, 165–196.

Yadav, N., Parveen, S., Chakravarty, S., & Banerjee, M. (2019). Skin anatomy and morphology. In A. Dwivedi, N. Agarwal, L. Ray, & A. K. Tripathi (Eds.), *Skin aging & cancer: Ambient UV-R exposure* (pp. 1–10). Springer Singapore. https://doi.org/10.1007/978-981-13-2541-0_1.

Yano, S., Tachibana, H., & Yamada, K. (2005). Flavones suppress the expression of the high-affinity IgE receptor FcepsilonRI in human basophilic KU812 cells. *Journal of Agricultural and Food Chemistry*, *53*(5), 1812–1817. Available from https://doi.org/10.1021/jf047929d.

Zhang, M., Zhou, J., Wang, L., Li, B., Guo, J., Guan, X., Han, Q., & Zhang, H. (2014). Caffeic acid reduces cutaneous tumor necrosis factor alpha (TNF-α), IL-6 and IL-1β levels and ameliorates skin edema in acute and chronic model of cutaneous inflammation in mice. *Biological and Pharmaceutical Bulletin*, *37*(3), 347–354. Available from https://doi.org/10.1248/bpb.b13-00459.

Zhang, Z., Li, X., Sang, S., McClements, D. J., Chen, L., Long, J., Jiao, A., Jin, Z., & Qiu, C. (2022). Polyphenols as plant-based nutraceuticals: Health effects, encapsulation, nano-delivery, and application. *Foods*, *11*(15), 2189. Available from https://doi.org/10.3390/foods11152189.

Zlabiene, U., Baranauskaite, J., Kopustinskiene, D. M., & Bernatoniene, J. (2021). In vitro and clinical safety assessment of the multiple w/o/w emulsion based on the active ingredients from *Rosmarinus officinalis* L., *Avena sativa* L. and *Linum usitatissimum* L. *Pharmaceutics*, *13*(5). Available from https://doi.org/10.3390/pharmaceutics13050732. https://www.mdpi.com/1999-4923/13/5/732/pdf.

Index

Note: Page numbers followed by "*f*" and "*t*" refer to figures and tables, respectively.

J

K

L

T

W

X

Z